Paternity in Primates: Genetic Tests and Theories

2nd Schultz-Biegert Symposium, Kartause Ittingen, Switzerland,
September 16–20, 1991

Paternity in Primates: Genetic Tests and Theories

Implications of Human DNA Fingerprinting

Editors
R.D. Martin, Zürich
A.F. Dixson; E.J. Wickings, Franceville

51 figures and 35 tables, 1992

Basel · München · Paris · London · New York · New Delhi · Bangkok · Singapore · Tokyo · Sydney

Library of Congress Cataloging-in-Publication Data
Schultz-Biegert Symposium (2nd: 1991: Kartause Ittingen)
Paternity in primates: tests and theories / 2nd Schultz-Biegert Symposium, Kartause Ittingen, Switzerland, September 16–20, 1991; editors, R.D. Martin, A.F. Dixson, E.J. Wickings.
Includes bibliographical references and index. (alk. paper)
1. Primates – Genetics – Congresses. 2. Primates – Behavior – Congresses.
3. Paternity testing – Congresses. 4. DNA fingerprints – Congresses.
5. Sexual behavior in animals – Congresses. I. Martin, R.D. (Robert D.), 1942–.
II. Dixson, A.F. III. Wickings, E.J. IV. Title.
[DNLM: 1. DNA Fingerprinting – congresses. 2. Paternity – congresses.
3. Primates – genetics – congresses. W 791 S387p 1991]
QL737.P9S36 1991 599.8′0415 -- dc20
ISBN 3-8055-5494-X

Printed in Switzerland on acid-free paper by Thür AG Offsetdruck, Pratteln
ISBN 3-8055-5494-X

Contents

Paternity in Primates

An Introduction

Following the successful launch of the series in 1989, the second Schultz-Biegert Symposium was held in the conference centre at the former Charterhouse of Ittingen (near Zürich) from 16th to 20th September 1991. On this occasion, the topic *Paternity in Primates: Genetic Tests and Theories* was selected and the symposium brought together 21 contributors from a broad international field, with participation from France, Gabon, Germany, Holland, Japan, Switzerland and the USA. Discussions were also enriched by a number of special guests: Prof. Hans Kummer (Zürich), Dr. Esther Signer (Leicester, UK) and Dr. Claude Schelling (Zürich).

The symposium was focussed on applications of recent developments in molecular genetics that permit reliable identification of individuals and detailed analysis of degrees of genetic relatedness. New techniques, most notably DNA fingerprinting, have opened up revolutionary new possibilities for the interpretation of primate behaviour in relation to reproductive success. In particular, paternity resulting from competitive interaction between males can now be determined directly, rather than being inferred from observed copulatory behaviour.

The starting-point for the new developments that led to the symposium is surprisingly recent. The technique of DNA fingerprinting for humans, discovered by Prof. Alec Jeffreys in 1984 and reported in *Nature* in 1985, rapidly led to a wide range of applications involving the identification of individuals and of genetic relationships between them. Among other things, the technique has been used to resolve disputed paternity cases, to check familial relationships in the context of immigration controls and to identify individuals involved in crimes. Progress in forensic applications of DNA fingerprints since 1985 has indeed been so rapid that a crime story in which this technique figured prominently has already been published: Joseph Wambaugh's *The Blooding* (1989). It was quickly realised that use of the technique of DNA fingerprinting is not restricted to humans, so applications to birds, non-human primates and other mammals soon followed.

The primary aim of the symposium was to bring together a group of scientists directly involved in the first studies in which the use of DNA fingerprinting and similar techniques is being extended to non-human primates. As the developments concerned are still at an early stage, it was not expected that the symposium would yield definitive results. Rather, the aim was to assess progress to date and to achieve a preliminary overview as a guide to future investigations. At this stage of research, pooling of information and experience is invaluable for channelling future work in the most promising directions and as an aid to newcomers. It is also important at this point to consider the kinds of questions that can be answered with the new methods and for this reason the symposium included a number of contributions dealing with relevant aspects of primate social behaviour. For those concerned with the interpretation of primate social behaviour, the potential of the new methods for determining paternity is of primary interest. Nevertheless, it is important to remember that new possibilities have also been opened up for the reliable identification of individuals over time, for checking pedigrees (including mother-infant relationships) and for investigation of population genetics. All of these aspects are, among other things, of great importance for maintenance of captive breeding colonies and for long-term measures furthering primate conservation.

Much discussion at the symposium centred around the new methods used for genetic characterization of primates and one outcome is the Appendix (pp. 275–280) summarizing different approaches and their advantages and disadvantages. Despite the relative recency of the developments concerned, a considerable variety of methods is being used for DNA typing in primates. Although some authors have used the original multilocus probes 33.6 and 33.15 developed by Jeffreys for human fingerprinting, others are using alternative DNA sequences or artificially synthesized oligonucleotide probes. Oligonucleotide probes may be constructed to replicate features of the Jeffreys probes, but they can also consist of limited repeats of simple sequences that bear no resemblance to the original probes. In all cases, informative banding patterns can be obtained, though the number of bands may be relatively small with certain oligonucleotide probes.

One important issue that was discussed was the emerging balance between classical approaches based on electrophoresis of blood proteins or restriction fragment length polymorphisms (RFLPs) and applications of new methods such as DNA fingerprinting. Although the impression is sometimes given that the new methods might rapidly replace the old, this is unlikely for several reasons. In the first place, it is both faster and less expensive to use electrophoresis of blood proteins as a preliminary screening measure in studies of genetic relatedness. Secondly, a number of the studies reported have shown that a combination of protein electrophoresis, RFLP analysis and DNA fingerprinting is more powerful than one of these methods alone. Thirdly, the genetic interpretation of electrophoretic patterns from blood proteins is relatively straightforward, whereas this is not true of DNA fingerprinting patterns. The high mutation rate underlying DNA fingerprint patterns generated with multi-locus probes is also a potential problem that has yet to be documented properly for non-human primates. It therefore seems that, for the forseeable future, a mixed strategy involving both classical techniques and DNA fingerprinting is to be recommended.

It is clear from the results obtained so far that the use of multi-locus probes for DNA fingerprinting has brought great benefits for the interpretation of genetic relationships in primate populations. Nevertheless, some reservations must be expressed. There is, for instance, a problem with reproducibility of DNA fingerprints, in the sense that the pattern for a given sample may vary to some extent between plates (e.g. in band intensity and hence observable band number). Further, there are the limitations that there is no simple genetic basis for the pattern obtained and that mutation may alter the band pattern often enough to cause problems in interpretation. For this reason, attention has been turning to single-locus probes specific to minisatellite and microsatellite DNA. Patterns obtained with single-locus probes have the drawback that the number of bands is strictly limited, but they are easily reproducible, open to straightforward genetic interpretation and barely affected by mutation. The amount of material required for each analysis is also markedly smaller. Given the availability of a set of different single-locus probes, it is possible to obtain sufficient information to match the band complexity of a muti-locus fingerprint while avoiding the major disadvantages.

An exciting extension of the strategy based on single-locus probes resides in the use of the polymerase chain reaction (PCR) to amplify specific sequences from the genome prior to genetic analysis. PCR is itself a relatively new technique, invented in 1983, and its application to microsatellite DNA has great potential. Using this combination, it is possible to take minute quantities of material (e.g. cells from hair follicles) for genetic analysis. This could reduce or eliminate the present need for blood samples, which usually require refrigeration, and holds promise for non-invasive sample collection. It must be emphasized, however, that multi-locus probes are unsuitable here and that the use of single-locus probes requires considerable development work for each new species studied, in order to identify appropriate sequences of DNA for investigation.

Another promising development that has already been applied in some cases is the use of non-radioactive methods for labelling of probes, thus eliminating the need for access to a registered isotope laboratory.

The papers presented reveal that the new methods have already been applied to a wide variety of non-human primate species. Although macaques *(Macaca fascicularis, M. fuscata, M. mulatta, M. sylvanus)* are disproportionately represented, preliminary studies have also been conducted on great apes (chimpanzee, gorilla), baboons, guenons, patas monkeys, colobus monkeys, marmosets and even some prosimians (dwarf bushbabies, ringtail lemurs). Most of the studies covered have, for obvious reasons, been conducted in captivity; but some results of extensions to field studies have also been reported (Barbary macaques, Japanese macaques, long-tailed macaques and common marmosets).

Despite the fact that research is at such an early stage, a number of important findings have already emerged. In many cases, the use of genetic typing has permitted reliable identification of many individuals and at least partial reconstruction of pedigree relationships, although the results for males have generally permitted only exclusion from paternity rather than positive proof of paternity. The preliminary results are good enough to show that male rank and copulatory frequency are not necessarily reliable guides to reproductive success and that there seem to be marked differences in this respect between species. One problem that has emerged is that, particularly with small and relatively inbred populations, variability in multi-locus DNA patterns may not be as marked as in humans. It is therefore a moot point whether the term 'fingerprint' is at present really justifiable for some non-human primate species. Further research is necessary to reach the level of reliable identification of individuals that is currently possible for humans. Another important finding from several studies is that, although mating between close relatives may occur at least in captivity, there is some kind of mechanism that leads to avoidance of inbreeding within matrilines.

Results already available bear witness to the enormous contribution that DNA profiling techniques can make to our understanding of the organization of primate societies and mating systems. The emerging information is beginning to clarify the relationship between mating behaviour and mating success and to throw new light on the question of sperm competition. Increased cooperation between laboratory geneticists and observers of primate behaviour, which the symposium was designed to foster, will now yield major new insights into the evolution of primate social organization.

As with its predecessor, the second Schultz-Biegert Symposium was preceded by a long period of advance preparation. Authors were requested to submit draft papers well in advance and galley proofs were hence available for discussion at the meeting. Various modifications were made subsequently following discussions during the symposium, but the advance availability of proofs has allowed us to publish this book within three months of the end of the meeting. In the fast-moving field of genetic profiling, this is an important bonus.

The symposium was organised in conjuction with *Folia Primatologica* and S. Karger AG, Basel. Financial support for the symposium was provided by grants from the A.H.-Schultz-Stiftung (Zürich), from the Julius-Klaus-Stiftung (Zürich) and from the Schweizerischer Nationalfonds zur Förderung der wissenschaftlichen Forschung. The symposium, which was co-organised with Dr. A.F. Dixson and Dr. E.J. Wickings of the Centre International de Recherches Médicales de Franceville (Gabon), depended heavily on the valuable assistance provided by the following people from the Anthropological Institute in Zürich: Elsbeth Rüegg, Dominique Cueni, Viktoria Langadakis, Krisztina Vasarhelyi, Franziska von Segesser and Dr. Zdenka Nechvátalová, who also helped prepare the manuscripts for publication. The logistic support and encouragement from Dr. Thomas Karger, Peter Schäfer and Marie-Louise de Stachelski of S. Karger publishers is also gratefully acknowledged.

23rd September 1991

Prof. Dr. R.D. Martin
University of Zürich

Martin RD, Dixson AF, Wickings EJ (eds): Paternity in Primates:
Genetic Tests and Theories. Basel, Karger, 1992, pp 1–2

Some Impacts of Paternity Studies on Primate Ethology

Hans Kummer

Ethologie und Wildforschung, Universität Zürich-Irchel, Zürich, Switzerland

This symposium volume summarizes the state of the art of determining paternity among primates in captivity and in the wild. With the same methods, we are told, we could identify siblings and estimate coefficients of relatedness, data that rank equally high on the list of a field worker's wishes.

Ethological field data on primates require thousands of observation hours. One great impact of the new methods of biochemical identification is that they will enhance, perhaps even multiply, the information to be drawn from the same behavioral data. To mention a few examples:

Since the rise of sociobiology, primatologists have searched for ways to test kin selection theory. As they could determine only matrilineal kinship, they knew just a part of the kin relationships at work. Masses of interactional data were not fully exploited.

Primate field workers have collected years of data in order to grasp the effects of male dominance on male fitness. With paternity unknown, their analyses ended with the unsatisfactory substitute measure of copulation frequency. The extensive studies on variations of male reproductive strategies were little more than interesting descriptions of primate behavioral versatility; now, researchers can measure and compare their effects on reproductive success.

Our own longitudinal study of Ethiopian hamadryas baboons in the seventies revealed behavioral mechanisms and a social structure pointing to an ability of harem males to monopolize their females at the cost of considerable effort. As the outbreak of the war with Somalia prevented further study and the paternity tests that had already been approved, the final piece of evidence that would have confirmed or rejected our interpretations is still missing. Even if we had been able to return after a few years, we would not have been able to recognize all previous individuals with certainty, and the descendency of the new generation would have been lost. With the methods described in this volume, individuals still alive can be recognized genetically after an interruption in observations and the pedigrees reconstructed. Had we thought of collecting a dozen hairs of each troop member from the beginning, some of the damage to our project plans could most probably be mended even now.

The new biochemical methods of kinship analysis will also promote the ethological analysis of kin recognition by the primates themselves. When a monkey or ape 'recognizes' her matrilineal kin she may simply be giving differential treatment to her early associates. Recognition of patrilineal kin in promiscuous societies is more demanding and may operate, for example, by one primate matching the phenotypes of other group members against its own.

The new tools described in this book, if put to work where they can have the greatest impact, will furnish primate social ethology with some hard answers where so far we have had to be content with weak correlative evidence.

Prof. Dr. Hans Kummer, Ethologie und Wildforschung, Universität Zürich-Irchel, Winterthurerstrasse 190, CH–8057 Zürich (Switzerland)

Martin RD, Dixson AF, Wickings EJ (eds): Paternity in Primates: Genetic Tests and Theories. Basel, Karger, 1992, pp 3–17

DNA Typing in Human Parentage Testing using Multilocus and Single-Locus Probes

Robert E. Lewis, Jr., Julius M. Cruse

Department of Pathology, University of Mississippi Medical Center, Jackson, Miss., USA

The determination of parentage has been a problem since biblical times. The first significant scientific advance in this field came at the turn of the century with Landsteiner's description of the ABO blood group system in 1900. Since that time, other genetic marker systems used in cases of disputed parentage have included additional red cell antigen systems, notably Rh and the MNSs, red cell isoenzymes, serum proteins and in more recent years the human leukocyte antigen (HLA) system [1]. Testing for these markers may assist in the exclusion of a man falsely accused of paternity, or they may reveal a probability that a particular man is a biological parent.

Whereas exclusion of a man falsely accused for paternity remains a primary reason for paternity testing, statistical analysis has yielded valuable information from test results when exclusion has not been established. Although paternal inclusion cannot be proven conclusively, it is possible to provide a mathematical estimate of the likelihood of paternity which has been accepted by the legal profession.

With the advent of recombinant deoxyribonucleic acid (DNA) technology, it is possible to identify inherited DNA markers through the use of genetic probes. These DNA probes are now used to establish identity with certainty, just as digital fingerprints were used in the past. Segregation of restriction fragment length polymorphisms (RFLPs) provides the most sophisticated test available for paternity. This technique can provide irrefutable evidence of family relationships. This report describes the use of both multilocus (MLP) and single-locus probes (SLP) in DNA typing for paternity. The unique features, advantages and limitations of each are presented.

Traditional tests for paternity take advantage of the multiple polymorphic genetic marker systems found in both red blood cell and lymphocyte (HLA) antigen systems. When both red blood cell and HLA typing are used, the mean exclusion capability, i.e. probability of excluding a falsely accused man in a paternity suit, is 91–97% [2]. For a man who is not excluded, such tests serve only for a statistical estimate of paternal inclusion.

The well-known complementarity of the two strands of alternating deoxyribose and phosphate molecules comprising the double helix of DNA is the basis for the high degree of specificity shown by DNA strands hybridizing with one another. This makes it possible to detect a specific sequence despite a great excess of unrelated sequences. A specific sequence may be detected even among the three billion bases that comprise the human genome. This is accomplished through the use of restriction endonucleases that cut DNA at specific sites to yield pieces or fragments of specific sizes. Thus, these enzymes slice DNA in a sequence-specific manner producing a large number of fragments of various sizes which are separated by electrophoresis in agarose gel. Thereafter, the DNA fragments are transferred to a Nylon membrane according to the method of Southern [3]. This membrane containing the transferred DNA fragments is then incubated with a labeled nucleic acid probe which hybridizes only with complementary sequences and will reveal their location. As the DNA is the same in all cells of the body, nucleated cells from any tissue of a person may be used. It is even possible to recover DNA for typing from tissue samples that are thousands of years old [4]. Extremely small amounts of DNA may be amplified by use of the recently developed polymerase chain reaction (PCR) [5–9]. Wyman and White [10] found that in using a particular probe there was variation in DNA fragment length derived from persons who were unrelated. They had discovered a hypervariable region (HVR). HVRs are also known as minisatellites or variable number of tandem repeats (VNTR). They are comprised of core tandem repeat sequences. Hypervariability is a consequence of alterations in the number of repeat units, as shown in figure 1. Hundreds to thousands of these loci are distributed throughout the genome of each animal. Since the discovery of HVRs, other such RFLPs have been found [11–16]. Whether or not a specific restriction endonuclease cleavage site is present can lead to the production of RFLPs, as two possible sizes of fragments (alleles) at that locus may be produced with that enzyme [17, 18]. Alternatively, DNA polymorphisms may be a consequence of the number of times a brief sequence is repeated between

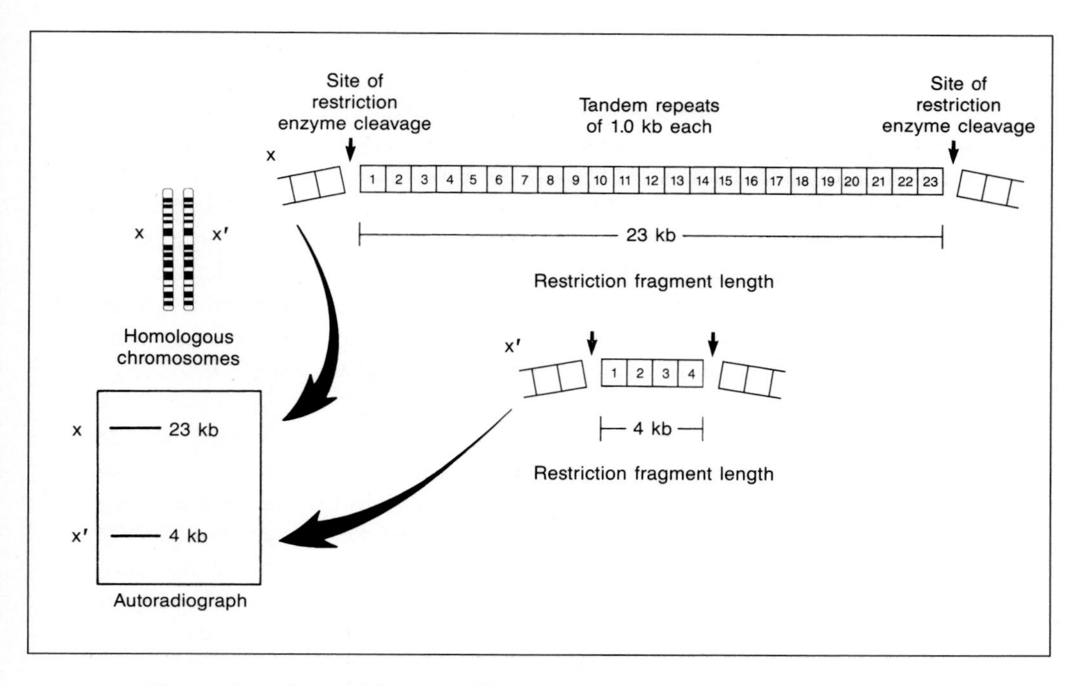

Fig. 1. Alleles (X and X′) at a specific locus on a homologous pair of chromosomes. Allele X consists of 23 tandem repeats and allele X′ consists of 4 tandem repeats. The restriction enzyme cleaves the DNA at fixed sites in the flanking regions of the alleles to generate 23-kb and 4-kb fragments.

sites cleaved by a particular restriction enzyme. The different numbers of tandem repeats (VNTRs) of core DNA sequences at a particular locus provide a distribution of fragment lengths. Estimation of the sizes of these fragment lengths identified in each person signifies that individual's genotype for a specific probe.

A tandem repeat refers to the repetitious end-to-end duplication of a core DNA sequence at a particular locus. Minisatellite refers to DNA regions comprised of tandem repeats of DNA short sequences. RFLP bands demonstrated with probes to minisatellite HVRs are inherited in a stable manner and segregate according to a Mendelian pattern [15]. Allele frequencies can be derived from population data on fragment sizes. Thus, HVR probes have tremendous potential in both forensic science and parentage testing [19–22].

Multilocus Probes

DNA typing makes use of two types of repeat regions. VNTR families expressing HVRs were studied first. Selected probes identify multiple related sequences distributed throughout each person's genome. These are termed multilocus probes (MLPs). They may reveal as many as 20 separate alleles. Because of this multiplicity of alleles, there is only a remote possibility that two unrelated persons would share the same pattern, i.e. about 1 in 30 billion. There is, however, a problem in deciphering the miltibanded arrangement of minisatellite RFLPs as it is difficult to ascertain which bands are allelic. Mutation rates of minisatellite HVRs remain to be demonstrated, but are recognized occasionally [23, 24].

DNA assays developed to solve cases of disputed parentage include multilocus minisatellite probes developed by Jeffreys et al. [15, 19, 25] and others [26–28] and single-locus minisatellite or VNTR probes pioneered by Nakamura et al. [29] and Wong et al. [24]. The recently developed PCR for DNA amplification may be applied to minisatellites [8, 30], microsatellites (simple dinucleotide repeats) [31–33], HLA loci [34, 35] and mt (DNA) [36]. This novel and innovative technology offers the possibility of analysis at the individual cell level [7, 37].

Jeffreys et al. [25] reviewed 1,702 cases of disputed paternity analyzed with the 33.6 and 33.15 probes. They claim that the multilocus DNA fingerprint probes [15, 19] provide the greatest individual specificity of all marker systems. These probes are comprised of a minisatellite found on chromosomes lcen-q24 and 7q31.3-qter, respectively [38]. The hybridization of these probes to multiple variable minisatellite fragments in the DNA of humans yields a DNA fingerprint that is individual-specific [15, 19]. Probes 33.6 and 33.15 showed little or no overlap in hypervariable loci in families of two large sibships tested. Most DNA fingerprints identified by each probe assorted independently in offspring [39]. Similar results in other studies confirm these observations. Whereas numerous studies confirm the analytical power of DNA fingerprints, the large sample analyses of Jeffreys et al. [25] have facilitated understanding of such other matters of concern in parentage testing as: (1) definition of the level of independence of DNA fingerprints identified with probes 33.6 and 33.15, (2) verification of the concept of band independence in DNA fingerprints, (3) solving the question of individual variation in number of bands and invariant human subpopulations, and (4) minisatellite mutation.

Jeffreys' minisatellite system with probes 33.15 and 33.6 identify alleles at multiple autosomal loci. Profiles of good quality using 33.15 and 33.6 probes can be produced for cats, dogs, fish and birds. The MLP of α-globin 3′HVR can identify DNA fragments in horses, cows, and dogs as well as in deer, elk and moose.

The DNA fingerprinting method using the two multilocus DNA probes yields a genetically determined, individual-specific band pattern. The probes identify multiple hypervariable regions present in an individual's DNA and exclude all males other than the true father, with the exception of an identical twin.

In this method, treatment of 700 μl of peripheral blood with phenol-chloroform and precipitation with ethanol permits the isolation of DNA. 1–10 μg of DNA is treated with restriction endonuclease HinfI to yield DNA fragments that are separated according to size using agarose gel electrophoresis on two separate gels. Following electrophoresis, the DNA fragments are transferred to Nylon membranes by Southern blotting and the multilocus DNA probes 33.6 and 33.15, derived from human minisatellites and labeled with phosphorus-32, are incubated with the membranes to permit hybridization. DNA band patterns are then developed by autoradiography.

DNA banding patterns from mother, child and alleged father are compared. For probe 33.6, all bands in the 3- to 12-kb molecular weight range are scored. For probe 33.15, all bands between the 3- and 25-kb molecular weight range are scored. After ascertaining all bands shared in common between the mother and the child's DNA fingerprints, the remainder of the child's bands must have been contributed by the true biological father. It is then only necessary to determine whether or not the alleged father possesses the paternal bands in the child's pattern. As shown in the examples (fig. 2), the sharing of bands between the alleged father and the child permits the establishment of paternity, whereas an alleged father who does not share most of the child's paternal bands is excluded as the biological father. In addition, occasional bands present in neither the father nor the mother but appearing in the child's pattern may be expected as a consequence of hypervariable region mutations [15, 23, 39].

DNA banding patterns using each of the two multilocus probes are analyzed separately and serve to counterbalance one another. The chance that an unrelated male might share an equal number of bands with a child as his true biological father has been ascertained to be 0.25^n. The 0.25 is the random band-sharing frequency, and n represents the number of paternal bands which the child shares with the alleged father [15, 19, 41]. Also considered in the calculations is the probability of an unassigned band, present in neither the mother's nor the father's pattern, appearing as a consequence of mutation, based on previously determined mutation rates [15, 23, 39].

Markowicz et al. [40] compared use of traditional paternity testing, including 6 red blood cell antigen systems and HLA phenotyping, with DNA fingerprinting analysis using two multilocus DNA probes. They analyzed 28 mother-child-alleged father trios. DNA fingerprinting confirmed the results obtained with red blood cell and HLA typing in 27 of the 28 cases.

The study is especially important, as it reveals that the combined use of red cell antigens and HLA was unable to exclude paternity in 19 of the 28 cases. Probabilities of paternities in these subjects varied from 94.0 to 99.9%, assuming a 0.5 prior probability. In 4 of the 19 cases, red cell isoenzyme and serum protein marker assays were also necessary to attain a probability of paternity value greater than or equal to 95%. In contrast to these traditional methods, DNA fingerprinting established paternity conclusively in all 19 of these cases. Using the two MLPs mentioned above, the number of bands which the child shared with the father varied from 12 to 29 with a mean value of 20. The statistical probability of a child sharing 20 bands with a man who is not his biological father is 0.25^{20} or about 1 in 1 trillion [15, 41].

When both red blood cell and HLA typing are combined, the average exclusion varies between 91 and 97% [2]. However, there is still a chance that a random male in the population may have the same phenotypic markers as the true biological father, making it impossible to positively prove paternity no matter how many genetic markers are investigated. By contrast, MLPs can identify simultaneously numerous highly variable genetic loci in human DNA [15, 19]. DNA fragments which these probes identify follow Mendelian patterns of inheritance [15, 19, 24, 39]. With the exception of identical twins, this technique permits the exclusion of males other than the true biological father in cases of disputed paternity.

In the 19 cases where traditional methods of RBC and HLA paternity testing could not exclude the alleged father, standard Bayesian techniques were used to calculate the probability of paternity [42]. Even though there was agreement between DNA fingerprinting analysis and these results, classical probability of paternity calculations are not applicable when using MLPs. Especially attractive is the elimination of the prior probability assumption with DNA fingerprinting and elimination of the need for sup-

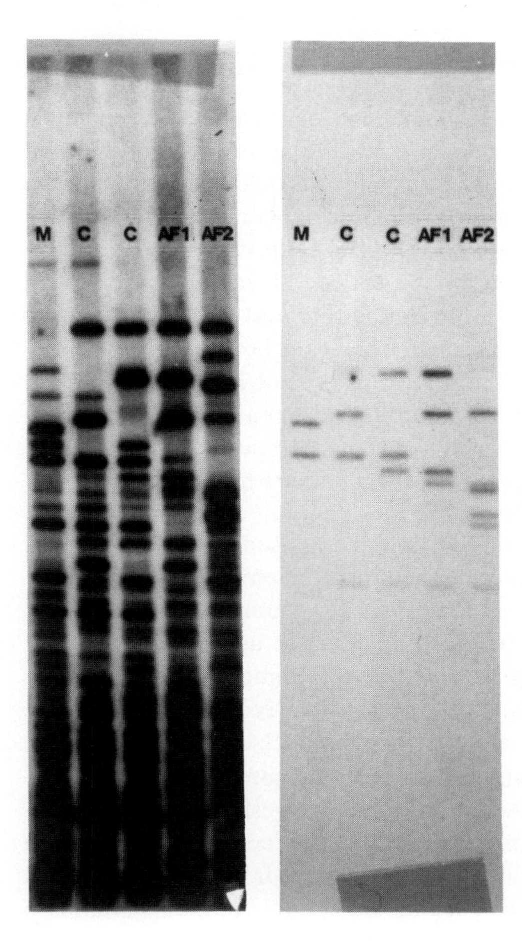

Fig. 2. A case of 2 alleged fathers known to be brothers. Conventional testing methods (red cell antigen, HLA, serum protein and enzymes) yielded high probabilities of paternity for both men (98.1% and 99.7% for AF1 and AF2, respectively). MLP analysis (33.15 shown here on the left) revealed that AF1 is the biological father of both children. Jeffreys' multilocus probes 33.15 and 33.6 were both used. The pattern on the right was obtained with a mixture of 4 single-locus probes used in combination to identify DNA fragments inherited from both parents.

plementary genetic testing, which is required when inclusionary statistics do not approach judicial standards.

Furthermore, DNA fingerprinting has a greater power of exclusion than 6 red cell antigens and HLA tests combined. Very small samples of nucleated cells yield sufficient DNA for testing. In addition, the DNA molecule is stable and can even be months or years old. Even a tiny amount of blood obtained from a heel stick in an infant is sufficient for testing when paternity of a newborn is in question. Figure 2 illustrates the power of MLPs in resolving difficult cases of disputed paternity.

Single-Locus Probes

Further probes which hybridize at only one locus have also been developed. Probes that identify a single locus of VNTRs have already been used successfully and show great promise for the future. They permit detection of a region of DNA repeats found in the genome only once and located at a unique site on a certain chromosome. Therefore, an individual can have only 2 alleles that SLPs will identify, as each cell of the body will have 2 copies of each chromosome, one from the mother and the other from the father [21]. When the lengths of related alleles on homologous chromosomes are the same, there will be only a single band in the DNA typing pattern. Therefore, the use of an SLP may yield either a single- or a double-band result for each individual. Single-locus markers such as the pYNH24 probe developed by White [3] may detect loci that are highly polymorphic, exceeding 30 alleles and 95% heterozygosity.

Akane et al. [43] point to the limitations imposed on RFLP markers by possessing only 2 alleles, rendering them insufficient to establish family pedigree analyses. DNA marker systems discovered during the past decade that have highly variable, multiallelic RFLPs arise as a consequence of differences in the number of tandem repeats of oligonucleotides at these loci. The minisatellite DNA sequences developed by Jeffreys et al. [15] (see above) have a highly complex and polymorphic DNA pattern. Thus, the DNA fingerprint is comprised of alleles from multiple loci that show partial homology with the probe. Akane et al. [43] suggested that the spontaneous mutation rates of these minisatellite loci are too great, approaching 5%, to permit their use in paternity testing. They favor the SLPs developed by Nakamura et al. [29] in which each of the variable number of tandem repeat (VNTR) markers may identify a single locus signifying a multiallelic RFLP. Although some Jeffreys probes have high mutation rates approaching 5%, others have lower rates which would make them more appropriate for paternity testing. They have properties similar to those of the Nakamura probes [23].

The different numbers of tandemly repeating sequence copies in VNTR loci make it possible for VNTR markers to identify multiple alleles. The ability of a single probe to detect one specific VNTR locus permits personal identity to be established at each locus, whereas the complex DNA fingerprint obtained with MLPs represents the sum of multiple RFLPs at hypervariable loci. The use of a multi-allelic RFLP yields allelic frequencies that are so low that a high probability of pater-

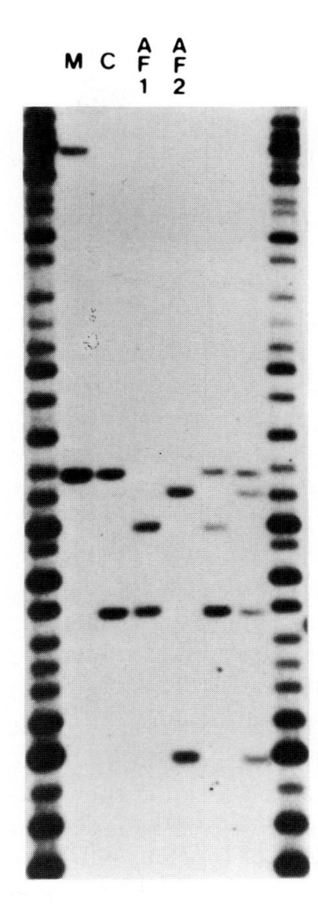

Fig. 3. An autoradiograph of a paternity case involving a mother (M), her child (C), and 2 alleged fathers (AF1 and AF2). DNA was digested with Pst I, size-separated and hybridized to a SLP which recognizes the highly polymorphic region D18S27. Size markers flank the samples. The paternal allele of the child (lower band) is present in AF1 but not AF2. Mixture lanes with C plus AF1 and C plus AF2 confirm this. These results show that AF2 is not the biological father, while AF1 is included as the possible biological father.

nity can be determined or the alleged father can be readily excluded. Figure 3 illustrates the versatility of SLPs in resolving cases of disputed paternity.

Comparison of Single-Locus Probes with Multi-Locus Probes

SLPs and MLPs both have advantages and disadvantages. For example, there is an advantage in that SLPs confine their recognition to two alleles for any single person, simplifying the assignment of a genotype.

SLPs also have the further advantage of strong signal strength. They are able to define a single VNTR and avoid the disadvantage experienced with MLPs, which identify many related yet degenerated sequences. This feature renders SLPs ten times more sensitive than MLPs. The development of single-locus VNTRs with greater than 90% allelic heterozygosity renders them capable of providing exclusionary and inclusionary data far greater than that generated when both HLA and isoenzyme typing are used for paternity testing. Data of this quality require two SLPs which define and identify VNTR loci that are assorting independently. In other words, results with at least two SLPs are needed. SLPs are gaining wide acceptance for paternity and forensic testing because of the unique characteristics cited above. Although it is laborious to prepare new SLPs, recent research has taken advantage of the fact that selected SLPs share certain sequences in common. Oligonucleotides of similar sequence have been synthesized and employed to identify novel RFLPs.

The principal strength of the MLP is the great number of fragments profiled, whereas the main power of the SLP resides in its ability to aid the analyst in determining allele frequencies in different populations. Whereas forensic analysis uses the SLPs almost exclusively, parentage testing is facilitated by both single-locus and multilocus approaches. The principal disadvantages of MLPs include band resolution and difficulties in interpretation of profiles and the absence of population allele frequency data bases.

The target DNA for both SLPs and MLPs is comprised of VNTR minisatellite alleles. There is a minimum requirement of 10–50 ng of DNA per digest for single-locus analysis and at least 0.5–1.0 μg of DNA per digest for the MLP. Whereas there are only one or two fragment bands per hybridization with SLPs, yielding clear band resolution, there are over 15 bands per hybridization for MLP, thereby diminishing band resolution. With single locus probes, 500-bp size bands may be interpreted, although probes may be directed to 0.5- to 24-kb fragments. MLPs permit interpretation of bands to a size of approximately 4 kb. Mixed samples are readily apparent when SLPs are used, as more than 2 bands could be detected. By contrast, interpretation of mixed samples is difficult with MLPs. With SLPs, probability calculations depend upon population allele frequency data. With MLPs, the probability calculations do not depend on population allele frequencies because loci are not defined. SLPs are generally species-specific, whereas MLPs frequently cross-react with a broad spectrum of species. SLPs employ blocking DNA and high-stringency hybridization washing conditions, which together lead to close matching in hybridization of probe and target DNA. By contrast, MLPs do not use blocking DNA. They require low-stringency hybridization washing conditions, which lead to a 30–40% probe-target DNA mismatch rate [3].

Current Status and Future Directions

Since 1989, estimates of the probability of paternity based on RFLP analysis have become increasingly conservative [44], after Lander [45] and Morris et al. [46] emphasized measurement variances. When SLPs are used individually, the PE varies from 70 to 90% [47]. By contrast, the use of several SLPs simultaneously yields a combined PE that may be in excess of 99%.

There is increasing emphasis on use of the PCR as interest focuses on retrospective DNA analysis of fixed tissue obtained from postmortem specimens. PCR methodology holds great potential in paternity testing with increasing use of improved techniques to evaluate amplified DNA products [7–9, 35, 48, 51]. Specifically, PCR may be used to amplify the variable number of tandem repeats (VNTR) loci together with electrophoretic identification of polymorphic bands in a procedure that does not require Southern blotting and hybridization with specific ^{32}P-labeled on nonradioactive probes [7–9]. The technique offers the advantages of both VNTR loci analysis and PCR, yielding high sensitivity, specificity and speed. Results obtained with this emerging technology must be compared with and confirmed by accepted methods of parentage testing that include HLA and red blood cell assays as well as RFLP Southern analysis.

A novel and innovative development is postmortem paternity testing in which PCR technology is used to amplify DNA from fixed tissue. In one case, analysis of the amplified DNA products led to derivation of an alleged father's genotype. In a fascinating report, Howard et al. [48] used HLA-DQA genotyping for exclusion. Although challenged in court, use of this technique has been supported by such authorities as Erlich et al. [51], who consider DQA typing reliable and reproducible. Strict control and careful evaluation must accompany the use of PCR amplification in paternity tests employing either fresh cells or fixed tissues. The high sensitivity of PCR constitutes both a strength and a weakness. The problem of contamination by minute quantities of DNA carryover, cross-contamination, nucleotide misincorporation, polymorphism of primer binding and variations in reaction temperatures may lead to both false-positive and false-negative results. All of these potential problems point once again to the necessity of confirming results obtained with this developing methodology by standard methods for paternity testing. Thus, the future holds bright prospects for DNA parentage testing as innovations in technology, exemplified by the development of PCR, open new vistas for scientific exploration and application in medical practice.

Acknowledgments

Figure 2 was provided through the courtesy of Cellmark Diagnostics of Germantown, Md., USA. Figure 3 was supplied through the kindness of Lifecodes Corporation of Valhalla, N.Y., USA. The authors are grateful to Ms. Berlinda Harris for exceptional secretarial assistance in preparing the manuscript for publication.

References

1 Silver H: Paternity testing. Crit Rev Clin Lab Sci 1989;27:391–408.
2 Hilderson Y, Henry MR: Scientific testing for paternity establishment. Paternity Establishment, ed 2. Washington, US Department of Health and Human Services, 1985, p 15.
3 Kirby LT: DNA Fingerprinting: An Introduction. New York, Stockton Press, 1990, pp 135–147.
4 Paabo S: Molecular cloning of ancient Egyptian mummy DNA. Nature 1985;314: 644.
5 Mullis K, Faloona F, Scharf S, Saiki R, Horn G, Erlich H: Specific enzymatic amplification of DNA in vitro: The polymerase chain reaction. Cold Spring Harbor Symp Quant Biol 1986;51:263–273.
6 Keller GH, Manak MM: DNA Probes. New York, Stockton Press, 1987, pp 215–231.
7 Jeffreys AJ, Wilson V, Neumann R, Kyte J: Amplification of human minisatellites by the polymerase chain reaction: Towards DNA fingerprinting of single cells. Nucleic Acids Res 1988;16:10953–10971.
8 Horn GT, Richards B, Klinger KW: Amplification of a highly polymorphic VNTR segment by the polymerase chain reaction. Nucleic Acids Res 1989;17:2140–2145.
9 Kasai K, Nakamura Y, White R: Amplification of a variable number of tandem repeats (VNTR) locus (pMCT 118) by the polymerase chain reaction (PCR) and its application to forensic science. J Forensic Sci 1990;35:1196–1200.
10 Wyman AR, White R: A highly polymorphic locus in human DNA. Proc Natl Acad Sci USA 1980;11:6754–6758.
11 Bell GI, Selby MJ, Ruther WJ: The highly polymorphic region near the human

insulin gene is composed of simple tandemly repeating sequences. Nature 1982;295: 31–35.
12 Proudfoot NJ, Gil A, Maniatis T: The structure of human zeta-globulin gene and a closely linked, nearly identical pseudogene. Cell 1982;31:553–563.
13 Promega Product Catalog. Madison, Promega Corporation, 1991.
14 Capon DJ, Chen EY, Levinson AD, Seeburg PH, Goeddel DV: Complete nucleotide sequences of the T24 human bladder carcinoma oncogene and its normal homologue. Nature 1983;302:33–35.
15 Jeffreys AJ, Wilson V, Thein SL: Hypervariable minisatellite regions in human DNA. Nature 1985;314:67–73.
16 Schumm JW, Knowlton R, Braman J, Barker D, Vorius G, Akots G, Brown V, Gravins C, Helms C, Hsiao K, Rediker K, Thurston J, Botstein D, Donis-Keller H: Detection of more than 500 single copy RFLPs by random screening. Cytogenet Cell Genet 1985;40:739.
17 Cooper DN, Schmidke J: DNA restriction fragment length polymorphisms and heterozygosity in the human genome. Hum Genet 1984;66:1.
18 Cooper DN, Youssoufian H. The CpG dinucleotide and human genetic disease. Hum Genet 1984;78:151.
19 Jeffreys AJ, Wilson V, Thein SL: Individual-specific 'fingerprints' of human DNA. Nature 1985;316:76–79.
20 Gill P, Jeffreys AJ, Werrett DJ: Forensic application of DNA 'fingerprints.' Nature 1985;318:577–579.
21 Jeffreys AJ, Brookfield JFY, Semeonoff R: Positive identification of an immigration test case using human DNA fingerprints. Nature 1985;317:818–819.
22 Odelberg SJ, Demers D, Westin E, Hossaini A: Establishing paternity using minisatellite DNA probes when the putative father is unavailable for testing. J Forensic Sci 1988;33:921–928.
23 Jeffreys AJ, Royle NJ, Wilson V, Wong Z: Spontaneous mutation rates to new length alleles at tandem repetitive hypervariable loci in human DNA. Nature 1988;332: 278–281.
24 Wong Z, Wilson V, Patel I, Povey S, Jeffreys AJ: Characterization of a panel of highly variable minisatellites cloned from human DNA. Ann Hum Genet 1987;51: 269–288.
25 Jeffreys AJ, Turner M, Debenham P: The efficiency of multilocus DNA fingerprint probes for individualization and establishment of family relationships, determined from extensive casework. Am J Hum Genet 1991;48:824–840.
26 Vassart G, Georges M, Monsieur R, Brocas H, Lequarre AS, Christophe D: A sequence in M13 phage detects hypervariable minisatellites in human and animal DNA. Science 1987;235:683–684.
27 Fowler SJ, Gill P, Werrett DJ, Higgs DR: Individual specific DNA fingerprints from a hypervariable region probe: Alpha-globin 3'HVR. Hum Genet 1988;79:142–146.
28 Georges M, Lequarre AS, Castelli M, Hanset R, Vassart G: DNA fingerprinting in domestic animals using four different minisatellite probes. Cytogenet Cell Genet 1988;47:127–131.
29 Nakamura Y, Leppert M, O'Connell P, et al: Variable number of tandem repeat (VNTR) markers for human gene mapping. Science 1987;235:1616–1622.

30 Jeffreys AJ, Neumann R, Wilson V: Repeat unit sequence variation in minisatellites: A novel source of DNA polymorphism for studying variation and mutation by single molecule analysis. Cell 1990;60:473–485.
31 Litt M, Luty JA: A hypervariable microsatellite revealed by in vitro amplification of a dinucleotide repeat within the cardiac muscle actin gene. Am J Hum Genet 1989; 44:397–401.
32 Tautz D: Hypervariability of simple sequences as a general source for polymorphic DNA markers. Nucleic Acid Res 1989;17:6463–6471.
33 Weber JL, May PE: Abundant class of human DNA polymorphisms which can be typed using the polymerase chain reaction. Am J Hum Genet 1989;44:388–396.
34 Saiki RK, Bugawan TL, Horn GT, Mullis KB, Erlich HA: Analysis of enzymatically amplified beta-globulin and HLA-DQalpha DNA with allele-specific oligonucleotide probes. Nature 1986;324:163–166.
35 Gyllensten UB, Erlich HA: Generation of single-stranded DNA by the polymerase chain reaction and its application to direct sequencing of the HLA-DQA locus. Proc Natl Acad Sci USA 1988;85:7652–7656.
36 Wrischnik LA, Higuchi RG, Stoneking M, Erlich HA, Arnheim N, Wilson AC: Length mutation in human mitochondrial DNA: Direct sequencing of enzymatically amplified DNA. Nucleic Acids Res 1987;15:529–542.
37 Li H, Gyllensten UB, Cui X, Saiki RK, Erlich HA, Arnheim N: Amplification and analysis of DNA sequences in single human sperm and diploid cells. Nature 1988; 335:414–417.
38 Jeffreys AJ, Macleod A, Neumann R, Povey S, Royle NJ: 'Major minisatellite loci' detected by minisatellite clones 33.5 and 33.15 correspond to the cognate loci D1S111 and D7S437. Genomics 1990;70:449–452.
39 Jeffreys AJ, Wilson V, Thein SL, Weatherall DJ, Ponder BAJ: DNA 'fingerprints' and segregation analysis of multiple markers in human pedigrees. Am J Hum Genet 1986;39:11–24.
40 Markowicz KR, Tonelli LA, Anderson MB, Green DJ, Herrin GL, Cotton RW, Gottschall JL, Garner DD: Use of deoxyribonucleic acid (DNA) fingerprints for identity determination: Comparison with traditional paternity testing methods. Part II. J Forensic Sci 1990;35:1270–1276.
41 Jeffreys AJ: Highly variable minisatellites and DNA fingerprints. Biochem Soc Trans 1987;15:309–317.
42 Walker RH: Probability in the analysis of paternity test results; in Silver H (ed): Paternity Testing. Washington, American Association of Blood Banks, 1978, pp 69–135.
43 Akane A, Matsubara K, Shiono H, et al: Paternity testing: Blood group systems and DNA analysis by variable number of tandem repeat markers. J Forensic Sci 1990;35: 1217–1225.
44 Walker RH, Crisan D: DNA technology: The fourth generation in parentage testing. Transfusion 1991;31:383–385.
45 Lander ES: DNA fingerprinting on trial. Nature 1989;339:501–505.
46 Morris JW, Sanda AI, Glassberg J: Biostatistical evaluation of evidence from continuous allele frequency distribution deoxyribonucleic acid (DNA) probes in reference to disputed paternity and identify. J Forensic Sci 1989;34:1311–1317.

47 Allen RW, Wallhermfechtel J, Miller WV: The application of restriction fragment length polymorphism mapping to parentage testing. Transfusion 1990;30:552–564.
48 Howard FLL, Collins CC, Heintz NH: Polymerase chain reaction and allele-specific oligonucleotides in paternity testing of the deceased. Transfusion 1991;31:441–442.
49 Said RK, Walsh PS, Levenson CH, Erlich HA: Genetic analysis of amplified DNA with immobilized sequence-specific oligonucleotide probes. Proc Natl Acad Sci 1988;85:7652–7656.
50 Helmuth R, Fildes N, Blake E, et al: HLA-DQa allele and genotype frequencies in various human populations, determined by using enzymatic amplification and oligonucleotide probes. Am J Hum Genet 1990;47:515–523.
51 Erlich HA, Higuchi R, Lichtenwalter K, Reynolds R, Sensabaugh G: Reliability of the HLA-DQa PCR-based oligonucleotide typing system (letter). J Forensic Sci 1990;35:1017–1019.

Robert E. Lewis, Jr., PhD, Department of Pathology,
University of Mississippi Medical Center, Jackson, MS 39216-4505 (USA)

Martin RD, Dixson AF, Wickings EJ (eds): Paternity in Primates:
Genetic Tests and Theories. Basel, Karger, 1992, pp 18–31

Biochemical Markers and Restriction Fragment Length Polymorphisms in Baboons: Their Power for Paternity Exclusion

John L. VandeBerg

Department of Genetics, Southwest Foundation for Biomedical Research,
San Antonio, Tex., USA

Identification of biochemical genetic markers in baboons (genus *Papio*) began shortly after current electrophoretic and staining methodologies for blood proteins were developed. Initially, the objectives were to compare genetic variation in baboons with that in humans and to establish markers for research on natural baboon populations [1–3]. Many markers have been developed subsequently during the course of research on the natural population structure of baboons [for example, see ref. 4].

The development and expansion of biomedical research programs in the 1960s and 1970s led to increased use of baboons as model organisms, and emphasis on genetics in the 1980s has stimulated the development of additional genetic markers as tools for research and for genetic management of captive breeding colonies [5]. Baboons are now used extensively as models for a wide variety of genetically mediated metabolic processes (e.g., metabolism of cholesterol [6], lipoproteins [7] and alcohol [8]) and diseases (e.g., atherosclerosis [9], alcoholic cirrhosis [10], epilepsy [11], lymphoma [12], hypertension [13] and hyaline membrane disease [14]). As powerful new genetic technologies have emerged, they have increasingly been applied to elucidate the mechanisms by which hereditary and environmental factors result in individual differences in these metabolic processes and in susceptibility to specific diseases.

Baboons are used extensively in research, as reflected by the results of a BIOSIS PREVIEWS search (via DIALOG) for publications concerning baboons, using the search terms [(BABOON?) or (PAPIO)]. The results revealed 213 publications in 1970, 448 in 1980, and 456 indexed to date for 1990 [15]. The number of these that dealt with genetic topics was determined using the search terms [(GENE or GENES) or (GENET?) or (BIOSIS concept code CC03502: general genetics and cytogenetics) or (BIOSIS concept code CC03506: animal genetics and cytogenetics)], combined with [(BABOON?) or (PAPIO)]. The results revealed 11 publications in 1970 (5.2% of the total publications that year), 39 (8.3% of the total) in 1980, and 54 (11.8% of the total) in 1990. These figures indicate a substantial increase in the role of genetics in research involving baboons.

Research on natural population structure and research on biomedical topics have a common need for methods to assess parentage. In the case of captive colonies, accurate records of paternity and maternity are required whenever research goals include analysis of inherited physiological characteristics and disease susceptibilities. They are also required for implementation of effective genetic management strategies to reduce inbreeding and to preserve genetic variation. Even in instances of single-male breeding groups, pedigree records require monitoring with genetic markers to ensure accuracy. Further, with multi-male breeding groups, paternity can be determined only by the use of markers to exclude all but one possible father.

In addition to their utilitarian value, genetic markers have also provided insight into the phylogenetic relationships among morphological subtypes of baboons [5, 16]. Genetic distance analyses clearly indicated that subtypes of common baboons merit classification as subspecies rather than as distinct species. According to the taxonomic rules of priority for nomenclature, the correct species name is *Papio hamadryas* Linnaeus, 1758. (Linnaeus actually designated baboons as *Papiones* in 1758, but the establishment of a binomial nomenclature led to the conventional use of *Papio* as the genus name. The authority for the name *Papio* is usually ascribed to Erxelben, 1777, but occasionally to Muller, 1773.) There are at least 5 subspecies: sacred baboons (*P. h. hamadryas* Linnaeus, 1758), yellow baboons (*P. h. cynocephalus* Linnaeus, 1766), chacma baboons (*P. h. ursinus* Kerr, 1792), red baboons (*P. h. papio* Desmarest, 1820), and olive baboons (*P. h. anubis* Lesson, 1827).

This communication describes the genetic markers that have been developed in baboons and their use for monitoring the paternity and maternity records of a large captive breeding colony. Based on the results, some theoretical considerations of the power and the limitations of these markers are discussed.

Materials and Methods

Most of the data presented in this paper were derived from baboons maintained at the Southwest Foundation for Biomedical Research (SFBR). The baboon population currently maintained at SFBR includes approximately 2,000 *P. h. anubis,* 100 *P. h. cynocephalus,* 200 *P. h. hamadryas,* 70 *P. h. papio,* 5 *P. h. ursinus,* and 900 hybrids, mostly between *P. h. anubis* and *P. h. cynocephalus.* Many of the breeding groups are maintained in outdoor cages, each of which contains 1 adult male and approximately 10–30 adult females, along with infants and juveniles. This caging arrangement permits the construction of pedigree records, which can be verified via the use of genetic markers.

Another group of approximately 50 adult males, 200 adult females, and 250 infants and juveniles, all *P. h. anubis,* is maintained in a 6-acre corral for the economic production of baboons for research that does not depend on pedigree structure. However, as methods of determining paternity become more powerful, it will be possible to establish accurate pedigrees for some of the progeny, thereby increasing their value for some research purposes.

To develop genetic markers for a variety of research projects, blood was withdrawn from the femoral veins of baboons immobilized by intramuscular injection of ketamine hydrochloride (10 mg/kg body weight). Sera were obtained from coagulated blood and placed in Tygon tubing, which was heat-sealed in segments containing approximately 60 μl each [17] and stored at –80 °C. This procedure ensures that each of the aliquots is subjected to only a single freeze-thaw cycle prior to analysis. Because the tubing is sealed, the procedure eliminates risk of desiccation or oxidation of samples over prolonged periods of time.

Erythrocytes in clots were also stored at –80 °C in Tygon tubing segments [18]. The preservation of erythrocytes within the structure of a clot allows superior retention of enzyme activity (unpublished data) by comparison with erythrocytes preserved for many years in ethylene glycol [19] or glycerol [18] solutions.

Protein variation was assessed by standard electrophoretic and isoelectric focusing methods using hydrolyzed starch, polyacrylamide, cellulose acetate, and agarose as gel media. Many of the electrophoretic systems have been described elsewhere [20] or in the references cited in table 1.

Genomic DNA was isolated from leukocytes obtained from buffy coats of uncoagulated blood by standard techniques, and digested with one or more restriction endonucleases. The resulting DNA fragments were size-fractionated by electrophoresis in agarose and transferred to Nylon membranes by Southern blotting. The baboon DNA samples were screened for restriction fragment length polymorphisms (RFLPs) using human cDNA and genomic clones as probes in standard hybridization reactions. Further details are provided in the references cited in table 2.

Biochemical markers and RFLPs have been used routinely at SFBR to verify the pedigree records derived from single-male breeding groups containing *P. h. anubis, P. h.*

cynocephalus, and their hybrids. When discrepant parent-offspring combinations were observed for any marker, the following sequence of steps was taken: (1) The samples from the recorded dam, sire, and offspring were retyped to check for error in the laboratory. (2) If the results were not due to laboratory error, fresh blood samples were collected from the offspring and the parents to check for mislabelling of any of the initial blood samples. (3) If the results from both sets of blood samples were consistent, then the case was considered a pedigree error, and the relevant link was broken in the computerized data base.

Results

Table 1 provides a summary of polymorphic biochemical markers that have been identified in baboons and can be typed using serum or red blood cells. This summary is a composite of results from baboons at SFBR and other populations examined by other investigators. The table includes markers whose genetic basis has been confirmed with family data and markers for which phenotypic and population data suggest a high likelihood of Mendelian inheritance. Two of the markers (G6PD and TBG) are X-linked in baboons (as they are in other mammals), and all of the others are autosomal. Some electrophoretic variations described in the literature were omitted from this table because the published data were too limited to inspire a high level of confidence in the genetic basis of those variations. In instances where the data for a single marker were derived at more than one laboratory, an attempt was made to standardize allelic designations for the same alleles observed at different laboratories. However, some inaccuracies might exist because the allelic isozymes observed at various laboratories were not actually cross-standardized experimentally, and some assumptions were made about the equivalence of markers based on the gene frequencies observed.

Table 2 provides a summary of RFLPs identified in baboons. All of these polymorphisms are inherited as autosomal codominant systems.

Table 3 summarizes the characteristics of 6 unlinked markers that have been used for monitoring the pedigrees of 1,128 family triads (sire, dam, offspring). Three of these markers, apoAIV, PLG, and MPI, were in Hardy-Weinberg proportions in the parents, whereas the other 3 were not. Of the 3 markers in Hardy-Weinberg proportions, only 1 (MPI) was typed in triads that had not been preselected on the basis of whether discrepancies had been detected at one or more of the other marker loci. Therefore, the data for MPI provide the most valid estimate of the actual number of

Table 1. Polymorphic biochemical markers in baboons

Protein or polypeptide	Abbreviation		Blood component	Alleles identified[b]					Reference
	protein	gene[a]		*P.h. anubis*	*P.h. cynocephalus*	*P.h. hamadryas*	*P.h. papio*	*P.h. ursinus*	
1 Adenine phosphoribosyl transferase	APRT	*APRT*	RBC	a,b	a,b	a	a	a	15, 16
2 Adenosine deaminase	ADA	*ADA*	RBC	a,b	a,b	a,b	b,c	a,b	4, 16
3 Adenylate kinase 1	AK-1	*AK1*	RBC	a,b	–	a,b	a	a	4, 21, 22
4 Albumin	ALB	*ALB*	serum	a,b	a	–	–	–	15
5 α_1-Antitrypsin (protease inhibitor)	PI	*PI*	serum	a	a	a,b,c	a,b	a,c	4,15,22
6 Apolipoprotein(a)	apo(a)	*LPA*	serum	a,b,c,d,e, g,h,i,n	a,b,c,d, e,f,k,n	a,b,c,f, g,h,i,n	b,f,g,l,n	c,d,n	23, 24
7 Apolipoprotein A-IV	apoAIV	*APOA4*	serum	a,c,d,e,g,h,i	c,d,e,f,g,i,j	a,e,g,i	a	b,c,d,e,g	15, 25
8 Carbonic anhydrase I	CA-I	*CA1*	RBC	a,b	a,b	a,b	b	b,c	4, 16
9 Carbonic anhydrase II	CA-II	*CA2*	RBC	a,c	b,c	c	c	c	16
10 Catalase	CAT	*CAT*	RBC	a,b	a,b	a,b	b	b	15
11 Complement component 3	C3	*C3*	serum	a,b,c,d,e,f, g,h,i,j,k	b,d,e,f,g,h, i,j,k	f,h,i	i	e	16
12 Complement component 6	C6	*C6*	serum	a,b	a,b	–	–	–	15
13 Complement component 7	C7	*C7*	serum	c	a,b,c	–	–	–	15
14 Complement component 9	C9	*C9*	serum	a,b	a,b	–	–	–	15
15 Glucose-6-phosphate dehydrogenase	G6PD	*G6PD*	RBC	b	b	b	a,b	b	5,15, 26

16 Glucose-phosphate isomerase	GPI	*GPI*	RBC	a	a	a	b	a	16
17 Group specific component	GC	*GC*	serum	a,b,c	a,b,c	a,c	–	–	15, 27
18 Malate dehydrogenase 1	MDH-1	*MDH1*	RBC	a	a	a	a	a,b	15, 22
19 Mannose-phosphate isomerase	MPI	*MPI*	RBC	a,c,d	c,d	b,c,d	c	c	15,28
20 Nucleoside phosphorylase	NP	*NP*	RBC	a,b	a,b	–	–	–	15,29
21 Peptidase B	PEP-B	*PEPB*	RBC	b	b	b	b	a,b	16
22 Phosphoglucomutase 1	PGM-1	*PGM1*	RBC	a,b	b	b	b	b	4,15
23 Phosphoglucomutase 2	PGM-2	*PGM2*	RBC	a,b	b	a,b	b	a,b	4, 15, 22
24 Phosphogluconate dehydrogenase	6PGD	*PGD*	RBC	a,b	b	b,c	b	a,b	16, 22
25 Phosphoglycerate kinase 1	PGK-1	*PGK1*	RBC	b	b	b	b	a,b	5, 15, 22
26 Plasminogen	PLG	*PLG*	serum	a,b,c,d	b,c,d	–	–	–	15, 30
27 Prealbumin esterase	PAES	–	serum	b	–	a,b	–	–	4
28 Prealbumin	PALB	*PALB*	serum	a,b	–	a,b,c	–	–	4
29 Properdin factor B	BF	*BF*	serum	a,c	–	a,b,c	–	–	15, 31
30 Thyroxin-binding globulin	TBG	*TBG*	serum	a,b	–	a,b	–	–	4, 15, 32
31 Transferrin	TF	*TF*	serum	b,c,f,g	b,c,d,f,g	a,b,c,e	c	c,d	15

[a] Gene designations are as per reference 39.

[b] All alleles except those for apo(a) are designated in alphabetical order of decreasing negative charge on the protein surface; apo(a) alleles are designated in alphabetical order of decreasing molecular size.

errors in the pedigree records. To estimate the total number of errors in the 1,128 triads, the power of the MPI polymorphism to detect discrepancies was calculated using the formula presented in Chakravarti and Li [40]. Under the assumption that one true parent is known, this calculation determines the probability that a randomly chosen animal would be detected as discrepant if assigned as the other parent. Given the allele frequencies observed in the SFBR pedigrees, 7% of incorrectly assigned parents will be detected using MPI genotypes alone. Four errors were detected by MPI typings of 1,040 triads. Therefore, the best estimate of the number of errors that actually exist in the 1,128 pedigree records is $(4/0.07) \times (1,128/1,040) = 62$.

Although 3 of the markers were not in Hardy-Weinberg proportions, the probability of detecting an existing error was calculated for each of the 6 markers on the basis of the observed gene frequencies and under the assumption that they were in Hardy-Weinberg proportions. Because markers not in Hardy-Weinberg proportions can have higher or lower power to

Table 2. RFLPs in baboons

Protein or polypeptide	Abbreviation		Restriction enzymes that detect polymorphisms[a]		Reference
	protein	gene[b]	*P.h. anubis*	*P.h. cynocephalus*	
Antithrombin III	AT-III	*AT3*		4	33, 34
Apolipoprotein A-I	apoAI	*APOA1*	2	2	35
Apolipoprotein B	apoB	*APOB*	4	4[c]	33, 34, 36
Arginase	ARG-1	*ARG1*		1	33, 34
Hexosaminidase B	HEX-B	*HEXB*		4	33, 34
Lecithin:cholesterol acyltransferase	LCAT	*LCAT*	1	1	37
Low-density lipoprotein receptor	LDLR	*LDLR*	1	1	38
Renin	REN	*REN*		3[c]	33, 34

[a] All polymorphisms are 2-allele systems, except for those indicated by footnote c.
[b] Genes are designated following reference 39.
[c] Some of these restriction enzymes detect three alleles, see reference 42.

detect errors than those in Hardy-Weinberg proportions, depending on the actual distribution of genotypes, the values (ranging from 0.07 to 0.30, table 3) are considered to be reasonable estimates of the power of these markers to detect errors in the existing pedigrees. The combined power of all 6 markers to detect pedigree errors was estimated to be 0.74. Because the population is randomly mated with respect to all markers, the genotypic frequencies of existing progeny, which are being used as parents to produce the next generation, are expected to be in Hardy-Weinberg proportions, and indeed they are in Hardy-Weinberg proportions for all of the markers except CA-I. Therefore, the calculated probabilities should be an even better estimate of the actual power of these markers to detect pedigree errors in subsequent generations.

Between 677 and 1,127 triads were typed for any one marker, revealing 22 pedigree errors, including 18 mistaken paternities and 4 mistaken maternities. One marker, C3, which was typed for 1,127 of the triads, revealed 15 of the 22 errors detected by all markers combined.

Table 3. Characteristics of unlinked markers that were used for pedigree monitoring in a mixed colony of *P.h. anubis* and *P.h. cynocephalus*

	Locus					
	APOA4	*PLG*	*C3*	*CA1*	*APRT*	*MPI*
Number of alleles	8	4	11	2	2	3
Frequency of most common allele among parents	0.51	0.53	0.71	0.84	0.84	0.93
Average heterozygosity of parents	0.58	0.51	0.45	0.32	0.23	0.13
P(E)	0.30	0.27	0.30	0.12	0.12	0.07
1-P(E)	0.70	0.73	0.70	0.88	0.88	0.93
Cumulative probability (across the columns) of not detecting an existing error	0.70	0.51	0.36	0.31	0.28	0.26
Cumulative probability (across the columns) of detecting an existing error	0.30	0.49	0.64	0.69	0.72	0.74

P(E) = Probability of detecting an existing error.

Discussion

The total of 39 loci with defined genetic markers provides substantial power for parentage determination in captive colonies and wild populations of baboons. As a consequence of ongoing biomedical research programs, including an initiative to develop a linkage map of the baboon genome [29, 30], the number of polymorphic markers can be expected to increase substantially in future years. However, the total number of polymorphic markers is deceptive because (1) each subspecies is polymorphic for only a subset of the markers, (2) any population of a subspecies is likely to have less variation than the subspecies as whole, and (3) some of the markers have low average heterozygosities, so they are useful only in rare instances. Despite these limitations, however, practical applications and theoretical considerations have demonstrated the utility of these markers.

The best estimate of existing error rate, derived from the MPI typings, is 62 errors in 1,128 triads (5.5%). Because 18 of the 22 detected errors were in the records of paternity, it is estimated that 51 (4.5%) of the 1,128 pedigrees have mistaken fathers and 11 (1.0%) have mistaken mothers. This error rate would reduce the power of genetic analyses (e.g. complex segregation analysis and linkage analysis) and of genetic management strategies. However, having detected 22 errors, we anticipate that the number of remaining errors in the 1,128 pedigrees is 40 (3.6%). If all triads are tested for all 6 markers, we expect that only 16 errors (1.4%) will remain (62×0.26, from table 3).

In the instances presented in table 3, the 3 most polymorphic markers (apoAIV, PLG, C3) are capable of detecting an estimated 64% (1–0.36) of existing pedigree errors in this population. If 3 more loci with discriminating power equivalent to those 3 were used with them, 87% ($1-0.36^2$) of pedigree errors could be detected. If 3 additional loci with equivalent power were added, 95% ($1-0.36^3$) of existing errors would be detected.

Although all 6 of the markers used for these analyses were biochemical markers, the implications of the results are equally pertinent to RFLPs. In theory, it would be possible to identify a large number of RFLPs exhibiting 2 alleles with nearly equal gene frequencies. The power of such a polymorphism (where $p = 0.5$ and $q = 0.5$) to detect pedigree errors in a population in Hardy-Weinberg proportions is 0.19 [calculated according to ref. 40]; 81.3% of pedigree errors would escape detection by a single such RFLP. Five idealized RFLPs would have a power to detect pedigree errors ($1-0.813^5 = 0.65$) equal to that of apoAIV, PLG, and C3 combined (0.64, table 3). It would require 14 such idealized RFLPs to detect 95% (1–0.813^{14}) of existing pedigree errors. We are beginning to type the triads for the RFLPs listed in table 2. Although the number of triads typed for any RFLP is considerably lower than the number typed for each biochemical marker, several additional discrepant triads have been detected, confirming the utility of the RFLPs for this purpose.

The detected pedigree errors were not a consequence of mistakes in labeling samples, in interpreting marker data, or in entering marker data into the computerized data base. All such mistakes were detected by resampling of every family for which initial results were inconsistent with Mendelian expectations. Investigation of the pedigree errors, using colony records and genetic marker data on additional animals, as well as consideration of animal management practices, suggested several possible contributing factors. Some paternity errors may have been a consequence of a juvenile male impregnating a female before he was believed to be capable of doing so. Others may have been a consequence of mating of a female with a male in an adjacent cage, separated by a single panel of cyclone fence; such matings have been observed. Some maternity errors may have been a consequence of inadvertent exchange of identification numbers by maintenance staff as the permanent identification numbers were assigned, or of infant swapping between mothers that gave birth at about the same time. Infant swapping has been well documented with genetic markers in squirrel monkeys [20]. Another likely cause of error for mothers or fathers is mistakes that occur at the time the infant is born in recording the correct identification number of the two parents into the permanent colony records. The results of this investigation confirm the importance of paternity and maternity verification in captive colonies of nonhuman primates, even when they are produced in harems believed to have only a single male capable of breeding.

Quite different from the problem of verifying pedigrees that are believed to be correct on the basis of colony records is the problem of paternity exclusion in multi-male breeding groups. This latter problem is presented in an extreme form in the case of the SFBR population of *P. h. anubis* maintained in a 6-acre corral. We have conducted a simulation with a subset of the polymorphic markers to explore the power of paternity exclusion in this population [41]. The simulation was based on the population as it existed in 1982. During that year, 115 of the 252 breeding-age females each gave birth to an offspring, for which there were 47 possible fathers. The simulation used actual gene frequency data for six loci (*APRT,*

ADA, CA1, C3, MPI and *PGD*) and assumed gene frequencies at two additional sets of 6 hypothetical loci equal to the frequencies at the actual 6 loci. Given the particular demographic structure of this population and under the assumption of random mating, it was concluded that (1) an average of 39 possible fathers could be excluded for each offspring, leaving 8 possible fathers, and (2) all but 1 father could be excluded for as many as 33% of the offspring. It was also concluded that reducing the number of males of breeding age by 50% would increase the proportion of progeny with known fathers to 44%. However, this improvement in paternity assignment was considered to be insufficient to justify the loss of genetic variability that would result from reducing the effective population size.

The power of paternity exclusion will continue to increase as more biochemical genetic markers and RFLPs are identified in baboons. Also, the cost-effectiveness of these markers for pedigree verification or paternity exclusion will increase as additional markers with high levels of heterozygosity are discovered and used to replace markers with lower levels of heterozygosity.

At the present time, there are no established methods for typing hypervariable regions of baboon DNA. The development of these technologies will undoubtedly increase the power of paternity exclusion for baboons, but it will be important to evaluate the power and cost-effectiveness of these technologies by comparison with biochemical markers and RFLPs that are already well characterized in this species. In the many instances where the typing of single-locus DNA and biochemical markers is required for a specific research application (e.g. population genetics, disease associations, gene mapping), the typing of hypervariable DNA markers might add only marginally to the parentage information available from the conventional markers.

Acknowledgments

I thank Dr. Candace Kammerer, Mary Jo Aivaliotis, Jim Bridges, and Mary Sparks for assistance in the pedigree monitoring process and in retrieving data from the computerized data base; Ruth Brooks for conducting the BIOSIS search; Dr. Jeffrey Rogers for helpful discussions throughout the preparation of this manuscript; and the many associates and assistants who contributed to acquisition of the data and their management in the computerized data base. Critical comments on the penultimate draft of the manuscript were provided by Drs. Candace Kammerer and Jean MacCluer. Acquisition of the data presented in this manuscript and preparation of the manuscript were supported by NIH grants HL28972, HL39890, HG00336, and NIH contract N015303.

References

1 Buettner-Janusch J: Hemoglobins and transferrins of baboons. Folia Primatol 1963; 1:73–87.
2 Barnicot NA, Jolly CJ, Huehns ER, et al: Red cell and serum protein variants in baboons; in Vagtborg H (ed): 1st International Symposium on the Baboon and Its Use as an Experimental Animal, San Antonio, 1963. Proceedings: The Baboon in Medical Research. Austin, University of Texas Press, 1965, vol 1, pp 323–338.
3 Kitchin FD, Barnicot NA, Jolly CJ: Variations in the group-specific (Gc) component and other blood proteins of baboons; in Vagtborg H (ed): 2nd International Symposium on the Baboon and Its Use as an Experimental Animal, San Antonio, 1965. Proceedings: The Baboon in Medical Research. Austin, University of Texas Press, 1967, vol 2, pp 637–657.
4 Shotake T, Nozawa K, Tanabe Y: Blood protein variations in baboons. I. Gene exchange and genetic distance between *Papio anubis, Papio hamadryas* and their hybrid. Jpn J Genet 1977;52:223–238.
5 VandeBerg JL, Cheng M-L: The genetics of baboons in biomedical research; in Else JG, Lee PC (eds): Primate Evolution. Selected Proceedings, 10th Congress, International Primatological Society, Nairobi, 1984, vol 1. Cambridge, Cambridge University Press, 1986, pp 317–327.
6 Dell RB, Mott GE, Jackson EM, et al: Whole body and tissue cholesterol turnover in the baboon. J Lipid Res 1985;26:327–337.
7 McGill HC Jr, Kushwaha R: Development and utilization of genetic dyslipoproteinemias in baboons; in Crepaldi G, Gotto AM, Manzato E, et al (eds): 8th International Symposium on Atherosclerosis, Rome, 1988; Atherosclerosis VIII. Amsterdam, Excerpta Medica, 1989, pp 145–148.
8 Holmes RS, Meyer J, VandeBerg JL: Baboon alcohol dehydrogenase isozymes: Purification and properties of liver Class I ADH. Moderate alcohol consumption reduces liver Class I and Class II ADH activities. Prog Clin Biol Res 1990;340:819–841.
9 McGill HC Jr, McMahan CA, Kruski AW, et al: Relationship of lipoprotein cholesterol concentrations to experimental atherosclerosis in baboons. Arteriosclerosis 1981;1:3–12.
10 Lieber CS, DeCarli LM, Rubin E: Sequential production of fatty liver, hepatitis, and cirrhosis in sub-human primates fed ethanol with adequate diets. Proc Natl Acad Sci USA 1975;72:437–441.
11 Ehlers CL, Killam EK: Circadian periodicity of brain activity and urinary excretion in the epileptic baboon. Am J Physiol 1980;239:R35–R41.
12 Crawford MH, Devor EJ, O'Rourke DH, et al: Genetics of primate lymphomas in a baboon *(Papio hamadryas)* colony of Sukhumi, USSR. Genetica 1987;73:145–157.
13 McGill HC Jr, Carey KD, Marinez YN: Cardiovascular responses to experimental hypertension in the baboon; in Dal Palu C, Ross R (eds): Hypertension and Atherosclerosis. Pathophysiology, Primary and Secondary Prevention. Amsterdam, Excerpta Medica, 1989, pp 87–97.
14 Coalson JJ, Kuehl TJ, Escobedo MB, et al: A baboon model of bronchopulmonary dysplasia. II. Pathologic features. Exp Mol Pathol 1982;36:335–350.
15 VandeBerg JL: Unpublished data.

16 Williams-Blangero S, VandeBerg JL, Blangero J, et al: Genetic differentiation between baboon subspecies: Relevance for biomedical research. Am J Primatol 1990;20:67–81.

17 Cheng M-L, Woodford SC, Hilburn JL, et al: A novel system for storage of sera frozen in small aliquots. J Biochem Biophys Methods 1986;13:47–51.

18 Cheng M-L, VandeBerg JL: Cryopreservation of erythrocytes in small aliquots for isozyme electrophoresis. Biochem Genet 1987;25:535–541.

19 VandeBerg JL, Johnston PG: A simple technique for long-term storage of erythrocytes for enzyme electrophoresis. Biochem Genet 1977;15:213–215.

20 VandeBerg JL, Aivaliotis MJ, Williams LE, et al: Biochemical genetic markers of squirrel monkeys and their use for pedigree validation. Biochem Genet 1990;28: 41–56.

21 Vergnes H, Cambefort Y, Gherardi M: Red cell and serum enzymes of Guinea baboon *(Papio papio)*. J Med Primatol 1978;7:8–18.

22 McDermid EM, Vos GH, Downing HJ: Blood groups, red cell enzymes and serum proteins of baboons and vervets. Folia Primatol 1973;19:312–326.

23 Rainwater DL, Manis GS, VandeBerg JL: Hereditary and dietary effects on apolipoprotein(a) isoforms and Lp(a) in baboons. J Lipid Res 1989;30:549–558.

24 Williams-Blangero S, Rainwater DL: Variation in Lp(a) levels and apo(a) isoform frequencies in five baboon subspecies. Hum Biol 1991;63:65–76.

25 Ferrell RE, Sepehrnia B, Kamboh MI, et al: Highly polymorphic apolipoprotein A-IV locus in baboon. J Lipid Res 1990;31:131–135.

26 Gourdin D, Vergnes H, Bouloux C, et al: Polymorphism of erythrocyte G-6-PD in the baboon. Am J Phys Anthropol 1972;37:281–288.

27 Dykes D, Crawford M, Polesky H: Gc subtypes identified by isoelectric focusing in baboons (*Papio hamadryas*). Hum Genet 1985;69:89–90.

28 VandeBerg JL, Aivaliotis MJ: Mannose-6-phosphate isomerase polymorphism in baboon erythrocytes. Biochem Genet 1990;28:495–501.

29 van Oorschot RAH, VandeBerg JL: Tight linkage between MPI and NP in baboons. Genomics 1991;9:783–785.

30 VandeBerg JL, Weitkamp L, Kammerer CM, et al: Linkage of plasminogen (PLG) and apolipoprotein(a) (LPA) in baboons. Genomics, in press.

31 Dykes DD, Crawford MH, Polesky HF: Genetic variants of properdin factor B (Bf) in *Papio hamadryas* baboons. Folia Primatol 1981;36:226–231.

32 Lockwood DH, Coppenhaver DH, Ferrell RE, et al: X-linked, polymorphic genetic variation of thyroxin-binding globulin (TBG) in baboons and screening of additional primates. Biochem Genet 1984;22:81–88.

33 Rogers J, Kidd KK: Population genetic structure in Tanzanian yellow baboons as measured by RFLPs in nuclear DNA. Am J Phys Anthropol 1989;78:291–292.

34 Rogers J: Personal communication (Southwest Foundation for Biomedical Research, San Antonio, Tex.).

35 Hixson JE, Borenstein S, Cox LA, et al: The baboon gene for apolipoprotein A-I: Characterization of a cDNA clone and identification of DNA polymorphism for genetic studies of cholesterol metabolism. Gene 1988;74:483–490.

36 Rogers J, Kidd JR, Murphy PD, et al: RFLPs for APOB in two subspecies of savanna baboons *(Papio hamadryas cynocephalus and Papio hamadryas anubis).*

Human Gene Mapping 9 (1987): Ninth International Workshop on Human Gene Mapping. Cytogenet Cell Genet 1987;46:683.
37 Hixson JE, Borenstein S, Cox LA: PvuII RFLP for the lecithin-cholesterol acyltransferase gene (LCAT) in baboons. Nucleic Acids Res 1990;18:384.
38 Hixson JE, Kammerer CM, Cox LA, et al: Identification of LDL receptor gene marker associated with altered levels of LDL cholesterol and apolipoprotein B in baboons. Arteriosclerosis 1989;9:829–835.
39 Human Gene Mapping 10 (1989): Tenth International Workshop on Human Gene Mapping. Cytogenet Cell Genet 1989;52:1–1148.
40 Chakravarti A, Li CC: The effect of linkage on paternity calculation; in Walker RH (ed): Inclusion Probabilities in Parentage Testing. Arlington, Virginia, American Association of Blood Banks, 1983, pp 411–422.
41 Dyke B, Williams-Blangero S, Dyer TD, et al: Use of isozymes in genetic management of nonhuman primate colonies. Prog Clin Biol Res 1990;344:563–574.
42 Rogers J: Nuclear DNA polymorphisms in hominoids and cercopithecoids: Applications to paternity testing; in Martin RD, Dixson AF, Wickings EJ (eds): Paternity in Primates: Genetic Tests and Theories. Basel, Karger, 1992, pp 82–95.

John L. VandeBerg, PhD, Department of Genetics, Southwest Foundation for Biomedical Research, PO Box 28147, San Antonio, TX 78228–0147 (USA)

Martin RD, Dixson AF, Wickings EJ (eds): Paternity in Primates: Genetic Tests and Theories. Basel, Karger, 1992, pp 32–52

A Comparison of the Success of Electrophoretic Methods and DNA Fingerprinting for Paternity Testing in Captive Groups of Rhesus Macaques

David Glenn Smith, Becky Rolfs, Joseph Lorenz

Department of Anthropology and California Primate Research Center, University of California, Davis, Calif., USA

The demand for experimental subjects suitable as animal models for human biomedical research [1, 2] and limitations on the importation of nonhuman primates to the USA [3] have led to rapid expansion of domestic nonhuman primate breeding programs during the last decade. The recent initiation of efforts to establish Specific Pathogen Free (SPF) colonies of rhesus macaques imposes serious constraints upon management practices. These constraints limit management flexibility in forming breeding groups, introducing new breeding stock into the groups and facilitating gene flow among breeding groups. They therefore necessitate a sound program for monitoring the genetic and genealogical structure of breeding groups. We have studied the genetic structure of several captive breeding groups at the California Regional Primate Research Center (CRPRC) since 1977 and provided recommendations designed to maximize genetic heterogeneity and minimize the opportunity for inbreeding in these groups. In this contribution, we describe our use of paternity exclusion analysis (PEA) to study the breeding history and genealogical structure of one such group, NC 7, and compare the efficiency of conventional methods for conducting PEA (i.e. electrophoretic analyses of blood proteins) with an approach using DNA fingerprinting.

NC 7 was established in 1976 with 3 adult male and 46 young adult female rhesus macaques *(Macaca mulatta),* all of which were of unknown ancestry, but which were presumed to be unrelated to one another. During the early fall of 1977, the 3 males were replaced with 4 unrelated adult males. Beginning in 1984, several infant male and female rhesus were introduced annually by foster-infant exchange between mothers, or cross-fostering [4], involving NC 7 and other breeding groups at the CRPRC. The oldest of these foster males introduced to NC 7 are presumed to have first become sexually mature in 1988, during which breeding season the offspring born during the spring of 1989 were conceived. Thus, all offspring born before 1989 in NC 7 are descended from 1 (or more) of the 7 founding males. We will hereafter refer to the descent of any particular animal in NC 7 with reference to its 'patrilines' (i.e. the founding males in its ancestry) and its 'matrilines' (i.e. the founding females in its ancestry), and its ancestry will be assessed from the proportionate contribution to its genome made by 1 or more of the 7 founding males and 1 or more of the 46 founding females.

Paternity Exclusion Analysis

Eight electrophoretically defined blood protein polymorphisms (table 1) were employed to determine the fathers of all offspring born during each year since 1976 by the method of paternity exclusion [5]. The electrophoretic methods used for this purpose have been reported elsewhere [6].

Table 1. Change in gene diversity (h_i) in group NC 7

$$h_i = 2n\left(1 - \sum_{i=1}^{m} p^2\right)/(2n-1) \qquad \overline{H} = \sum_{j=1}^{r} h_j / r_j$$

Electrophoretic system	Founders	1977–1981	1982–1987	1988–1990
Transferrin	0.745	0.671	0.740	0.757
Carbonic anhydrase II	0.297	0.245	0.331	0.300
Albumin	0.468	0.463	0.356	0.416
Prealbumin	0.077	0.058	0.034	0.071
6-Phosphogluconate dehydrogenase	0.389	0.348	0.450	0.364
Glucose phosphate isomerase	0.204	0.221	0.199	0.054
Diaphorase	0.460	0.490	0.477	0.438
Vitamin D binding protein	0.014	0.026	0.065	0.071
$\overline{H}$	0.332	0.315	0.331	0.309

Although a much larger number of genetic polymorphisms can be characterized using serological or electrophoretic techniques, these 8 have proven to be informative for PEA, are less expensive and easier to characterize than other markers, and are the only markers screened for *all* offspring born in NC 7 since 1976. All males that were at least 4 years of age and living in NC 7 when a given offspring was conceived were regarded as possible fathers of that offspring. Males below the age of 4 were regarded as sexually immature because our previous studies have shown that young rhesus males living in captive groups at the CRPRC rarely, if ever, father offspring before reaching this age [7].

1,702 exclusions were required to identify the fathers of 270 offspring that were born in NC 7 since 1976 and from which blood samples were

Table 2. Success of PEA in group NC 7 between 1977 and 1990 using eight electrophoretically defined polymorphisms

	Adult males	Unrelated patrilines	Offspring	Exclusions %	Paternity (and patrilineal affiliation) identified, %	Potentially inbred offspring	Actually inbred	Additional number possibly inbred
1977	3[1]	3[1]	24	83	67 (67)	0	–	–
1978	4	4	14	79	36 (36)	0	–	–
1979	4	4	25	77	44 (44)	0	–	–
1980	4	4	25	79	52 (52)	0	–	–
1981	4	4	23	86	56 (56)	1	0	0
1982	8	6	20	79	45 (50)	14	0	0
1983	9	6	26	91	52 (57)	20	2	0
1984	9	6	23	89	50 (60)	17	3	2
1985	6	4	14	93	71 (100)	9	3	0
1986	6	4	18	96	78 (100)	16	2	1
1987	6	4	8	88	50 (63)	8	2	3
1988	9	4	13	93	69 (77)	13	7	4
1989	14	9[2]	18	76	6 (17)	18	1	11
1990	18	9[2]	19	82	21 (21)	18	1	13
Total	30	12[2]	270	85[3]	50 (56)	134	21	34

[1] These 3 males were replaced with 4 unrelated males during the summer of 1977.
[2] Five of these patrilines represent newly sexually mature males introduced into group NC 7 from separate groups by cross fostering.
[3] Represents 1,439 of the 1,702 exclusions required.

obtained. 1,439 (or about 85%) of these exclusions could be successfully made using these eight genetic polymorphisms, resulting in the identification of the fathers of about half of the 270 offspring (table 2). The number (n) of adult males in NC 7 increased from 3 in 1977 to 18 in 1990. As n–1 exclusions are required to identify the father of any given offspring, the increase in the number of adult males has significantly reduced our success of paternity identification in recent years. The proportion of offspring whose fathers are identified declined more sharply after 1988 than did the number of successful exclusions. This is partly because, as the number of males increased, a larger proportion of them were closely related to (and therefore shared genetic markers in common with) at least one other breeding male in NC 7. However, the founding paternal ancestor of about 6% of all offspring could be identified even though its father could not be determined. Thus, there is a significant negative correlation between the proportion of offspring for which paternity could be identified ($r = -0.62$, $p < 0.01$) and the number of adult males in NC 7, but not for the proportion of all required exclusions that could be made ($r = -0.06$). The number of possible fathers for offspring born in NC 7 during 1989 and 1990 increased dramatically and the fathers of only 6% and 21% of the offspring born during those years, respectively, could be identified. When the data for these 2 years are excluded from the correlation, neither the success of exclusion nor paternity identification appears significantly reduced by the increasing number of adult males. In fact, both correlations ($r = 0.53$ and $r = 0.08$, respectively) were positive and the increasing number of adult males is statistically significantly correlated, at the 0.05 level of probability, with a *greater* success in making exclusions. This in part results from management policies (e.g. culling practices) designed to maintain the success of PEA. Nevertheless, the eight genetic markers employed provide for reasonably successful PEA when the number of adult males is fewer than 10 but are rather ineffective for identifying paternity when the number of adult males is greater.

Inbreeding

After 1981, increasing numbers of sexually mature females in NC 7 were closely related to adult males in the group and their own offspring were therefore potentially inbred. By 1986, virtually all offspring born in

NC 7 were potentially inbred (i.e. their mothers were related to at least 1 adult male in the group), as shown in table 2. As inbred offspring can be identified, so long as all males unrelated to them can be excluded from being their fathers, unsuccessful paternity identification does not necessarily preclude the identification of inbred offspring, although inbreeding coefficients cannot always be estimated. The first inbred offspring were born in 1983, when 10% of all potentially inbred offspring (and 8% of all offspring born) were in fact inbred. By 1988, between 54% and 85% of all offspring born were inbred. In addition, with each passing generation, some matrilines and patrilines are lost and the approximately randomly mating young males and females come to descend from a greater proportion of all of the few remaining founding matrilines and patrilines. Thus, without intervention, within two or three additional generations all adult males in NC 7 would be related to each other and, necessarily, to the mothers of all offspring born in NC 7 all of which, therefore, will be at least remotely inbred.

Distribution of Reproductive Success

PEA permits the estimation of the reproductive success (RS) of the adult males in NC 7. Rates of inbreeding and loss of genetic heterogeneity are greatest when variance in RS among adult males (or females) is high [8], because the effective population size is low under this condition. For this reason, it is desirable to estimate and monitor the RS of each male and founding lineage.

To enhance the success of PEA during some years, particularly before 1987, all rhesus macaques in NC 7 were typed with antisera to identify phenotypes for 19 polymorphic blood group systems [9]. The use of these serologically defined polymorphisms increased the proportion of offspring whose fathers could be identified to over 80% [8]. Most offspring whose fathers could not be identified could be shown to have been fathered by one of only 2 or 3 adult males. By assigning equal fractions of such offspring to all nonexcluded males, estimates of the number of offspring fathered by each male during each year were made, as given in table 3. This estimate for the 7 founding males (which sired about 133 offspring) ranged from 2 to 47.7, with a variance of 185.0 (or 148.9 taking only the 4 males that replaced the original set of 3 males during 1977). In sharp contrast, the variance in reproductive success of the 46 founding females in NC 7, which have produced between 0 and 12 offspring each (a total of 143 offspring) since 1977 was only 7.1 (data not shown). These estimates of variance in RS of adult males and adult females in NC 7 are slightly higher and lower, respectively, than values reported earlier [10], which were based on offspring born before 1985. This high variance in RS of adult males significantly reduces the effective population size of NC 7, increases the rate of random inbreeding and jeopardizes the survival of less reproductively successful patrilines.

Although the variance in RS generally increases in years when there are fewer adult males in NC 7, as illustrated in table 3 for variance in estimates of *relative* RS, the correlation is not quite statistically significant at the 0.05 level of probability ($r = -0.44$, $p < 0.06$). When the relative RS values of the sons and grandsons of the founding males are pooled with those of their founding patriarch, the variance in RS among the 7 patrilines varies erratically from year to year. The absence of adult males of patrilines A and C between 1978 and 1982 led to dramatic increases in the cumulative RS of two other patrilines (E and F). While the successful reproduction of a son of the patriarch of patriline C has since increased the representation of that patriline among offspring produced in NC 7 to near average levels, one of the original 3 founding patrilines has since become extinct.

In contrast to the yearly variance in RS among patrilines, the cumulative RS of the 7 patrilines has not substantially changed since about 1980, largely due to concerted colony management efforts to maximize homogeneity of RS (i.e. minimize variance in RS) among all founding patrilines by favoring, for retention as future breeding males, young males whose fathers and grandfathers experienced relatively low RS. For the same reason, the cumulative RS of the 7 patrilines is highly correlated ($r = 0.95$, $p < 0.0005$) with the lineage composition of the current adult population of NC 7, as shown in table 4. The variance in representation among the 6 surviving patrilines in the current adult population (0.013) is significantly greater ($F = 4.15$ with 5 and 14 degrees of freedom for the numerator and denominator, respectively, $p < 0.02$) than that among the 15 surviving matrilines (0.003), but this difference is far less, due to selective culling, than the difference in variance in RS between founding patriarchs and matriarchs themselves. The lower variance among patrilineal representation in males, compared to females, now living in NC 7 is not statistically

Table 3. Estimated RS of sexually mature males (i.e. ≥ 4 years of age) in group NC 7 between 1977 and 1990 during every year each male was living in the group and their patriline (in parentheses)[1]

Male ID No. (and patriline)	1977	1978	1979	1980	1981	1982	1983	1984	1985	1986	1987	1988	1989	1990	Total
7618(A)	13														13.0
17466(A)						2	3.8	2							7.8
7836(B)	2														2
7838(C)	9														9
17179(C)						5	6	4	1	3	3	1.0	1.6	1.1	25.7
17226(C)						0	1.5	1							2.5
21041(C)												0.8	0.6	0	1.4
21213(C)												0.3	1.1	0.3	1.7
11(D)		5	8.4	4.2	5.5	0.5	1	1							25.6
18257(D)							1	2	2.5	8					13.5
19045(D)									1.5	2	1	0.2	0.7	1.3	6.7
21079(D)												0.6	1.0	1.0	2.6
22444(D)														1.0	1.0
1243(E)		3	3	6.8	5										17.8
17700(E)						0	5.4	8.5	5.5	1.5	1.4	0	1.7	1.0	25.0
19036(E)									3.5	3.5					7.0
21165(E)												3.4	1.8	1.0	6.2
21806(E)													1.8	2.6	4.4
22387(E)														0.5	0.5
22605(E)														1.4	1.4

16888(F)		5	7.8	12.3	7.5	8.5	3.3	2.5	0	0	0.8				47.7
17502(F)						3	1	1.5							5.5
20127(F)											1	6.5	1.1	3.3	11.9
20197(F)											0.8	0.2	0.8	0.3	2.1
16898(G)		1	5.8	1.7	5.0	1	3	0.5							18.0
21802(X1)[2]													1.5	1.5	3.0
21814(X2)[2]													0.1	0.1	0.2
21872(X3)[2]													0.9	0.3	1.2
21876(X4)[2]													3.3	2.3	5.6
22330(X5)[2]														0	0
Totals	24	14	25	25	23	20	26	23	14	18	8	13	18	19	270.0
Variance among patrilines															
Yearly	0.036	0.014	0.007	0.024	0.002	0.040	0.006	0.012	0.063	0.041	0.011	0.024	0.007	0.010	
Cumulative	0.036	0.011	0.004	0.006	0.007	0.008	0.007	0.006	0.006	0.006	0.006	0.007	0.008	0.008	

[1] Equal portions of offspring whose fathers could not be identified were assigned to all nonexcluded sexually mature males living in the group.

[2] X1, X2, X3, X4 and X5 are five different males born in four different groups other than NC 7. Four of these males were introduced into NC 7 as newborns by foster infant exchange during the spring of 1984 and the fifth was similarly introduced into NC 7 during the spring of 1985. Although 21814 and 21872, both born in group NC 2, share no parents in common, they share one grandmother and one grandfather in common and thus are related by a kinship coefficient of 0.0625 (as full first cousins).

significant at the 0.05 level of probability ($F = 1.70$ with 5 degrees of freedom for both the numerator and denominator, $p < 0.3$); but this probably results from the focus of management upon manipulating patrilineal representation by managing the adult male membership of NC 7 rather than adult female membership. The rationale behind this strategy is that the RS of any single male is far more influential than that of any given female on the lineage composition of NC 7.

Table 4. Lineage composition[1] of current adult population (i.e. $>$ 3 years of age) of group NC 7 by sex

Patriline	Male	%	Female	%	Total, %	Matriline	Male	%	Female	%	Total, %
7618	0.875	0.132	1.125	0.069	0.087	16334	0.125	0.020	0.750	0.042	0.037
7838	0.750	0.113	1.083	0.067	0.080	16343	0	0.000	0.583	0.033	0.024
11	1.750	0.265	3.208	0.197	0.217	16348	0.250	0.039	0.708	0.040	0.040
1243	1.500	0.226	4.834	0.297	0.277	16349	0.500	0.078	1.000	0.056	0.062
16888	1.625	0.245	5.250	0.324	0.301	16352	0.500	0.078	2.250	0.127	0.114
16898	0.125	0.019	0.750	0.046	0.038	16354	0.250	0.039	1.500	0.085	0.072
						16356	0	0.000	0.375	0.021	0.015
Total	6.625		16.25			16361	1.500	0.236	3.750	0.210	0.218
						16363	0.125	0.020	1.375	0.078	0.062
						16365	0.500	0.078	1.750	0.099	0.093
						16586	0.750	0.118	0.250	0.014	0.041
						16587	0.250	0.039	0.250	0.014	0.021
						16588	0.625	0.098	0	0.000	0.026
						16595	0.250	0.039	0.584	0.033	0.035
						16601	0.750	0.118	2.625	0.148	0.140
						Total	6.375		17.750		
SD[2]		0.095		0.124	0.112			0.061		0.059	0.055

[1] Lineage composition was estimated as the proportion of all founding lineages represented by each animal above 3 years of age in group NC 7. For example, an animal whose mother is a founder and whose father was born in the group contributes 0.5 to its mother's matriline and 0.25 each to its father's patriline and matriline. Thus, the number of male and female animals included in these calculations is given by summing the numbers for male patriline and matriline composition and female patriline and matriline composition, respectively (i.e., 13 males and 34 females).

[2] Four males and 4 females over 3 years of age, all unrelated to each other and to all other animals in group NC 7, were introduced to NC 7 from other groups as foster infants and were omitted from this estimate. Founding patriline 7836(B) which is now extinct was omitted from those estimates, as were all 31 founding matrilines which have become extinct in NC 7 since 1977.

Loss of Gene and Lineage Diversity

It is noteworthy that a much greater number of founding matrilines (i.e. 31 of the original 46) than founding patrilines (only 1 of the original 7) no longer survive in NC 7. Nine of these 31 matriarchs died without producing any offspring in NC 7, a larger number produced only 3 or fewer offspring, which themselves produced no surviving offspring, and the remnants of the nonsurviving matrilines were removed intact to inhibit aggression directed by or against those particular matrilines. Thus, it is clear that the heterogeneity of lineage composition in closed groups of limited size can decline very rapidly. This process, together with the increase in the incidence of inbreeding, has already led to a perceptible reduction in genetic diversity of NC 7. Based on gene frequencies associated with the eight electrophoretically defined polymorphisms employed for the PEA (table 2), Nei's index of gene diversity [11] for offspring born between 1988 and 1990 is about 7% lower than that for the founders of NC 7, as shown in table 1.

To counteract the loss of genetic diversity, an infant cross-fostering program was initiated in 1983 to provide both male and female gene flow among all captive groups at the CRPRC [4]. Since that year, 13 infants born in other groups (8 males and 5 females) have been successfully fostered into NC 7, 11 of which (6 males and 5 females ranging from 1 to 8 years of age) still remain in NC 7 and some of which have already produced offspring. As there are increasing numbers of foster individuals among the adult males of NC 7, the loss of genetic variation and the incidence of inbreeding can be maintained at acceptable levels and genetic monitoring provided by PEA will be facilitated by the reduction in the number of related adult males. However, the relatively low success of PEA during the last few years has prevented us from determining whether cross-fostered males will breed as successfully in their adoptive groups as do natal males.

DNA Fingerprinting

In an effort to develop a more effective protocol for conducting PEA, we have been screening rhesus monkeys in NC 7 to identify DNA polymorphisms associated with multiple alleles. The level of heterozygosity for most, but not all [12], RFLPs is relatively low. The recent identification of regions in genomic DNA of mammals where variation occurs in the number of tandemly repeated short sequences (or minisatellites), so-called VNTR loci [13, 14], provide a promising focus for improving PEA. Jeffreys et al. [15] recently cloned such a sequence from the human myoglobin gene and employed it to probe enzymatically digested genomic DNA of randomly selected individuals. They found that the resulting 'genetic fingerprint', associated with tens of loci scattered throughout the genome [16], could uniquely identify an individual [17]. Moreover, probing the same genomic DNA digested with two different restriction enzymes generates two genetic fingerprints associated with loci which segregate independently of each other. Weiss et al. [18] and others [19–21] recently showed that these human probes reveal informative DNA fingerprints for several species of the genus *Macaca.*

We have used an oligonucleotide (the 22mer, TGG AGG AGG GCT GGA GGA GGG C; Molecular Biosystems, Inc.), corresponding to the core sequence of one of these clones (the 33.6 minisatellite), conjugated with the enzyme alkaline phosphatase [22] to generate 'fingerprints' for offspring recently born (during 1989 and 1990) in the macaque group NC 7, along with their mothers and all adult males. As there were 14 and 18 possible fathers of the offspring born in 1989 and 1990, respectively, with each of them consanguineously related, on average, to 3.3 ± 2.1 other adult males in the group, this exercise provides a rigorous test of the utility of fingerprinting for PEA in captive groups of rhesus monkeys. Nine of the 37 animals born during 1989 or 1990 (6 born in 1989 and 3 born in 1990) whose electrophoretic markers had been previously studied were no longer available for bleeding and were excluded from this part of this study.

Methods

Between 2 and 5 ml blood was collected from each of the 28 remaining animals, their mothers and all possible fathers into EDTA tubes. Buffy coats were removed and frozen for several months. Samples from the possible fathers were thawed and washed several times with RBC lysing solution [23]. The buffy coats were then suspended in 3 ml of TEN buffer (10 m*M* Tris-HCl/1 m*M* EDTA/10 m*M* NaCl, pH 8.0). 100 μl of proteinase K (20 mg/ml) and 100 μl of 20% SDS were added to the samples, which were incubated at 37 °C overnight. The DNA was extracted using a modification of a high-salt extraction method [23], collected on a sterile pasteur pipette, washed in 70% EtOH, air-dried, redissolved in TE buffer (10 m*M* Tris-HCl/1 m*M* EDTA, pH 8.0) then treated with 6 μl of RNase (500 μg/ml, Boehringer-Mannheim) for 1 h at 37 °C. The samples were then reprecipitated using 1/10 vol 3 *M* Na acetate and 2 vol 100% EtOH, washed in 70% EtOH, dried and redissolved in 1 ml of TE. The buffy coats of the mothers and infants were not washed with RBC lysing solution prior to the salt extraction procedure. After the DNA had been obtained and resuspended (as described above), each sample was extracted twice with phenol/chloroform to remove excess protein and once with chloroform to remove excess phenol. The samples were then treated as described above.

DNA yields and purity were quantified by measuring the absorbance of each sample at 260 and 280 nm on a spectrophotometer (Hitachi-Perkin-Elmer Model 139). Approximately 10 μg of DNA from each sample was digested overnight with 50 units of Hae III and Hinf I (separately) at 37 °C. The digests were precipitated with ethanol, washed, dried and redissolved in 12 μl TE; 4 μl of loading solution containing dye markers (xylene cyanol and bromophenol blue) was then added. The samples were loaded onto a 1% SeaKem (FMC) gel and run for 24 h at 40 V. The first and last lanes contained molecular weight standards ranging between 1 and 12 kb. The sample of each offspring was run adjacent to its mother's sample and those for the 18 possible fathers were run together on the same gel.

The DNA digests were transferred to Nylon membranes (Boehringer-Mannheim) by alkaline Southern blotting (0.4 *M* NaOH/0.6 *M* NaCl) for 3 h. The membrane was rinsed for 15 min in 2 × SSC and vacuum-dried at 120 °C overnight.

Each Nylon membrane was prehybridized for 30 min at 50 °C in 10 ml 5 × SSC, 0.5% BSA and 1% SDS, pH 7.0, in a Robbins hybridization incubator, then incubated for at least 15 min at 50 °C in 10 ml of the prehybridization buffer containing the 33.6 alkaline-phosphatase-conjugated oligonucleotide probe (Molecular Biosystems). The Nylon membrane was washed twice successively at 50 °C for 5 min in 1 × SSC, 1% SDS, then twice more in 1 × SSC, 1% Triton X-100, and finally washed twice successively at room temperature for 5 min in 1 × SSC. Bands were visualized colorimetrically by incubating the Nylon membrane (sealed in a plastic bag, then wrapped in aluminium foil) at 37 °C for at least 4 h in a Tris-HCl solution (10 ml 1.0 *M* Tris-HCl, pH 8.5, 2 ml 5 *M* NaCl, 1.01 g $MgCl_2 \cdot 6\ H_2O$ and 100 μl 0.1 *M* $ZnCl_2$, raised to 100 ml with dH_2O) containing 80 μl each of a solution (Gibco) containing NBT (75 mg/ml) and BCIP (50 mg/ml). Band color continued to intensify (without significant interference from background staining) as membranes incubated for several days. When the membranes were sufficiently developed the stain solution was replaced with dH_2O and xerox copies were produced for analysis.

Fingerprinting profiles for paternally inherited bands of each offspring were characterized by assigning approximate molecular weights to bands that were present in the offspring but absent in the mother. For this purpose, a ruler connecting the corresponding molecular weight standards in the first and last lanes of each Nylon membrane was employed, progressing from the largest to the smallest standard fragment lengths. The profile of approximate molecular weight estimates for all paternally inherited bands of each offspring was then compared with those, developed on a single membrane, of all 18 fathers flanked (in the first and last lanes) by the molecular weight standards. All adult males lacking a clearly resolved band identified as a 'paternal marker' for a particular offspring were excluded as fathers of that offspring. When all but a single male had been excluded as the father of a particular offspring, the remaining male was regarded as the father. These results were compared with the results of PEA based on the eight electrophoretically defined polymorphisms and the comparative efficiency of each method was evaluated.

Results

The electrophoretically defined phenotypes of the 14 possible fathers of offspring born during 1989 and the additional 4 males that are possible fathers of offspring born in 1990 are given in table 5. The number of males heterozygous or homozygous for relatively rare alleles at diallelic loci varies from 2 (for PA and Gc) to 11 (for 6-PGD), with a mean value of 5.3 ± 3.6. An inordinately high incidence of heterozygosity at the Dia and PGD loci limits the usefulness of these two polymorphisms for PEA. Six alleles

Table 5. Electrophoretically defined phenotypes of 18 adult males in group NC 7 for eight protein-coding loci

ID No.	Locus							
	Tf	PA	Al	CA II	Gc	GPI	PGD	Dia
17179	BH	2-2	2-2	2-1	2-2	1-1	AA	CD
17700	CG	2-2	2-2	1-1	2-2	1-1	AB	CC
19045	CG	2-2	1-1	1-1	2-2	2-1	AA	CD
20127	D′H	2-2	2-1	1-1	2-2	1-1	AB	CC
20192	CC	2-2	2-2	1-1	2-2	1-1	AA	CD
21041	BC	2-2	2-1	1-1	2-2	1-1	AA	DD
21079	CC	2-2	2-1	1-1	2-2	2-1	AB	CD
21165	GG	2-2	2-2	1-1	2-1	1-1	AB	CD
21213	GH	2-2	2-2	1-1	2-2	2-1	AB	DD
21802	CD′	2-2	1-1	1-1	2-2	2-1	AB	CD
21806	AG	2-2	2-2	1-1	2-2	1-1	AB	CD
21814	CC	2-2	2-1	1-1	2-1	2-1	AA	DD
21872	CC	2-2	2-2	2-2	2-2	1-1	AA	CD
21876	CG	2-1	2-1	2-1	2-2	1-1	AB	CD
22330[1]	DF	2-2	2-2	2-1	2-2	1-1	AA	CD
22387[1]	CC	2-2	2-2	1-1	2-2	1-1	AB	CD
22444[1]	CC	2-2	2-1	1-1	2-2	1-1	AB	CD
22605[1]	CG	2-1	2-2	1-1	2-2	1-1	AB	CD

[1] These males were not judged to be sexually mature until the fall of 1989 and are therefore regarded as possible fathers of offspring born in 1990, but not 1989.
Tf = Transferrin; PA = prealbumin; Al = albumin; CA II = carbonic anhydrase II; Gc = group-specific component (vitamin D binding protein); GPI = glucose phosphate isomerase (phosphohexose isomerase); PGD = 6-phosphogluconate-dehydrogenase; Dia = NADH diaphorase.

which segregate at the transferrin locus occur in the 14 males that are possible fathers of offspring born in 1989, while eight Tf alleles occur in the 18 fathers of offspring born in 1990. The number of males that carry at least one copy of the same transferrin allele varies from 1 (for Tf^{A}) to 12 (for TF^{C}, the most common Tf allele in rhesus monkeys in NC 7), with an average of 3.6 ± 3.9.

The success of PEA for the 28 offspring born during 1989 and 1990 for which genetic fingerprints were also available is illustrated in table 6. The Tf locus alone provided 65% of the total number of exclusions (i.e. 428) required to identify paternity of all 28 offspring, but identified the fathers of only two of these offspring. Three polymorphisms, for each of which several males carry rare alleles (PA, Gc and GPI) provided no exclusions which could not be made using one of the other polymorphisms, and the remaining 4 polymorphisms provided only an additional 72 of the remaining 150 exclusions not made using the Tf polymorphism. Thus, the electrophoretic markers provided for almost 82% of the required number of exclusions, but revealed the fathers of only 6 (21.4%) of the 28 offspring.

Genetic fingerprints of the 18 adult males that are possible fathers of offspring born in 1990, generated using the 33.6 minisatellite probe after digesting the target DNA with the Hinf I and the Hae III restriction

Table 6. Number of successful exclusions for 28 offspring born in 1989 and 1990 using eight electrophoretic markers

Electrophoretic marker	Number exclusions provided	Additional exclusions provided	Cumulative number of exclusions (fathers identified)	Percent of required exclusions (fathers identified)
Transferrin (Tf)	278	278	278 (2)	65.0 (7.4)
Albumin (Al)	100	34	312 (3)	72.9 (10.7)
NADH diaphorase (Dia)	61	24	336 (5)	78.5 (17.9)
Carbonic anhydrase II (CA II)	44	12	348 (5)	81.3 (17.9)
6-Phosphogluconate dehydrogenase (6-PGD)	27	2	350 (6)	81.8 (21.4)
Prealbumin (PA)	0	–	350 (6)	81.8 (21.4)
Vitamin D binding protein (nGc)	0	–	350 (6)	81.8 (21.4)
Glucose phosphate isomerase (GPI)	0	–	350 (6)	81.8 (21.4)

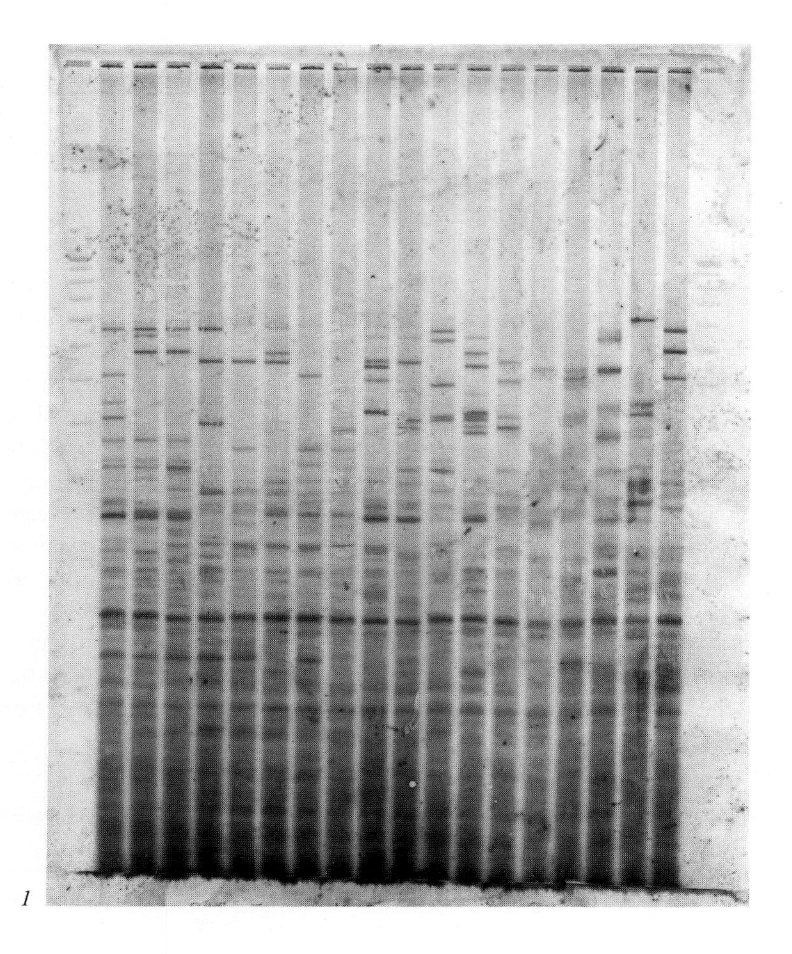

Fig. 1–3. Genetic fingerprint of 18 adult males that are possible fathers of infants born during 1990 in group NC 7. Fourteen of these males are all possible fathers of infants born in 1989. Samples from related males appear in adjacent lanes to illustrate clustering of fingerprint patterns. After cutting genomic DNA with the Hinf I *(1)*, Hae III *(2)* or Alu I *(3)* restriction enzyme, electrophoretically separated fragments (cathode at top) were hybridized to the 33.6 minisatellite myoglobin probe covalently attached to the enzyme alkaline phosphatase. Molecular weight markers in the first and last lanes identify fragment lengths of unit kb size from 12 kb (most cathodal marker) to 2 kb (most anodal marker).

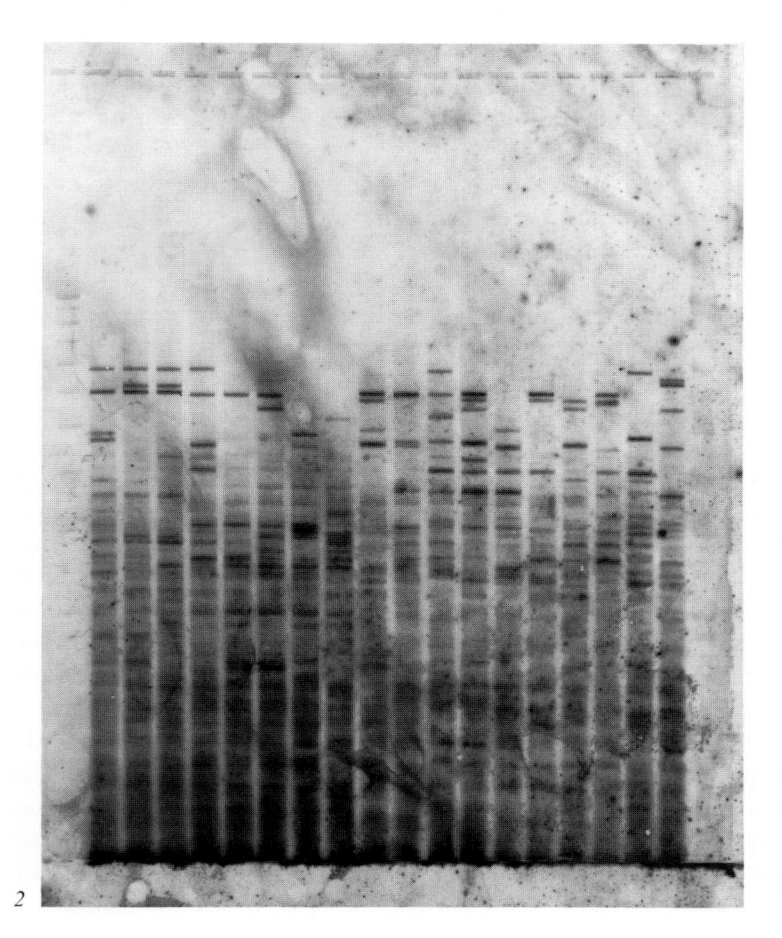

2

enzymes, are shown in figures 1 and 2, respectively. The most easily resolved of the 60–70 different bands detectable appeared in the high-molecular-weight region between 3 and 8 kb, but some fragments less than 2 kb in size were polymorphic. Differences in fragment lengths of all bands were better resolved in subsequent agarose gels of lower concentration (0.8%) and after a longer period of electrophoresis (30 h) at 40 V power. The number of males sharing each of these bands larger in size than 3 kb ranged between 1 and 14 for Hinf I (with an average value of 4.3 ± 2.9) and between 1 and 18 for Hae III (with an average value of 6.8 ± 4.8). The fingerprints of infants exhibited, on average, 5.5 ± 2.4 Hinf I bands and 4.7 ± 1.9 Hae III bands that were not present in their mothers and that were therefore regarded as inherited from their father. It was not always possible to identify unequivocally all 'paternal markers', especially those with low molecular weights or those represented by very faint bands, in the fathers ultimately assigned to individual offspring. Moreover, the intensity of some bands in some animals changed somewhat when exposed to different experimental conditions. However, in no case was a male designated the sire of an infant on the basis of electrophoretic markers but excluded as that sire on the basis of genetic fingerprinting when only clearly resolved bands were employed. Conversely, no sires assigned on the basis of genetic fingerprints were excluded on the basis of electrophoretic phenotypes. The 33.6 probe yielded 353 and 376 exclusions based on the Hinf I and Hae III fingerprints, respectively, giving a combined total of 412 exclusions, or 96.3% of those required. With this probe and the two restriction enzymes the fathers of 17 of the 28 infants were identified. When combined with the electrophoretic markers, the fathers of all but 2 of these 28 offspring could be identified and 426 of the total (428) exclusions required, representing 99.5%, could be made.

The estimates of RS of each patriline during 1989 and 1990 based on electrophoretic markers, which are given in table 3 (with all foster males combined and treated as one 'nonindigenous patriline'), were highly correlated with those based on both electrophoretic markers and DNA fingerprints ($r = 0.91$; $p \leqslant 0.02$). Nevertheless, the improved success of PEA provided valuable additional information. First, it is significant that, of the four of the original seven patrilines still represented by adult males in NC 7, none appears to be in danger of extinction. Each of the four patrilines produced between 3 and 10 offspring, with an average of 6.6 ± 2.8. Second, males introduced to NC 7 through cross-fostering produced at least 5 (possibly six) of the 28 offspring. Only 2 of these 5 cross-fostered males failed to produce any of the 28 offspring and 1 of these 2 was only 4 years old when the offspring born in 1990 were conceived. Thus, there is no indication that cross-fostered males will have a lower RS than natal males. Third, very young (i.e. 4-year-old) males can successfully compete with a much larger number of older males for access to mates. Although none of the 4 4-year-old males fathered offspring born in 1990, 3 of the 5 males that were 4 years old when the 1989 cohort of offspring were con-

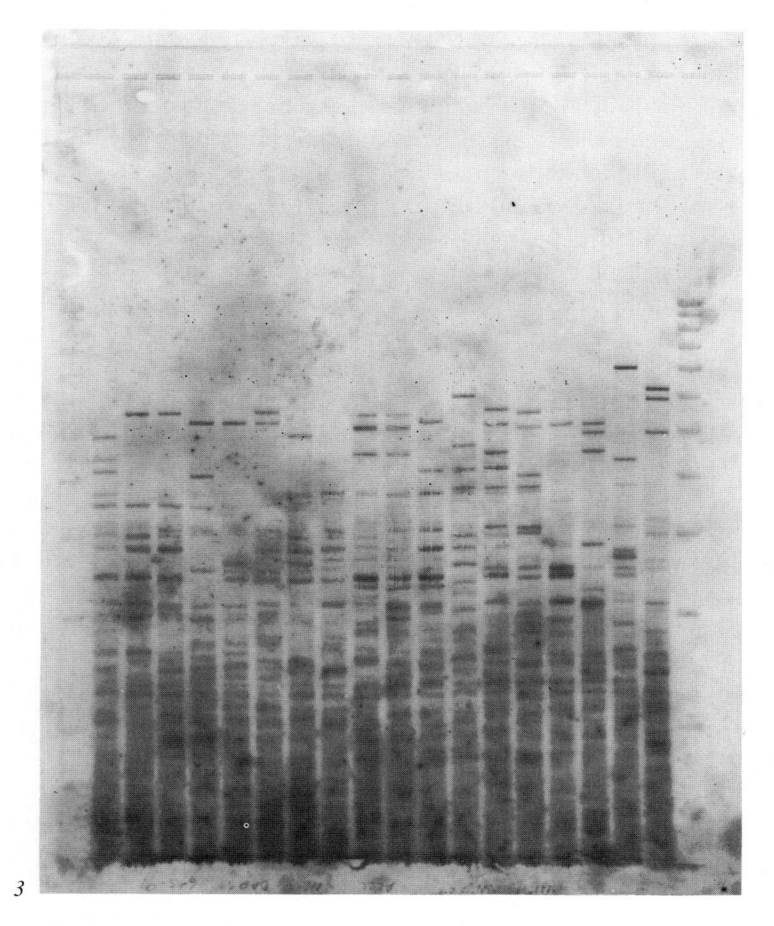

3

(For legend, see p. 46.)

ceived fathered at least one member of that cohort. Fourth, the inbreeding status of 19 of the 28 offspring could not be ascertained using electrophoretic markers alone. Genetic fingerprinting revealed that 10 of these 19 were in fact inbred and 8 were not inbred. The inbreeding status of 1 of the 19 – 1 of the two infants among the 28 studied whose father could not be determined (either of 2 different males could be its father) – remains in question.

It is clear that the methods of genetic fingerprinting employed in this study are a sufficient and necessary complement to the use of electrophoretic markers for PEA. More recent studies in our laboratory have shown that other restriction enzymes, such as Alu I, and other probes conjugated with alkaline phosphatase, such as the 33.15 myoglobin probe [15], provide additional markers useful for PEA (fig. 3). DNA fingerprints cannot replace electrophoretic methods for monitoring the demographic and genetic parameters of captive groups of rhesus macaques but should be considered a valuable, if not necessary, complement to these methods [25]. The low cost of characterizing phenotypes for electrophoretic markers, which provided 82% of all required exclusions, and the absence of ambiguity in their interpretation justify its use as a first resort for PEA. Unresolved cases can then be evaluated by genetic fingerprinting using DNA cut with the Hae III restriction enzyme, which yielded slightly higher success of PEA together with the electrophoretic markers, than when these markers are used together with the Hinf I fingerprints. The use, in this study, of a commercially available probe labeled with alkaline phosphatase has the advantage that it avoids the risk of exposure to radioactive materials and the costs of disposal of radioactive waste. The use of additional restriction enzymes and probes might be necessary in even more highly inbred captive groups or in free ranging groups for which a sufficient number of uniquely identifying paternal markers to provide for paternity inclusion analysis is required.

Acknowledgments

This study was supported by USPHS grants RR00169, HD32804 and R24RR05090. We are indebted to the California Regional Primate Research Center (CRPRC) colony management and animal care staff for providing for the collection of blood samples and supplying demographic records necessary for this study. Mr. Glen Byrnes and Dr. Dominico Bornoco provided technical advice and training without which this study would not have been possible.

References

1 Goodwin WJ, Augistine J: The primate research centers program of the National Institutes of Health. Fed Proc 1975;34:1641–1642.

2 Cornelius CD, Rosenberg DP: Spontaneous diseases in nonhuman primates. Am J Med 1983;74:169–171.

3 Held JR: Breeding and use of non-human primates in the USA. Int J Study Anim Prob 1980;2:27–37.
4 Smith S: Infant cross-fostering in captive rhesus monkeys *(Macaca mulatta)*. Am J Primatol 1986;11:229–237.
5 Smith DG: Paternity exclusion in six captive groups of rhesus monkeys *(Macaca mulatta)*. Am J Phys Anthropol 1980;53:243–249.
6 Smith DG: Use of genetic markers in the colony management of nonhuman primates: A review. Lab Anim Sci 1982;32:540–546.
7 Smith DG, Smith S: Paternal rank and reproductive success of natal rhesus males. Anim Behav 1988;36:554–562.
8 Smith DG: Incidence and consequences of inbreeding in three captive groups of rhesus macaques *(Macaca mulatta)*; in Benirschke K (ed): Primates: The Road to Self-Sustaining Populations. New York, Springer, 1986.
9 Sullivan PT, Blystad C, Stone WH: Immunogenetic studies on the rhesus monkey *(Macaca mulatta)*. XI. Use of blood groups in problems of parentage. Lab Anim Sci 1977;27:348–351.
10 Smith DG: Use of genetic markers in the management of captive groups of rhesus monkeys at the California Primate Research Center, USA. Jap J Med Sci Biol 1985; 38:44–48.
11 Nei M: Molecular Evolutionary Genetics. New York, Columbia University Press, 1987, pp 177–183.
12 Rogers J: Nuclear DNA polymorphisms in hominoids and cercopithecoids: applications to paternity testing; in Martin RD, Dixson AF, Wickings EJ (eds): Paternity in Primates: Genetic Tests and Theories. Basel, Karger, 1992, pp 82–95.
13 Wyman AR, White R: A highly polymorphic locus in human DNA. Proc Natn Acad Sci USA 1980;77:6754–6758.
14 Nakamura Y, Leppert M, O'Connell P, Wolff R, Holm T, Culver M, Martin C, Fujimoto E, Hoff M, Kumlin E, White R: Variable number of tandem repeat (VNTR) markers for human gene mapping. Science 1987;235:1616–1622.
15 Jeffreys AJ, Wilson V, Thein SL: Hypervariable 'minisatellite' regions in human DNA. Nature 1985;314:67–73.
16 Jeffreys AJ, Wilson V, Thein SL, Weatherall DJ, Ponder BAJ: DNA 'fingerprints' and segregation analysis of multiple markers in human pedigrees. Am J Hum Genet 1986;39:11–24.
17 Jeffreys AJ, Wilson V, Thein SL: Individual-specific 'fingerprints' of human DNA. Nature 1985;316:76–79.
18 Weiss ML, Wilson V, Chan C, Turner T, Jeffreys AJ: Application of DNA fingerprinting probes to Old World monkeys. Am J Primatol 1988;16:73–79.
19 Inoue M, Mitsunaga F, Ohsawa H, et al: Paternity testing in captive Japanese macaques *(Macaca fuscata)* using DNA fingerprinting; in Martin RD, Dixson AF, Wickings EJ (eds): Paternity in Primates: Genetic Tests and Theories. Basel, Karger, 1992, pp 131–140.
20 de Ruiter JR, Scheffrahn W; Trommelen GJJM, et al: Male social rank and reproductive success in wild long-tailed macaques. Paternity exclusions by blood protein analysis and DNA fingerprinting; in Martin RD, Dixson AF, Wickings EJ (eds): Paternity in Primates: Genetic Tests and Theories. Basel, Karger, 1992, pp 175–191.

21 Ménard, N, Scheffrahn W, Vallet D, et al: Application of blood protein electrophoresis and DNA fingerprinting to the analysis of paternity and social characteristics of wild barbary macaques; in Martin RD, Dixson AF, Wickings EJ (eds): Paternity in Primates: Genetic Tests and Theories. Basel, Karger, 1992, pp 155–174.
22 Edman JC, Evans-Holm ME, Marich JE, Ruth JL: Rapid DNA fingerprinting using alkaline phosphatase-conjugated oligonucleotides. Nucleic Acids Res 1988;16: 6235.
23 Miller SA, Dykes DD, Polesky HF: A simple salting out procedure for extracting DNA from human nucleated cells. Nucleic Acids Res 1988;16:1215.
24 Maniatis T, Fritsch EF, Sambrook J: Molecular Cloning: A Laboratory Manual. Cold Spring Laboratory, Cold Spring Harbor, New York, 1982.
25 Lanigan C: A short guide to DNA techniques for genealogy construction with special attention to Japanese Macaques; in Fedigan LM, Asquith PJ (eds): The Monkeys of Arashiyama: Thirty-five Years of Research in Japan and the West. Albany, State University of New York Press, 1991.

David Glenn Smith, PhD, Department of Anthropology and
California Primate Research Center, 330 Young Hall, University of California,
Davis, CA 95616 (USA)

Martin RD, Dixson AF, Wickings EJ (eds): Paternity in Primates: Genetic Tests and Theories. Basel, Karger, 1992, pp 53–62

Genetic Identification of Nonhuman Primates using Tandem-Repetitive DNA Sequences

Keiko Washio

Department of Legal Medicine, Institute of Community Medicine, University of Tsukuba, Ibaraki, Japan

The primate genome contains both unique (single-copy) and repetitive (multiple-copy) DNA sequences. There is a family of tandem-repetitive sequences and included in this family are tandem repeats of short sequence units (6–60 nucleotides), called either variable number of tandem repeat (VNTR) sequences or minisatellites [1, 2]. Because these DNA regions show substantial variabilities in length due to variations in the number of repeat units among individuals, they are used as informative genetic markers for linkage analyses, gene mapping and individual identification including paternity testing in humans [3]. In nonhuman primates, individual identification is of great significance for studies of sociobiology, demography, and ecology. However, conventional genetic markers such as blood groups and polymorphic proteins are of limited value because of their relatively low levels of heterozygosity. DNA typing using highly polymorphic tandem-repetitive sequences is now recognised as an informative method for achieving individual identification including paternity tests.

Jeffreys et al. [1] demonstrated the existence of tandem repeat probes which simultaneously hybridize with a number of these highly polymorphic DNA regions and developed a DNA fingerprinting system on this basis. The Southern blot hybridization patterns obtained with these probes consist of multiple hypervariable fragments and are specific to individuals in humans. With the confirmation of a relatively stable inheritance of the fragments, DNA fingerprinting emerged as a powerful method for individual identification of humans [4] and subsequently of various animals including birds, dogs and cats [5, 6]. Additional probes that detect multiple hypervariable DNA fragments have also been constructed and used for DNA fingerprinting [7, 8].

DNA typing has frequently been limited either by extremely small sample quantities or by inadequate field preservation conditions. These drawbacks limit the availability of adequate amounts of the high-molecular-weight (HMW) DNAs that are necessary for conventional Southern blot hybridization analyses. The polymerase chain reaction (PCR) technique developed by Saiki et al. [9] offsets these limitations, permitting specific amplification of a selected DNA segment in the presence of two oligonucleotide primers and the heat-resistant Taq polymerase. Sufficient quantities for DNA typing have been produced from trace DNAs extracted from blood stains, a single sperm, and from the small number of the cells contained in the roots or shafts of hairs [10, 11]. These DNAs do not necessarily have a high molecular weight, as the target sequences are restricted in length (<3–4 kb) because of a limitation on PCR amplification. The human genome contains 50,000–100,000 interspersed copies of tandem repeat blocks of a very short unit, (dT–dG)·(dC–dA)n, [(TG)n], with n being 15–30 [12]. For human populations, these dinucleotide repeats have been shown to be highly polymorphic in length, using PCR amplification and subsequent electrophoresis resolution in denaturation polyacrylamide gels [13, 14].

In the present study, we show the application of two informative DNA typing systems to Old World monkeys and apes. One is DNA fingerprinting using a 28-bp repeat downstream of the c-Ha-*ras*-1 gene as a probe and the other is typing of PCR-amplified dinucleotide repeats. The advantages and limitations of these two systems are discussed.

DNA Fingerprinting using a 28-bp Repeat

Methods

The species examined are 3 species of Old World monkeys and 1 ape species, the Japanese macaque (*Macaca fuscata*), the crab-eating macaque (*Macaca fascicularis*), the rhesus macaque (*Macaca mulatta*) and the chimpanzee (*Pan troglodytes*). HMW DNAs were prepared from peripheral blood lymphocytes using the standard method [15]. DNA (2 mg) was digested with the 4-bp restriction endonuclease Sau3AI, electrophoresed through 0.6% agarose gels, and blotted on to nitrocellulose filters. A 28-bp repeat probe was purified from plasmid pEJ and labelled with [α-^{32}P] dCT by nick-translation [8]. Hybridization was carried out in 1 *M* NaCl at 65 °C, as previously described [8, 16].

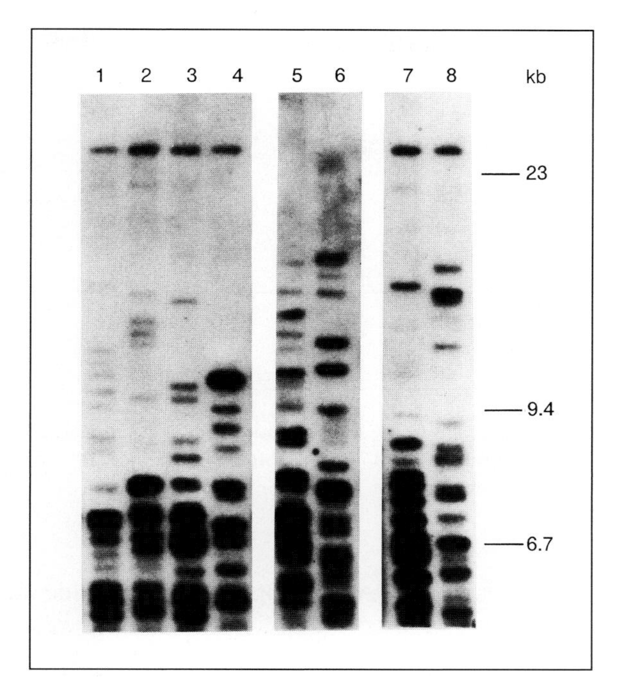

Fig. 1. DNA fingerprints using a 28-bp repeat probe obtained from 3 species of Old World monkeys; Japanese macaques (lanes 1–4), crab-eating macaques (lanes 5 and 6), and rhesus macaques (lanes 7 and 8) [reprinted from ref. 16].

Hypervariable Patterns in Nonhuman Primates

Using highly stringent washing conditions (0.015 *M* NaCl/0.0015 *M* sodium citrate/0.1% sodium dodecyl sulfate (SDS) at 65 °C), only 1 or 2 DNA fragments were observed in each subject. In less stringent conditions (0.15 *M* NaCl/0.015 *M* sodium citrate/0.1% SDS at 65 °C), however, all samples examined showed multiple DNA fragments that hybridized with the 28-bp repeat probe in both macaques (fig. 1) and chimpanzees (data not shown). DNA fragments less than 5 kb were allowed to run off the gels and the largest fragments were clearly resolved. The majority of them were found to be unique to each subject and only a few common bands were shared between individuals, although the number examined was small for each species. No identical patterns were observed in either intra- or inter-specific comparisons and all patterns were hence specific to individuals in

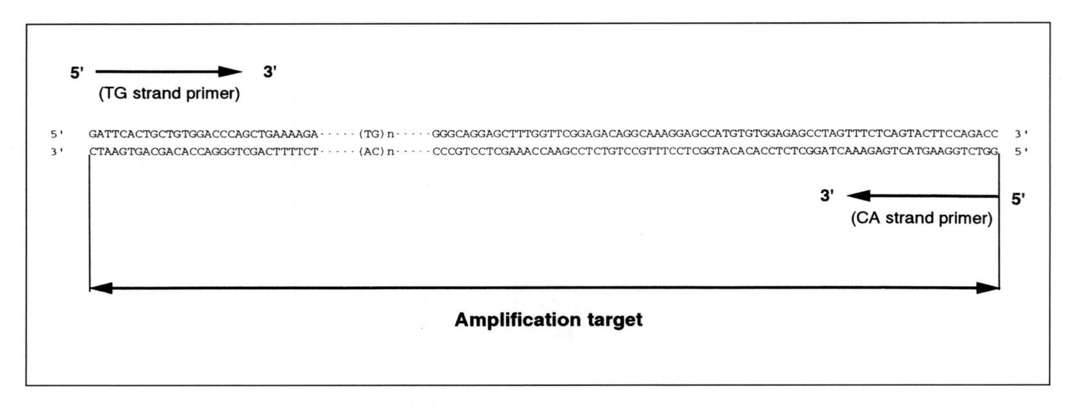

Fig. 2. Human nucleotide sequences of the dinucleotide repeat (TG repeat) and its flanking DNA regions at apolipoprotein AII [17]. The dotted region indicates the TG repeat and arrows indicate the primer positions.

all species. These results suggest the DNA fingerprinting can provide individual-specific patterns that can be used for individual identification in nonhuman primates as well as in humans [8, 16].

Typing of PCR-Amplified Dinucleotide Repeat

Methods

A dinucleotide repeat, (TG)n, at the apolipoprotein locus AII of chimpanzees and Japanese monkeys was amplified by PCR. The same primer sequences that flank the repeat block in humans were also used for chimpanzees and Japanese macaques (fig. 2) [14]. Two synthesized oligonucleotides, 5′-GATTCACTGCTGTGGACCCA-3′ (TG strand) and 5′-GGTCTGGAAGTACTGAGAAA-3′ (CA strand), were column-purified by Sephadex-25. Only the TG strand primer was end-labeled with [γ-^{32}P] ATP, using T4 polynucleotide kinase. The polymerase chain reaction was performed in a reaction mixture containing 400 n*M* of both primers, 250 μ*M* of dATP/dGTP/dCTP/dTTP, 67 m*M* Tris-HCl (pH 8.8), 6.7 m*M* $MgCl_2$, 16.6 m*M* $(NH_4)_2SO_4$, 10 m*M* 2-mercaptoethanol, 0.0067 m*M* Na_2 EDTA, 5% dimethylsulfoxide, and 1 unit of Taq polymerase. Both HMW and partially degraded DNAs (30–100 ng) from peripheral blood were used as templates. The size and quantity of template DNAs were confirmed after electrophoresis in agarose gels by ethidium bromide staining. Thirty reaction cycles were conducted with three steps; denaturation for 1 min at 92 °C, annealing for 1 min at 60 °C, and extension of a new strand for 2 min at 72 °C. The PCR product (1 μl) was mixed with formamide-dye solution (1 μl) and applied to a denaturation gel containing 8% polyacrylamide and 7 *M* urea. After electrophoresis, the gels were fixed, dried, and autoradiographed on X-ray films.

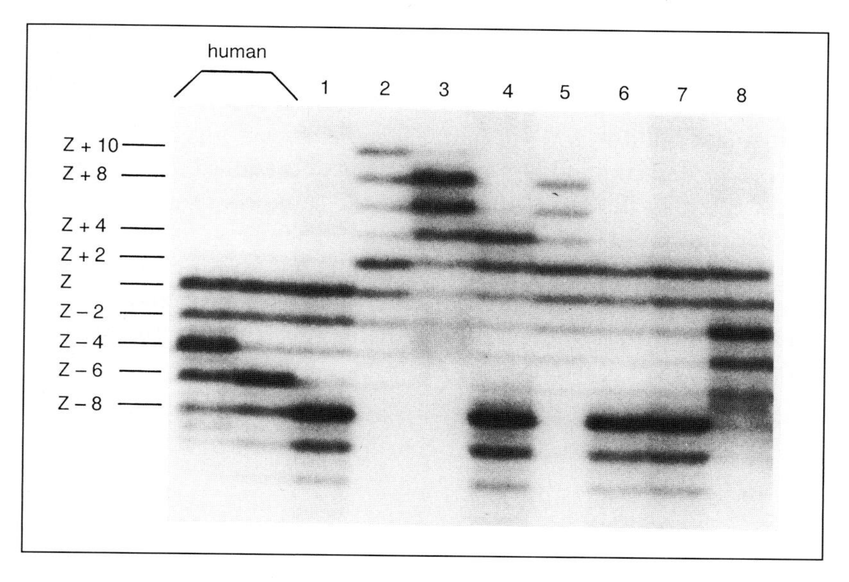

Fig. 3. Dinucleotide repeat polymorphism at the apolipoprotein AII detected by PCR following denaturation polyacrylamide gel electrophoresis. Allele names are indicated at the major band positions. Peripheral blood samples of chimpanzees are shown in lanes 1–8.

Highly Polymorphic Patterns in Chimpanzee

DNA templates from chimpanzees can be amplified using the primers developed from human sequences. Eight chimpanzees having unknown relatedness show polymorphic patterns (fig. 3). Each sample exhibits one or two intense major bands and a few minor bands which are found at intervals of every 2 nucleotides. Among 7 alleles identified, 5 alleles corresponded to human population alleles, i.e., Z+8, Z+4, Z+2, Z, and Z–2. Allele Z, having the 137-bp major fragment, is the most common allele in both the Caucasian [14] and Japanese human populations. Human polymorphism information contents [18] and heterozygosities were estimated as 0.65 and 0.74, respectively, for the Caucasian population [14], and 0.67 and 0.72 for the Japanese population [unpublished data], with 6 alleles being observed from either 41–45 (actual number indeterminate) Caucasian or 46 Japanese samples. Although the population size of chimpanzees examined was small, the number of detected alleles (7) compared favorably with those found with human populations. Furthermore, neither the largest nor the smallest allele (Z+10, Z−8) was detected in either the Caucasian or Japanese human population. These results indicate that the dinucleotide repeat at the apolipoprotein AII locus is highly polymorphic in chimpanzees, and is an informative genetic marker just as in human populations [14]. Template DNAs from 8 Japanese macaques could also be amplified using the same primers and the same conditions; however, all samples showed identical patterns (data not shown).

Partially degraded DNAs (size: 200–400 bp) from 2 chimpanzees were also examined under the same conditions (fig. 3, lanes 5 and 6). The PCR products obtained exhibited discrete major bands with an intensity similar to those of HMW bands.

Advantages and Limitations of the DNA Typing Systems

DNA Fingerprinting

The advantage of DNA fingerprinting is that one probe can detect multiple polymorphic loci simultaneously; therefore it has been demonstrated to be a powerful means for individual identification in humans. In nonhuman primates, including some macaques and chimpanzees, hypervariable patterns have been obtained using Jeffreys' probes and a 28-bp repeat probe. In Japanese macaques, protein analyses have failed to permit individual identification because of limited variability, whereas DNA fingerprinting exhibits hypervariable patterns between individuals and promises success in the discrimination of individuals.

On the other hand, the features of DNA fingerprints are so complicated that each phenotype cannot be determined, therefore the patterns can only be compared with each other on the same gel.

It is generally not known which DNA fingerprint fragments belong to which locus, unless a significant number of family samples are analysed, so locus identification and determination of allele frequencies are only remotely possible. In view of these circumstances, we note two considerations that must be taken into account when applying this method practically to captive/wild populations. Firstly, the degree of individual specificity of a DNA fingerprint cannot be estimated from allele frequencies. Jeffreys et al. [4] estimated the probability that 2 individuals show the same hybridization patterns using pattern analyses which counted the fragments shared by any 2 unrelated individuals instead of using allele frequencies.

As each primate population has a different degree of genetic variation reflecting its reproductive system and other factors, the probability for each primate population should be estimated before practical applications are attempted. Secondly, analyses of hypervariable VNTR sequences cloned from humans [19] and inbred mice [20] have shown that high heterozygosities are associated with high mutation rates [21–23]. Spontaneous mutation events producing new length alleles have also been observed in DNA fingerprints from human pedigree samples [1]. Because locus identification is very difficult in DNA fingerprinting, some attention must be given to the inclusion of these hypermutable fragments in the DNA fingerprints examined.

DNA fingerprinting can be also accomplished using mixed multiple minisatellite probes combined with knowledge concerning the loci and allele frequencies, although the detected loci are fewer than those found using single-probe DNA fingerprinting [19].

PCR Application to Individual Identification

It is known that PCR is a powerful method for individual identification in both forensic examinations and animal investigations, where a sufficient quantity of biological specimens is difficult to obtain. For this reason, a number of DNA regions have been amplified by PCR. Polymorphisms of amplified DNAs are detected as changes occurring in the nucleotide sequences [24]. Substitution of genome and/or mitochondria sequences were detected by differential hybridization to allele-specific oligoprobes, by the gain/loss of restriction endonuclease sites, by different electrophoretic mobility in polyacrylamide gel occurring due to single-strand conformational polymorphism, and also by direct sequence determination. Length divergences caused by gain/loss of a nucleotide block were observed after gel electrophoresis followed by either ethidium bromide staining or by Southern blot hybridization. Although VNTR sequences are highly informative makers, they frequently exceed the PCR amplification limits because of the wide range of repetition numbers among allleles, thus they cannot always be exactly amplified [23, 25]. Additionally, they are usually resolved in agarose gels or acrylamide gels, which do not permit resolution of small differences.

By contrast, dinucleotide repeat blocks for most alleles are within a few hundred nucleotides, as the repeat unit is only two nucleotides, so they can be amplified exactly. After amplification, PCR products are resolved in denaturation polyacrylamide gels, which can discriminate a single

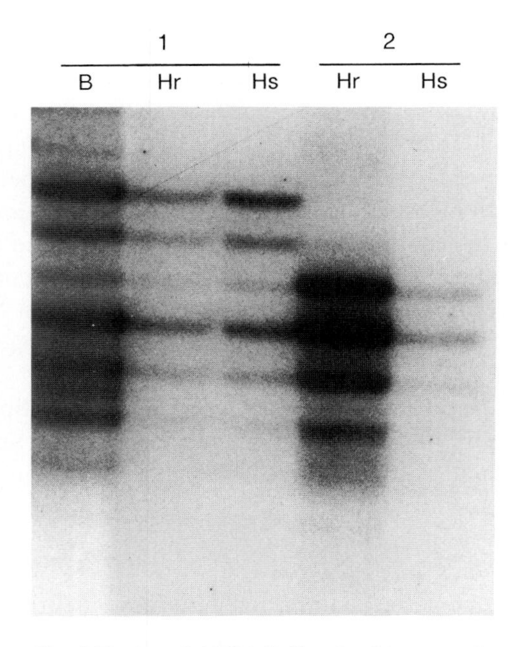

Fig. 4. Typing of APOA-2 dinucleotide repeat locus from human hair samples. Phenotypes of individuals No. 1 and 2 are Z, Z–6 and Z–4, Z–6, respectively. B = Blood; Hr = hair root; Hs = hair shaft.

nucleotide length difference [13, 14]. In this preliminary study, it has been shown that the (TG)n block at the apolipoprotein AII gene can be amplified and shown to be highly polymorphic in chimpanzees and that typing is possible even from partially degraded DNA samples. Additionally, typing was performed using a single hair obtained from 2 human individuals (fig. 4). The PCR products from both root and shaft of human hair showed identical patterns to those obtained from peripheral blood. These results indicate that dinucleotide repeat typing using PCR is a powerful new method, especially for field investigation. A sample of 8 Japanese macaques did not show polymorphism at this locus, but considering the large number of dinucleotide repeat polymorphisms at dispersed chromosome loci that have been reported in the human genome, combinations of selected dinucleotide repeat polymorphisms are expected to be the most informative genetic markers for individual identification in most wild nonhuman primate populations.

Acknowledgments

I wish to acknowledge that the study reported here was carried out in close collaboration with the research group of the Department of Anthropology, The University of Tokyo. I am grateful to Drs. S. Ueda and K. Kurosaki for technical support and helpful discussions, and to Dr. S. Misawa for helpful advice. I also thank Dr. O. Takenaka for providing blood samples from nonhuman primates. This study was supported by grants from the Ministry of Education, Science and Culture of Japan, and from the University of Tsukuba, Project Research.

References

1 Jeffreys AJ, Wilson V, Thein SL: Hypervariable 'minisatellite' regions in human DNA. Nature 1985;314:67–73.
2 Nakamura Y, Leppert M, O'Connell P, et al: Variable number of tandem repeat (VNTR) markers for human gene mapping. Science 1987;235:1616–1622.
3 Donis-Keller H, Green P, Helms P, et al: A genetic linkage map of the human genome. Cell 1987;51:319–337.
4 Jeffreys AJ, Wilson V, Thein SL: Individual-specific 'fingerprints' of human DNA. Nature 1985;316:76–79.
5 Wetton JH, Carter RE, Parkin DT, et al: Demographic study of a wild house sparrow population by DNA fingerprinting. Nature 1987;327:147–149.
6 Jeffreys AJ, Morton DB: DNA fingerprints of dogs and cats. Anim Genet 1987;18: 1–15.
7 Vassart G, George M, Monsieur R, et al: A sequence in M13 phage detects hypervariable minisatellites in human and animal DNA. Science 1987;235:683–684.
8 Washio K, Misawa S, Ueda S: Probe walking: Development of novel probes for DNA fingerprinting. Hum Genet 1989;83:223–226.
9 Saiki RK, Gelfand DH, Stoffel S, et al: Primer-directed enzymatic amplification of DNA with a thermostable DNA polymerase. Science 1988;239:487–491.
10 Li H, Gyllensten UB, Cui X, et al: Amplification and analysis of DNA sequences in single human sperm and diploid cells. Nature 1988;335:414–417.
11 Higuchi R, von Beroldingen CH, Sensabaugh GF, et al: DNA typing from single hairs. Nature 1988;332:543–546.
12 Hamada H, Petrino MG, Kakunaga T: A novel repeated element with Z-DNA-forming potential is widely found in evolutionarily diverse eukaryotic genomes. Proc Natl Acad Sci USA 1982;79:6465–6469.
13 Tautz D: Hypervariability of simple sequences as a general source for polymorphic DNA markers. Nucleic Acids Res 1989;17:6463–6471.
14 Weber JL, May PE: Abundant class of human DNA polymorphisms which can be typed using the polymerase chain reaction. Am J Hum Genet 1989;44:388–396.
15 Sambrook J, Fritsch EF, Maniatis T: Molecular Cloning: A Laboratory Manual, ed 2. New York, Cold Spring Harbor Laboratory Press, 1989.
16 Washio K, Misawa S, Ueda S: Individual identification of non-human primates using DNA fingerprinting. Primates 1989;30:217–221.

17 Knott TJ, Wallis SC, Robertson ME, et al: The human apolipoprotein AII gene: Structural organization and sites of expression. Nucleic Acids Res 1985;13:6387–6398.
18 Botstein D, White RL, Skolnick M, et al: Construction of a genetic linkage map in man using restriction fragment length polymorphisms. Am J Hum Genet 1980;32: 314–331.
19 Wong Z, Wilson V, Patel I, et al: Characterization of a panel of highly variable minisatellites cloned from human DNA. Ann Hum Genet 1987;51:269–288.
20 Jeffreys AJ, Wilson V, Kelly R, et al: Mouse DNA 'fingerprints': Analysis of chromosome localization and germ-line stability of hypervariable loci in recombinant inbred strains. Nucleic Acids Res 1987;15:2823–2836.
21 Jeffreys AJ, Royle NJ, Wilson V, et al: Spontaneous mutation rates to new length alleles at tandem-repetitive hypervariable loci in human DNA. Nature 1988;332: 278–281.
22 Kelly R, Bulfield G, Collick A, et al: Characterization of a highly unstable mouse minisatellite locus: Evidence for somatic mutation during early development. Genomics 1989;5:844–856.
23 Jeffreys AJ, Neumann R, Wilson V: Repeat unit sequence variation in minisatellites: A novel source of DNA polymorphism for studying variation and mutation by single molecular analysis. Cell 1990;60:473–485.
24 Innis MA, Gelfand H, Sninsky JJ, et al (eds): PCR protocols: A Guide to Methods and Applications. San Diego, Academic Press, 1990.
25 Jeffreys AJ, Wilson V, Neumann R, et al: Amplification of human minisatellites by the polymerase chain reaction: Towards DNA fingerprinting of single cells. Nucleic Acids Res 1988;16:10953–10970.

Dr. Keiko Washio, Department of Legal Medicine,
Institute of Community Medicine, University of Tsukuba, Ibaraki 305 (Japan)

Martin RD, Dixson AF, Wickings EJ (eds): Paternity in Primates: Genetic Tests and Theories. Basel, Karger, 1992, pp 63–81

Paternity Exclusion using Multiple Hypervariable Microsatellite Loci Amplified from Nuclear DNA of Hair Cells

Phillip A. Morin, David S. Woodruff

Department of Biology and Center for Molecular Genetics, University of California, San Diego, La Jolla, Calif., USA

Recent advances in biochemistry and genetics now permit unambiguous or statistically acceptable paternity exclusion in multi-male primate groups, at least in theory. In practice, however, secondary problems of applying the new molecular methodologies to natural populations remain. These include difficulties of acquiring tissue samples from free-ranging animals and of interpreting whole-genomic patterns called DNA fingerprints. In this paper we describe the successful development and application of a new approach to paternity assignment based on noninvasive tissue sampling and rapid characterization of individuals' genotypes at multiple hypervariable nuclear loci. This approach promises to facilitate the resolution of a number of hitherto intractable problems involving paternity assessment in captive and more natural social units.

Studies of free-ranging primates have historically focused on social and reproductive groups, as sociality has dominated primate evolution. Reproductive strategies, ecological limits, inclusive fitness, predator defense, kin and group selection arguments have all been used to explain the evolution of these social groups [1–3]. Implicit in these arguments is the assumption that the social unit is identical to the reproductive unit, or at the very least, strongly associated with it. As behavioral data have accumulated it has become clear, however, that this is not always the case [4–6]. Indeed, the social unit may, in some cases, be quite unrelated to the reproductive unit. Behavioral data reveal such unexpected patterns of social organization in several groups of primates, but the true genetic relationships in such groups, and therefore the reproductive units, remain unknown or poorly defined.

The characterization of relationships within and among groups of free-ranging primates will permit investigators to test numerous hypotheses regarding primate sociobiology and evolution. Specific points that may be addressed include: reproductive success of individuals, correlations of dominance status and behaviors with reproductive success, levels of relatedness and inbreeding, gene flow between communities and across potential physical barriers, effective population size, and intrapopulation variation. Although we concentrate on paternity assessment in this paper, pedigree data have much broader applicability. Retrospective sociobiological analyses, based on the establishment of paternities in long-studied groups, will be feasible and of great potential significance. It will also be possible to test various hypotheses regarding the known variation in reproductive strategies between sexes, populations and species [7; Morin, in preparation].

Genetic studies of free-ranging populations have long been hampered by tissue-sampling problems. To obtain enough protein or DNA for available analytical procedures, animals had to be captured and bled, biopsied, or killed. Tissue samples then had to be processed, frozen in liquid nitrogen or stored in sterile tissue culture media, and shipped immediately to the laboratory for analysis. Logistically, this was often very difficult at remote field sites and, at some sites, was not possible as the disturbance of the animals under study would have undone any habituation accomplished by behavioral researchers.

Several biochemical genetic methodologies have been developed for paternity assessment, but all have drawbacks:

(1) Allozymes were used to establish population levels of variability (proportion of loci that are polymorphic and individual heterozygosity) and structure, and patterns of geographic variation, but with a few notable exceptions [8, 71–74] have been of only limited use for resolving pedigree relationships.

(2) Mitochondrial DNA (mtDNA) restriction fragment length polymorphism (RFLP) analyses were important in establishing population genetic structure but are less useful at the primate family or social community level because mtDNA typically exhibits strict maternal inheritance [9, 10].

(3) DNA fingerprinting based on hypervariable nuclear sequences can be used to distinguish relationships among close relatives where all poten-

tial sires are known [11–13, 74–79]. Variation in these diverse loci, which are dispersed throughout the genome, is generally due to copy number of tandem repeats of the short minisatellite sequence. Such variable number tandem repeat (VNTR) patterns are analysed by digestion with restriction enzymes and transfer hybridization using minisatellite sequence probes to reveal variation at a large number of hypervariable loci simultaneously [14]. The resulting DNA fingerprint is a complex, multifragment pattern which is often unique to an individual. It provides a measure of genetically controlled variation but, because the individual fragments cannot be assigned to specific loci, does not permit formal genotyping. In some populations, the whole-genomic patterns are too complex for unambiguous interpretation and in others, involving small inbred populations, there may be insufficient marker 'alleles' to establish pedigree relationships [15]. The major technical and statistical problems inherent in DNA fingerprinting [16–18] and its restricted applicability to natural populations are discussed elsewhere [19, 20, 71, 73, 76–82].

(4) A preferable approach would be to establish an individual's genotype at specific hypervariable loci using synthetic oligonucleotides as probes. This approach is initially more laborious, but has the distinct advantage of providing genotypic data of the type required for pedigree analyses. Allele-specific oligonucleotide (ASO) probes have recently become available for human VNTR [21, 22] and single copy sequences [e.g. HLA-DQA1; ref. 23–25] and will undoubtedly resolve paternity questions in some situations. Unfortunately, such probe-based procedures require relatively large amounts (hundreds of nanograms to micrograms) of intact, high-molecular-weight DNA, and the probes themselves may have to be developed separately for each nonhuman primate species.

In 1989, with these limitations of existing methodologies in mind, we set out to develop a noninvasive, individual genotyping procedure based on specific genes and gene fragments.

Amplification of Hypervariable Microsatellite Loci

In 1983, a simple method for making unlimited copies of specific DNA fragments was rediscovered by Mullis [26]. The development of the polymerase chain reaction (PCR) for the in vitro enzymatic amplification of DNA [27, 28] has opened up vast opportunities for genetic investigation. This procedure permits the isolation and replication of several million copies of single, specific gene fragments from the mitochondrial or nuclear genomes. Large tissue samples, frozen at the time of collection, are no longer a prerequisite for many DNA level studies. Minute samples of tissues as diverse as saliva, semen, hair, bone, tooth, and (for birds) feather have provided enough DNA for amplification by the PCR, and even highly degraded DNA may still yield PCR products that can be sequenced or analyzed in other ways [29–31; Morin, unpubl. data]. For primate studies, the discovery of hair as a DNA source [23] has been especially important, as this tissue can be collected noninvasively from many species without disturbing animals or disrupting the habituation process.

By itself, the development of the PCR was not enough to permit us to undertake the types of genetic studies needed to elucidate the relationships among individuals in primate communities. Further progress awaited the discovery and characterization of hypervariable loci small enough to be efficiently amplified from partially degraded DNA. This breakthrough came in 1989 when Weber and May [32], Litt and Luty [33] and Tautz [34] published PCR primer sequences for several human loci, called microsatellites, which contained variable numbers of the dinucleotide tandem repeat (VNDR) $(dC\text{-}dA)_n$, where n is typically in the range of 6–30. These VNDR loci typically exhibit high levels of length polymorphism and their allelic variants follow Mendelian heritability patterns [32, 35]. They are two orders of magnitude shorter than the minisatellite VNTR loci employed in DNA fingerprinting and are consequently easier to amplify and interpret. There are 50,000–100,000 dinucleotide tandem repeats interspersed throughout the human genome, usually in introns of functional nuclear genes. Their functional significance is still unknown. Since those first publications in 1989, over 100 such loci have been identified and characterized in humans [J. Weber, Marshfield Medical Research Foundation, pers. commun.] and many also occur in other primates, including prosimians [A. Merenlender, Princeton University, pers. commun.; Morin, unpubl. data].

For establishing relationships among individuals in a primate community, these microsatellite loci offer several benefits over previously mentioned proteins and gene fragments. First, alleles differ in the number of repeats, and therefore in base pair (bp) length, rather than in nucleotide sequence. This means that variation is easily detected by separating alleles by size on a polyacrylamide gel and obviates the need to sequence the DNA. Second, allele lengths can vary substantially within a population and a large proportion of the individuals within a population may be heterozygous. As a consequence, heterozygotes with unique allele combinations may be relatively common in many communities. Third, the presence of many different, highly variable microsatellite loci enables us to prepare a multiple-locus characterization for each individual. The resulting multiple-locus designations differ from the complex whole-genome pattern of a minisatellite DNA fingerprint in that the Mendelian heritability of every allele can be established. Finally, DNA extraction, amplification, and allele identification are relatively fast procedures. Specific microsatellite loci can be extracted, amplified, and electrophoresed on a gel in a single working day; the subsequent visualization of the alleles by autoradiography typically takes an additional 24 h.

Protocols for the Characterization of Microsatellite Loci from Hair

Sample Collection and Storage

Six to ten hairs per individual are adequate for most pedigree studies. Although circumstances may dictate that hairs be picked up or plucked by hand, we have found that the collector's own DNA can become a PCR contaminant; ideally, hairs should be handled with forceps, hemostat or gloves. Hair samples must be unambiguously assignable to a single individual and labeled and packaged to ensure sample integrity. For amplification, the best hairs are freshly plucked from the animal, or 'groomed' from its coat. Such hairs typically have intact roots surrounded by sheath cells, and relatively large quantities of high-quality DNA. If the animals cannot be touched, then hairs may be collected from resting sites (e.g., sleeping nests). Such hairs may be identified as belonging to a known individual if these sites are freshly made and used by only one animal. Also, hairs may be collected from the ground after a solitary animal is observed self-grooming, but this is risky as one can rarely be certain of their origin. Hairs should be put into labeled plastic or paper envelopes and stored in a dry place until they can be shipped to a laboratory with freezer facilities. Although no studies of the effect of time on the degradation of DNA in hairs have been reported, samples studied in this laboratory have usually been held at ambient temperature for up to 3 months and then frozen at – 80 °C. One chimpanzee hair sample was 'stored' in a field notebook for 2 years, and yielded sufficiently high-quality DNA to enable us to amplify and sequence a 340-bp region of the mitochondrial cytochrome b gene.

DNA Extraction

DNA should be extracted from single hairs for each individual animal whenever possible. Two or more hairs may be extracted simultaneously only when their provenance is known with certainty. In our experience, 1–2 freshly plucked hairs are sufficient for the type of analyses described. A single plucked hair will yield more than 200 ng and shed hair more than 10 ng of DNA [23]; as the PCR will amplify a sequence from a single template, in theory at least, a hair yields enough DNA to type hundreds of microsatellite markers. In practice 15–25 loci per hair can be typed. Hair roots are washed with 90% ethanol and sterile water, and approximately 2–4 mm of the root end is cut off and placed into 200 μl of 5% Chelex 100 (BioRad). The samples are incubated at 56 °C for 20 min, vortexed, incubated at 100 °C for 8 min, vortexed again, and centrifuged at 14,000 *g* for 2 min. A section of the shaft of each hair should also be 'extracted' separately to ensure that the DNA one amplifies is not contaminating DNA from the surface of the hair.

Primer End-Labeling

One VNDR primer per locus is end-labeled with 2 μCi/μl of γ-ATP^{32} (3000 Ci/m*M*), in a final concentration of 1 μ*M*, using T4 polynucleotide kinase. End-labeling reactions are carried out at 37 °C for 30 min, and stopped by incubation at 100 °C for 2 min. In the next few years, technical advances should enable us to switch to nonradioactive (e.g., biotinylated) primers.

Polymerase Chain Reaction

The use of Taq DNA polymerase and gene-specific oligonucleotide primers allows repeated sequence replication and denaturation by differential incubation using a thermal cycler. The PCR conditions vary with each locus, but in general they involve primary denaturing at 94 °C for 3 min followed by 35 step-cycles of amplification under the following conditions: primer binding or annealing at 55 °C (60 °C for the last 15 cycles) for 5 s, DNA synthesis or extension at 74 °C for 30 s, and denaturation at 92 °C for 1 min. The reaction conditions for a final volume of 25 μl are: 10 m*M* Tris, pH 8.3; 50 m*M* KCl; 0.01% gelatin; 1.5 m*M* $MgCl_2$; 0.2 m*M* of each dNTP; 0.4 μ*M* unlabeled primer (+) (= 10 pmol); 0.4 μ*M* unlabeled primer (–) (= 10 pmol); 0.08 μ*M* end-labeled (g-^{32}P, 3,000 Ci/mmol) primer (–) (= 2 pmol); and 1 unit Amplitaq™ Taq polymerase (Perkin-Elmer Cetus). This regularly results in at least a 100,000-fold amplification of the selected microsatellite sequence.

Once conditions for the resolution and interpretation of a specific VNDR locus have been optimized, one may proceed to amplify additional microsatellite markers simultaneously [e.g. fig. 3 in ref. 32]. A limit to the number of loci that can be amplified in the same reaction is set by the need to keep their resulting electrophoretic patterns separate and interpretable on the gels and autoradiographs.

Electrophoresis

Amplified products are visualized by electrophoretic separation of the polymorphic size fragments. Loading dye (4 μl) is added to each PCR reaction tube, mixed, and 4 μl of product loaded in each well of a 20 × 40 cm, 8% polyacrylamide, 50% urea sequencing gel. Electrophoresis at 2000 V and 30 mA is carried out for 3–5 h depending on the size of the product. Gels are then either wrapped in plastic and exposed to autoradiography film for 12–48 h at – 80 °C, or preferably dried and exposed to the film at room temperature. Homozygotes produce a single band, heterozygotes two bands, representing alleles of the same or different length, respectively. Alleles that differ in length by only a single repeat (or two nucleotides) can be distinguished. Care must be taken with partially degraded samples to ensure that heterozygotes are not mis-scored as homozygotes, as shorter fragments may amplify preferentially to longer fragments [25]; it is important to carry the PCR reaction through enough step cycles to produce adequate quantities of the underrepresented longer fragment, if present.

Paternity Assessment

For paternity analysis, the probability of exclusion of a nonfather is proportional to the number and relative frequencies of alleles at each locus. Such data are not yet available for any microsatellite locus for any natural or captive population of nonhuman primates. In captive colonies, this information may be estimated by genotyping large numbers of individuals; but if the animals are of diverse geographic origin, or are inbred, this approach will give misleading results. For many species, therefore, large samples from wild populations will be needed to establish statistically significant frequencies for each allele at each locus. Until such data are available, one can proceed to calculate the probability of excluding nonfathers by assuming equal allele frequencies and making conservative estimates of the number of alleles. In this situation, we have followed Smouse and Chakraborty [36] in using the equation:

$$E_H(\text{K alleles}) = \frac{(K-1)(K^3 - K^2 - 2K + 3)}{K^4}$$

where E_H is the exclusion probability from Selvin [37]. If, in the future, the exact allele frequencies are determined for a locus in the species of interest, the exclusion probability for a given set of K alleles with frequencies denoted P_k: $k = 1, ..., K$, can be evaluated with Chakravarti's [38] equation:

$$E(\text{K alleles}) = a_1 - 2a_2 + a_3 + 3(a_2a_3 - a_5) - 2(a_2^2 - a_4),$$

where

$$a_i = \sum_{k=1}^{K} P_k{}^i$$

It is important to note here that these estimates assume the population in Hardy-Weinberg equilibrium and that exclusion probability is strongly influenced by allele frequencies. If the assumption of either panmixia or equal allele frequencies is violated, such estimates will be too high, and should be considered 'best case' estimates. This would be the case if there were close geneological relationships among potential mates in a population.

Applications

Pedigree Analysis in a Captive Colony of Chimpanzees, Pan troglodytes

Four microsatellites were used to characterize the 4 founders and 3 offspring among the chimpanzees at the Asheboro Zoo in North Carolina. Four loci [Mfd18, Mfd23, Mfd32, LL-1; ref. 32, 33, 39–41] were sufficient to establish unique individual-specific allele patterns in all animals in this small colony and to assign paternity for all 3 offspring (fig. 1). No other possible pairwise combinations of adult males and females, regardless of sex or individual age, could possibly have contributed the alleles found in any of the offspring.

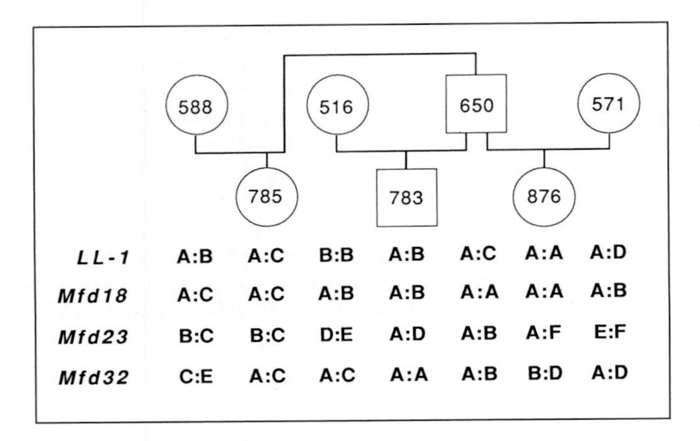

Fig. 1. Pedigree relationships among 7 members of the chimpanzee colony at the Asheboro Zoo, N.C., USA. Each individual was characterized at 4 microsatellite loci, and the alleles have been arbitrarily given letter designations, A–F. Each individual's multiple-locus genotype pattern is unique and alternative parentages can be excluded. Circles = Males; squares = females.

Pedigree Analysis in a Captive Colony of Bonobos, Pan paniscus

In the case of a subgroup of the Zoological Society of San Diego's bonobos, a more complicated pedigree was analyzed with the same 4 microsatellite loci (fig. 2). This group was founded by 3 individuals and is now in its second generation. These 4 loci were not sufficient to create unique individual-specific allele patterns or to exclude all alternate paternities. Additional loci will therefore be needed for more complex pedigrees of captive colonies and for free-ranging situations.

Genetic Variability and Paternity Assessment in Free-Ranging Chimpanzees

Preliminary results from our laboratory suggest that microsatellite loci are highly polymorphic and some alleles are widely distributed throughout the ranges of the three chimpanzee subspecies. In a survey of 25 *Pan troglodytes troglodytes* from Gabon, representing the central African subspecies, the Mfd18 locus exhibited 8 alleles, and more than 90% of the individuals were heterozygous. The LL-1 locus exhibited 12 alleles and greater than 90% heterozygosity for the same individuals. These preliminary data reveal no marked changes in allele frequencies across 500 km in Gabon,

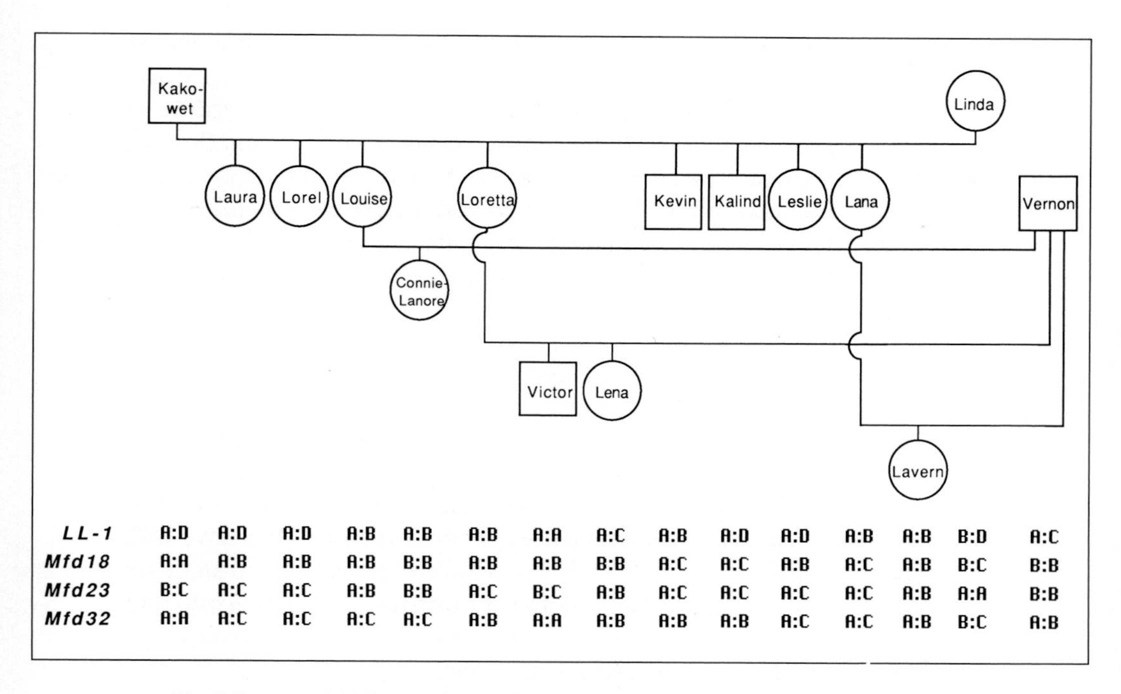

Fig. 2. Purported pedigree relationships based on behavioral observations among 15 members of the bonobo colony at the San Diego Wild Animal Park, Calif., USA. Each individual was genotyped at the same 4 microsatellite loci, as in figure 1. Lower levels of variability at these loci in this colony meant that not all individual multilocus genotype patterns are unique (e.g., Laura, Lorel and Leslie share one pattern) and that paternity exclusion was possible in only 30% of the cases. Circles = Males; squares = females.

indicating either a relatively high mutation rate for the number of tandem (CA) repeats, or more likely, significant gene flow across the subspecies range. There is no need to invoke strong natural selection to explain the maintenance of such high variability, as microsatellite loci have no known function and are thought to be selectively neutral [42].

Preliminary data from hair samples of eastern chimpanzees (*Pan t. schweinfurthii*; Kasakela community, Gombe Stream Reserve, Tanzania) indicate similar frequencies of alleles at these 2 loci: to date, we have detected 5 Mfd18 alleles in 7 individuals and four LL-1 alleles in 10 individuals.

The large number of alleles at the microsatellite loci should permit us to establish pedigree relationships of many individuals and detect the reproductive effects (gene flow) of successful dispersal in wild chimpanzee communities. For example, if we assume that all alleles are at equal frequency in the population and that there is insignificant inbreeding at Gombe, the probability of excluding nonfathers (P_e) with the Mfd18 locus is 74%. The higher allelic diversity of the LL-1 locus would provide a P_e of 83%. The combined use of these 2 loci would, therefore, provide a 96% probability of exclusion of nonfathers in a paternity case. Thus, with the addition of 2–3 more loci of similar diversity, we expect to raise P_e to greater than 99%.

Paternity Assessment in Other Species of Primates

We expect that similar levels of variability will be found in other nonhuman primates. DNA samples of captive lowland gorilla, *Gorilla g. gorilla*, were available and preliminary analyses suggest high levels of microsatellite variation: 6 or more alleles were found at each of 3 loci examined in 10–16 individuals, with greater than 80% heterozygosity. We have also succeeded in amplifying 1–5 VNDR loci with Weber's primers from hairs of orangutan *(Pongo pygmaeus)*, lion-tailed macaque *(Macaca silenus)* and white-handed gibbon *(Hylobates lar)*; in the latter, 4 individuals from a zoo in Thailand all possessed different Mfd23 alleles.

Discussion

The Cost of Microsatellite Genotyping

The advantages of the microsatellite-based genotyping system are noted above; its limitations are analogous to those found in other systems. The associated protocols are expensive and still require postgraduate levels of expertise in molecular biology and biochemistry. A working knowledge of sterile technique and the rigorous application of the specific precautions required to minimize contamination of the amplification reaction, with products of previous PCR reactions (product carryover) or exogenous DNA, are essential [25]. Accidental contamination of nonhuman primate samples with human DNA must be guarded against as transspecific homologies make it more difficult to detect. At this time, it is best to regard all PCR-based projects as pilot studies, and to expect that protocol optimization will take much longer (months) and be more expensive than expected.

Orrego [43] and Hillis et al. [44] provide useful notes on equipping and organizing a laboratory for this type of work. Once procedures are optimized for a particular locus, species, sample type, and laboratory, the reagents and supplies cost less than US$ 20 to process a single sample through the DNA extraction, PCR, electrophoresis, and autoradiography steps. These costs do not include the necessary laboratory and darkroom equipment (US$ 50,000) or the time and extensive work needed for primer optimization.

These costs notwithstanding, the microsatellite method is still less expensive than DNA fingerprinting and DNA sequencing. For comparison, the cost of minisatellite DNA fingerprinting has been estimated at US$ 50–100 per individual sample for supplies and US$ 100–200 when labor is factored in [45]. Similar costs would be incurred if the Cetus Corp. human HLA DQ1 typing kit were found to be useful for nonhuman primates. Recognizing that laboratory costs will be the major impediment to the now required use of the new technologies in many sociobiological studies, Weatherhead and Montgomerie [45] have called for the establishment of regional facilities to provide economic genotyping services.

Types of Microsatellites Available for Genotyping

Primers for VNDR regions have been published only for human DNA sequences. Although most of those we have tested appear to be useful for other primates, each primer amplification protocol must be optimized and tested for intrinsic variation in each study population. This involves repetitive experiments to establish the optimal annealing temperature; too low a temperature will allow product strands to re-anneal, often mismatched along the CA repeat region, and create product fragments at intervals of two nucleotides that may interfere with interpretation of the allele pattern [46]. If the match of primer and template DNA is perfect, an annealing temperature just below the calculated melting temperature of the primer-template DNA can be used [47]. If the match is not perfect, such a temperature may result in little or no PCR product, and a lower temperature will have to be used. To optimize this parameter, one may choose to derive empirically the best set of conditions for a given set of human-derived primers, or sequence the PCR product to see if sequences internal to the human primers may be used for the synthesis of species-specific primers. Additional nonhuman primate protocols are discussed by Washio [83].

In addition to the $(CA)_n$ repeats discussed above, length polymorphism has been described in other abundant simple-sequence tandem repeats, including $(GT)_n$ dinucleotide and poly(A) repeats [32, 48, 49]. Pools van Amstel et al. report two $(GT)_n$ dinucleotide repeats in the protein S alpha gene and protein S beta pseudogene in humans [50]. They found that their human primers amplified these regions in two chimpanzees *(Pan t. schweinfurthii)* and discovered size fragment length variation in the pS pseudogene GT repeat [Ploos van Amstel, pers. commun.]. $(GT)_n$ repeats are, of course, the complimentary strand of $(CA)_n$ repeat regions.

Recently, tandemly repeated tri- and tetra-nucleotide sequences have been identified and found to be highly polymorphic in humans [47, 51–56]. These microsatellites may also be useful in nonhuman primates. They have the added advantage of suffering less from strand-slippage problems than do the dinucleotide repeats. Ploos van Amstel et al. [57] analyzed two distinct highly polymorphic subregions of a tetra-nucleotide repeat region in the von Willebrand factor gene in humans. In each subregion, 6 size alleles were discovered among 24 Caucasians. The human primers will amplify homologous chimpanzee loci [Ploos van Amstel, pers. commun.] and it is reasonable to expect similar levels of variation in the other great apes. As more human primers are tested on other taxa, a set of 'universal' or generic primers may become available for paternity testing; until then, however, each new primer will have to be tested for both sequence conservation and product size diversity.

Noninvasive DNA Sampling

The noninvasive sampling techniques introduced here are more humane than techniques requiring animal restraint, bleeding or biopsy. Current procedures are inherently painful and stressful to both the individual animals and the social group, and typically require the services of trained veterinary personnel. Another advantage of using hair samples is that, in some cases, it will enable researchers to genotype animals too young or too sick to be bled.

Noninvasive DNA sampling is, of course, not limited to hairs; procedures employing other tissue types are under active development in this and other laboratories. DNA extracted from bones and teeth promises to be useful in retrospective studies of pedigree relationships where only skeletal material of deceased colony members is available [Morin, unpubl. data]. Buccal cells, which may be isolated from sugarcane wadges, have been successfully used as a DNA source in one study of relationships in a captive colony of chimpanzees [58]. The possibility of isolating host DNA from cells in feces has not yet received serious attention for nonhuman primates.

Paternity Assessment and Primate Conservation

The paternity assessment technique introduced here is important not just for studies of sociobiological issues but increasingly for the conservation of the genetic resources of threatened species in both captive and free-ranging situations. The methods of genetic management of captive and threatened populations are reviewed elsewhere [59–62] and include several practices which presuppose the availability of data on paternity. Paternity information is necessary in the design of sound breeding programs to avoid inbreeding depression, to retard the inevitable loss of innate genetic variability, and to identify sibships and other genetically comparable groups for experimental purposes. Pedigrees are a prerequisite for the identification and control of genetic disease. Establishing paternity permits managers to equalize founder contributions to maximize the genetic effective population size (N_e) of small and fragmented groups. Smith's [63, 71] proposals for rhesus macaque colony management by selective culling of genetically excess males and the simulation of gene flow by cross-fostering of infants would be facilitated by the genotyping procedures introduced here. In the wild, it will become increasingly necessary for managers to move males occasionally between now-isolated populations to simulate historical patterns of male emigration and gene flow and retard the loss of local allelic diversity in fragments of the metapopulation [64]. The success of such translocations can be assessed by subsequent noninvasive monitoring of the subpopulations.

One of the problems confronted by primate colony managers and conservationists is the relatively high infant mortality experienced in captive breeding situations [65]. Neonatal and first-year mortality rates of 15% in primates generally and over 20% in the great apes [66, 67] are a source of frustration. As improved management reduces the mortality attributable to inbreeding to insignificance in many colonies [68], attention turns to other aspects of animal care. One hitherto unquantified variable involves outbreeding depression. Outbreeding depression is an increase in gametic incompatibility, zygotic and embryonic mortality, stillbirths, decreased fertility, and increased mortality that may be manifest in the F1 or delayed until the F2 or backcross generations [69]. Unlike inbreeding depression, where the decline in fitness is attributed to interactions within loci (generally dominance effects), outbreeding problems are caused by interactions between loci: to a breakup of coadapted gene complexes or favorable epistatic relationships [70]. Some of the observed mortality in captive primates may be due to the inadvertent mating of individuals of diverse geographic origin representing genetically divergent races or subspecies.

The subspecies taxon has been all but abandoned by biologists, as it is rarely congruent with the evolutionary significant units of interest. Subspecific taxonomic categories are also notoriously poor guides for conservation management decisions [59–62]. Some morphologically or geographically defined subspecies in the genera *Ateles, Aotus* and *Pongo* are now recognized as genetically well-differentiated species. Although human races are remarkably similar genetically, the racial differentiation of many other species of primates may mask far greater underlying genetic differentiation. In chimpanzees, *P. troglodytes*, for example, no attempt has been made to manage the North American captive populations according to the three recognized subspecies. Although most captive apes are probable west African *P. t. verus*, some are undoubtedly central African *P. t. troglodytes*. The hybrids between these taxa are viable but of unknown fitness relative to the offspring of intrasubspecific matings. As we can now distinguish these two subspecies genetically [Morin and Woodruff, in preparation], we will soon be able to reassess the breeding records for evidence of outbreeding depression.

In the 4 years since multilocus minisatellite DNA fingerprinting was first used to determine patterns of parentage in birds, it has revolutionized our understanding of avian social and mating systems [11, 45, and references therein]. In addition to revealing new levels of complexity, it has opened the results of decades of behavioral ecology research on male and female reproductive strategies to reinterpretation. Hopefully, the noninvasive sampling and single-locus microsatellite genotyping technique of establishing pedigree relationships described here will enable investigators to undertake parallel studies of both free-ranging and managed primates.

Acknowledgments

We thank James Weber (Marshfield Medical Research Foundation, Marshfield, Wisc.) for providing us with microsatellite primers to test on nonhuman primates. The keepers and staff of the Center for the Reproduction of Endangered Species, Zoological Society of San Diego, and the Asheboro Zoo, Zoological Society of North Carolina, provided hair samples for the captive bonobo and chimpanzee studies, respectively. Jane Goodall, Jean Wickings, and Christophe and Hedwig Boesch contributed hair samples from African chimpanzees. Britta Becker, Heather Boyd, Katherine Nielsen, and especially Gayle Yamamoto contributed to the development of the protocols in our laboratory. This research was supported by grants from the US National Institutes of Health, the US National Science Foundation and the Academic Senate of the University of California.

References

1 Wrangham RW: Mutualism, kinship, and social evolution; in Kings College Sociobiology Group (eds): Current Problems in Sociobiology. Cambridge, Cambridge University Press, 1982, pp 269–289.

2 Smuts BB, Cheney DL, Seyfarth RM, et al (eds): Primate Societies. Chicago, University of Chicago Press, 1987 pp 1–578.

3 Symington MM: Fission-fusion social organization in *Ateles* and *Pan*. Int J Primatol 1990;11:47–61.

4 Silk JB: Social behavior in evolutionary perspective; in Smuts BB, Cheney DL, Seyfarth RM, et al (eds): Primate Societies. Chicago, University of Chicago Press, 1987, pp 318–329.

5 Smuts BB: Sexual competition and mate choice; in Smuts BB, Cheney DL, Seyfarth RM, et al (eds): Primate Societies. Chicago, University of Chicago Press, 1987, pp 385–399.

6 Cheney DL, Seyfarth RM, Smuts BB, et al: Future of primate research; in Smuts BB, Cheney DL, Seyfarth RM, et al (eds): Primate Societies. Chicago, University of Chicago Press, 1987, pp 491–497.

7 Tilson R: Primate mating systems and their consequences for captive management; in Benirschke K (ed): Primates: The Road to Self-Sustaining Populations. New York, Springer, 1986, pp 361–373.

8 Pope TR: The reproductive consequences of male cooperation in the red howler monkey: Paternity exclusion in multi-male and single-male troops using genetic markers. Behav Ecol Sociobiol 1990;27:439–446.

9 Avise JC, Arnold J, Ball RM, et al: Intraspecific phylogeography: The mitochondrial DNA bridge between population genetics and systematics. Annu Rev Ecol Syst 1987;18:489–522.

10 Honeycutt R, Wheeler W: Mitochondrial DNA: Variation in human and higher primates; in Dutta S, Winter W (ed): DNA Systematics: Human and Higher Primates. Boca Raton, The Chemical Rubber Company, 1989.

11 Burke T: DNA fingerprinting and other methods for the study of mating success. Trends Ecol Evol 1989;4:139–144.

12 Ely J, Ferrell RE: DNA 'fingerprints' and paternity ascertainment in chimpanzees (*Pan troglodytes*). Zoo Biol 1990;9:91–98.

13 Inoue M, Takenaka A, Tanaka S, et al: Paternity discrimination in a Japanese macaque troop by DNA fingerprinting. Primates 1990;31:563–570.

14 Dowling TE, Moritz C, Palmer JD: Nucleic acids. II. Restriction site analysis; in Hillis DM, Moritz C (eds): Molecular Systematics. Sunderland, Sinauer Associates 1990, pp 250–317.

15 Morin PA, Ryder OA: Founder contribution and pedigree inference in a captive breeding colony of lion-tailed macaques, using mitochondrial DNA and DNA fingerprint analyses. Zoo Biol 1991;10:341–352.

16 Jeffreys AJ, Turner M, Debenham P: The efficiency of multilocus DNA fingerprint probes for individualization and establishment of family relationships determined from extensive casework. Am J Hum Genet 1991;48:824–840.

17 Lander ES: DNA fingerprinting on trial. Nature 1989;339:501–505.

18 Lander ES: Research on DNA typing catching up with courtroom application. Am J Hum Genet 1991;48:819–823.

19 Lynch M: Estimation of relatedness by DNA fingerprinting. Mol Biol Evol 1988;5: 584–599.

20 Ely J, Alford P, Ferrell RE: DNA 'fingerprinting' and the genetic management of a captive chimpanzee population *(Pan troglodytes)*. Am J Primatol 1991;24:39–54.

21 Nakamura Y, Leppert M, O'Connel P, et al: Variable number of tandem repeat (VNTR) markers for human gene mapping. Science 1987;235:1616–1622.

22 Nakamura Y, Carlson M, Krapcho K, et al: New approach for isolation of VNTR markers. Am J Hum Genet 1988;43:854–859.

23 Higuchi R, von Beroldingen, CH, Sensabaugh GF, et al: DNA typing from single hairs. Nature 1988;332:543–546.

24 Saiki RK, Gelfand DH, Stoffel S, et al: Primer-directed enzymatic amplification of DNA with a thermostable DNA polymerase. Science 1988;239:487–491.

25 Erlich HA: Recent advances in the polymerase chain reaction. Science 1991;252: 1643–1651.

26 Mullis K: The unusual origin of the polymerase chain reaction. Sci Am 1990;262: 56–65.

27 Saiki RK, Scharf S, Faloona F, et al. Enzymatic amplification of β-globin genomic sequences and restriction site analysis for diagnosis of sickle cell anemia. Science 1985;230:1350–1354.

28 Innis MA, Gelfand DH, Sninsky JJ, et al (eds): PCR Protocols. A Guide to Methods and Applications. San Diego, Academic Press, 1990, pp 1–482.

29 von Beroldingen CH, Higuchi RG, Sensabaugh GF, et al: Analysis of enzymatically amplified HLA-DQα DNA from single human hairs. Am J Hum Genet 1987;41: 725.

30 von Beroldingen C: Applications of PCR to the analysis of biological evidence; in Erlich HA (ed): PCR Technology. New York, Stockton Press, 1989, pp 209–224.

31 Pääbo, S: Amplifying ancient DNA; in Innis MA, Gelfand DH, Sninsky JJ, et al (eds): PCR Protocols. A Guide to Methods and Applications. San Diego, Academic Press, 1990, pp 159–166.

32 Weber JL, May PE: Abundant class of human DNA polymorphisms which can be typed using the polymerase chain reaction. Am J Hum Genet 1989;44:388–396.

33 Litt M, Luty JA: A hypervariable microsatellite revealed by in vitro amplification of a dinucleotide repeat within the cardiac muscle actin gene. Am J Hum Genet 1989; 44:397–401.

34 Tautz D: Hypervariability of simple sequences is a general source for polymorphic markers. Nucleic Acids Res 1989;17:6436–6471.

35 Weber JL: Informativeness of human $(dC\text{-}dA)_n.(dT\text{-}dG)_n$ polymorphisms. Genomics 1990;7:524–530.

36 Smouse PE, Chakraborty R: The use of restriction fragment length polymorphisms in paternity analysis. Am J Hum Genet 1986;38:918–939.

37 Selvin S: Probability of non-paternity determined by multiple allele codominant systems. Am J Hum Genet 1980;32:276–278.

38 Chakravarti A, Li CC: The effect of linkage on paternity calculations; in Walker RH, Duquesnoy RJ, Jennings ER, et al (eds): Inclusion Probabilities in Parentage Testing. Arlington VA, American Association of Blood Banks, 1983, pp 411–420.

39 Weber JL, May PE: Dinucleotide repeat polymorphism at the D18S35 locus. Nucleic Acid Res 1990;18:6465.

40 Weber JL, Kwitek AE, May PE: Dinucleotide repeat polymorphisms at the D16S260, D16S261, D16S265, and D16S267 loci. Nucleic Acids Res 1990;18: 4034.
41 Weber JL, Kwitek AE, May PE, et al: Dinucleotide repeat polymorphism at the D8S85, D8S87, and D8S88 loci. Nucleic Acids Res 1990;18:4038.
42 Singer M, Berg P: Genes and Genomes: A Changing Perspective. Mill Valley, University Science Books, 1991, pp 1–929.
43 Orrego C: Organizing a laboratory for PCR work; in Innis MA, Gelfand DH, Sninsky JJ, et al (eds): PCR Protocols: A Guide to Methods and Applications. San Diego, Academic Press, 1990, pp 447–454.
44 Hillis DM, Larson A, Davis SK, et al: Nucleic acids. III. Sequencing; in Hillis DM, Moritz C (eds): Molecular Systematics. Sunderland, Sinauer Associates, 1990, pp 318–370.
45 Weatherhead PJ, Montgomerie RD: Good news and bad news about DNA fingerprinting. Trends Ecol Evol 1991;6:173–174.
46 Coggins LW, Oprey M: DNA tertiary structures formed in vitro by misaligned hybridization of multiple tandem repeat sequences. Nucleic Acids Res 1989;17: 7417–7426.
47 Luty JA, Guo Z, Willard HF, et al: Five polymorphic microsatellite VNTRs on the human X chromosome. Am J Hum Genet 1990;46:776–783.
48 Economou EP, Bergen AW, Warren AC, et al: The polydeoxyadenylate tract of the Alu repetitive elements is polymorphic in the human genome. Proc Natl Acad Sci USA 1990;87:2951–2954.
49 Peterson MB, Schinzel AA, Binkert F, et al: Use of short sequence repeat DNA polymorphisms after PCR amplification to detect the parental origin of the additional chromosome 21 in Downs syndrome. Am J Hum Genet 1991;48:65–71.
50 Ploos van Amstel HK, Reitsma PH, van der Logt CPE, et al: Intron-exon organization of the active human protein S gene PSα and its pseudogene PSβ: Duplication and silencing during primate evolution. Biochemistry 1990;29:7853–7861.
51 Polymeropoulos MH, Rath DS, Xiau H, et al: Trinucleotide repeat polymorphism at the human pancreatic phospholipase-A-2 gene (PLA2). Nucleic Acids Res 1990;18: 7468.
52 Polymeropoulos MH, Rath DS, Xiau H, et al: Trinucleotide repeat polymorphism at the human intestinal fatty acid binding protein gene (FABP2). Nucleic Acids Res 1990;18:7198.
53 Polymeropoulos MH, Xiau H, Rath DS, et al: Tetranucleotide repeat polymorphism at the human aromatase cytochrome-P-450 gene (CYP19). Nucleic Acids Res 1991; 19:195.
54 Zuliani G, Hobbs HH: Tetranucleotide repeat polymorphism in the LPL gene. Nucleic Acids Res 1990;18:4958.
55 Zuliani G, Hobbs HH: Tetranucleotide repeat polymorphism in the apolipoprotein-C-III gene. Nucleic Acids Res 1990;18:4299.
56 Zuliani G, Hobbs HH: Tetranucleotide repeat polymorphism in the apolipoprotein-B gene. Nucleic Acids Res 1990;18:4299.
57 Ploos van Amstel HK, Reitsma PH: Tetranucleotide repeat polymorphism in the vWB gene. Nucleic Acids Res 1991;18:4957.
58 Takasaki H, Takenaka O: Paternity testing in chimpanzees with DNA amplification

from hairs and buccal cells in wadges: A preliminary note; in Ehara A, et al (eds): Primatology Today. Proc 13th Congr Int Primatol Soc, Amsterdam, Elsevier, 1991, pp 613–616.

59 Woodruff DS: The problems of conserving genes and species; in Western D, Pearl M (eds): Conservation for the Twenty-First Century. Oxford, Oxford University Press, 1989, pp 76–88.

60 Woodruff DS: Genetically based measures of uniqueness: commentary; in Orians G, Brown G, Kunin W, et al (eds): The Preservation and Valuation of Biological Resources. Seattle, University of Washington Press, 1990, pp 199–132.

61 Woodruff DS: Genetics and demography in the conservation of biodiversity. J Sci Soc Thailand, in press.

62 Woodruff DS: Genetics and the conservation of animals in fragmented habitats. J Malay Nature Soc, in press.

63 Smith DG: Incidence and consequences of inbreeding in three captive groups of rhesus macaques *(Macaca mulatta);* in Benirschke K (ed): Primates: The Road to Self-Sustaining Populations. New York, Springer, 1986, pp 857–874.

64 Gilpin M: The genetic effective size of a metapopulation. Biol J Linn Soc 1991;42: 165–175.

65 Benirschke K (ed): Primates: The Road to Self-Sustaining Populations. New York, Springer, 1986, pp 1–1044.

66 Flesness NR: Captive status and genetic considerations; in Benirschke K (ed): Primates: The Road to Self-Sustaining Populations. New York, Springer, 1986, pp 845–856.

67 Seal US, Flesness NR: Captive chimpanzee populations – Past, present and future; in Benirschke K (ed): Primates: The Road to Self-Sustaining Populations. New York, Springer, 1986, pp 47–55.

68 Ralls K, Ballou JD, Templeton AR: Estimation of lethal equivalents and the cost of inbreeding in mammals. Cons Biol 1988;2:185–193.

69 Templeton AR, Hemmer H, Mace G, et al. Local adaptation, coadaptation, and population boundaries. Zoo Biol 1986;5:115–126.

70 Lynch M: The genetic interpretation of inbreeding depression and outbreeding depression. Evolution 1991;45:622–629.

71 Smith DG, Rolfs B, Lorenz J: A comparison of the success of electrophoretic methods and DNA fingerprinting for paternity testing in captive groups of rhesus macaques; in Martin RD, Dixson AF, Wickings EJ (eds): Paternity in Primates: Genetic Tests and Theories. Basel, Karger, 1992, pp 32–52.

72 VandeBerg JL: Biochemical markers and restriction fragment length polymorphisms in baboons: Their power for paternity exclusion; in Martin RD, Dixson AF, Wickings EJ (eds): Paternity in Primates: Genetic Tests and Theories. Basel, Karger, 1992, pp 18–31.

73 Ménard M, Scheffrahn W, Vallet D, Zidane C, Reber Ch: Application of blood protein electrophoresis and DNA fingerprinting to the analysis of paternity and social characteristics of wild barbary macaques; in Martin RD, Dixson AF, Wickings EJ (eds): Paternity in Primates: Genetic Tests and Theories. Basel, Karger, 1992, pp 155–174.

74 de Ruiter JR, Scheffrahn W, Trommelen GJJM, et al: Male social rank and reproductive success in wild long-tailed macaques. Paternity exclusions by blood protein analysis and DNA fingerprinting; in Martin RD, Dixson AF, Wickings EJ (eds): Paternity in Primates: Genetic Tests and Theories. Basel, Karger, 1992, pp 175–191.

75 Lewis RE, Cruse JM: DNA typing in human parentage testing using multilocus and single-locus probes; in Martin RD, Dixson AF, Wickings EJ (eds): Paternity in Primates: Genetic Tests and Theories. Basel, Karger, 1992, pp 3–17.
76 Wickings EJ, Dixson AF: Application of DNA fingerprinting to familial studies of Gabonese primates; in Martin RD, Dixson AF, Wickings EJ (eds): Paternity in Primates: Genetic Tests and Theories. Basel, Karger, 1992, pp 113–130.
77 Turner TR, Weiss ML, Pereira ME: DNA fingerprinting and paternity assessment in Old World monkeys and ringtailed lemurs; in Martin RD, Dixson AF, Wickings EJ (eds): Paternity in Primates: Genetic Tests and Theories. Basel, Karger, 1992, pp 96–112.
78 Inoue M, Mitsunaga F, Ohsawa H, et al: Paternity testing in captive Japanese macaques *(Macaca fuscata)* using DNA fingerprinting; in Martin RD, Dixson AF, Wickings EJ (eds): Paternity in Primates: Genetic Tests and Theories. Basel, Karger, 1992, pp 131–140.
79 Kuester, J, Paul A, Arnemann J: Paternity determination by oligonucleotide DNA fingerprinting in barbary macaques *(Macaca sylvanus);* in Martin RD, Dixson AF, Wickings EJ (eds): Paternity in Primates: Genetic Tests and Theories. Basel, Karger, 1992, pp 141–154.
80 Rogers J: Nuclear DNA polymorphisms in hominoids and cercopithecoids: Applications to paternity testing; in Martin RD, Dixson AF, Wickings EJ (eds): Paternity in Primates: Genetic Tests and Theories. Basel, Karger, 1992, pp 82–95.
81 Anzenberger G: Monogamous social systems and paternity in primates; in Martin RD, Dixson AF, Wickings EJ (eds): Paternity in Primates: Genetic Tests and Theories. Basel, Karger, 1992, pp 203–224.
82 Dixson AF, Anzenberger G, Monteiro Da Cruz MAO, et al: DNA fingerprinting of free-ranging groups of common marmosets *(Callithrix jacchus jacchus)* in NE Brazil; in Martin RD, Dixson AF, Wickings EJ (eds): Paternity in Primates: Genetic Tests and Theories. Basel, Karger, 1992, pp 192–202.
83 Washio K: Genetic identification of nonhuman primates using tandem-repetitive DNA sequences; in Martin RD, Dixson AF, Wickings EJ (eds): Paternity in Primates: Genetic Tests and Theories. Basel, Karger, 1992, pp 53–62.

Phillip A. Morin, PhD cand, Department of Biology, University of California, San Diego, La Jolla, CA 92093–0116 (USA)

Martin RD, Dixson AF, Wickings EJ (eds): Paternity in Primates: Genetic Tests and Theories. Basel, Karger, 1992, pp 82–95

Nuclear DNA Polymorphisms in Hominoids and Cercopithecoids: Applications to Paternity Testing

Jeffrey Rogers

Department of Genetics, Southwest Foundation for Biomedical Research, San Antonio, Tex., USA

Molecular genetic techniques now provide greater opportunities to examine genetic variation in populations of nonhuman primates than previous methods. Techniques developed in human genetics are making a major impact in studies of nonhuman species, particularly in the analysis of paternity. But the circumstances and constraints in studies of paternity among nonhuman primates are different from those in human cases. Furthermore, it is not necessary to judge all nonhuman paternity analyses by the same standards of success. This paper reviews the contexts within which paternity testing is conducted with nonhuman primates, and argues for a mixed strategy that combines the use of DNA polymorphisms, in both hypervariable repetitive sequences and single-copy unique sequences, with the continued use of traditional protein polymorphisms.

Context of Paternity Analysis

There are two basic contexts in which investigators undertake paternity testing in nonhuman primates, and the differences between them have important ramifications. Some studies are carried out with captive colonies, where infants are born in a closed population. In these situations, it is possible to compile a complete listing of all the possible fathers of a given infant. Furthermore, all these males may be available for genetic testing, or at least comprehensive testing of surviving group members is feasible.

Other paternity studies are conducted with wild populations. While studies of natural populations often attempt to include all potential fathers in analyses, this is more difficult than in captive colonies for several reasons. First, field investigators may know the identity of several potential fathers, but may be unable to obtain genetic data from all of them, due to inability to capture particular individuals or to the disappearance of potential fathers prior to genetic sampling. Second, not all potential fathers may be recognized as such. A male may mate in a group after a very short, unrecognized period of membership, or after a period of peripheral interaction with the group. Mating by extra-group males greatly complicates attempts at paternity analysis, and in fact casts uncertainty on all attempts to define the complete set of potential fathers for a given infant [see, for example, ref. 1].

The situations in which paternity testing is done also differ in their broader scientific context. Studies of natural populations will usually include demographic, ecological and/or behavioral investigations, possibly including documentation of consortships. On the other hand, most captive colonies are maintained for biomedical research. As a result, the captive animals for which paternity information is desired are likely to be used for detailed physiological, endocrinological, biochemical or behavioral investigations that are quite different from the studies of wild animals.

But field and captive studies also have important features in common. Genetic analysis of paternity is seldom if ever the only analysis undertaken on a given study population, wild or captive. In both situations, other data (some of them genetic) in addition to data specifically obtained for paternity testing will either be available or would be helpful to other projects. The design of paternity-testing strategies can be coordinated with other analyses for mutual benefit. For example, if genetic analyses are to be conducted on a specific wild population for the purpose of paternity testing, such testing can be designed to generate data that are also useful for other applications such as analyses of population genetic structure or genetic distances between this and other conspecific populations. In captive contexts, the immediate objective may be determining pedigree relationships efficiently, but only some types of genetic data will be useful for other research projects such as genetic linkage or genetic epidemiology. This issue will be discussed further in the final section of this paper.

The power of paternity analyses in both captive colonies and wild populations may suffer from a lack of genetic variability. Distinguishing among fathers and non-fathers will be more difficult whenever genetic variability among individuals is low. Sound management of captive colonies in order to retain as much variability as possible is of course beneficial for future attempts to resolve paternity, as well as for other reasons. In field studies, investigators can only endeavor to find additional highly polymorphic genetic markers for comparisons among individuals.

Finally, it should be acknowledged that accurate records of mother-infant relationships are a prerequisite for paternity testing wherever it is done. For natural populations, mother-infant relationships are commonly inferred from behavioral observations. Misidentification is a possibility in both field and captive situations. For example, VandeBerg et al. [2] examined pedigrees of captive squirrel monkeys (*Saimiri boliviensis* and *Saimiri sciureus*) and found 2 examples of mistaken maternity out of 79 infants tested. This can occur either through mistakes in records or actual swapping of infants by mothers that give birth at approximately the same time. Prior to any paternity testing, mother-infant pairs must be checked to detect any inaccuracies in these inferred relationships.

Criteria for Evaluating the Success of Paternity Analysis

It is not necessary to judge all analyses of paternity by the same criteria of success. It may be the desire of all investigators to identify the true fathers of all infants in a lengthy list. This is the most ambitious goal that can be proposed and in some circumstances it is an unrealistic one, especially in field studies. Nevertheless, paternity testing can provide valuable information, and hence a paternity analysis can be considered a success even if it falls short of this level of precision.

Several different levels of precision can be achieved, depending on the completeness of the data and the amount of time and resources invested in attempting to test and exclude candidate fathers. It is often possible to develop a set of probability statements that assign relative likelihoods of paternity to each of several potential fathers, including the unambiguous exclusion of a proportion of candidates [3, 4]. This result may not be acceptable in some circumstances, such as tests of certain sociobiological models. But in other circumstances a set of relative likelihoods will be sufficient. For example, if the intention is to estimate the number of males contributing to each generation in a captive colony or a free-ranging social group, then probability statements describing the likelihood of paternity for all ascertained males can be combined into a maximum likelihood estimate of the number of different contributing fathers.

In some research contexts, it may be sufficient to exclude some males as the father of a given infant without attempting to determine the identity of the true father. Consider an infant born in a captive social group containing several adult males, adult females and subadult males. The researcher may wish to know whether any subadult males actually breed and produce offspring in the group. It would be adequate to exclude either all the fully adult males or all the subadult males as father of the new infant without being concerned with determining precisely which male is the true father. Another possible example would be a field study with records of consortships. One might wish to determine the proportion of cases in which the male observed to consort with a cycling female can be excluded as father of her infant.

Depending on the reason for doing paternity testing, the researcher may set a very ambitious and difficult goal or may be able to accept a less precise result as adequate. This is significant because new molecular genetic techniques provide opportunities for detailed and extensive genetic testing. But this testing can be time-consuming and expensive, especially if it requires the development of new molecular probes (see below). While it may be possible in many circumstances to obtain data that lead to very precise answers to questions of paternity, it is advisable to determine the level of precision required prior to beginning the genetic experiments to generate those data. Considerable energy and resources may be saved.

Methods for Detecting Variation in the Nuclear Genome among Individual Nonhuman Primates

Over the past 10–15 years, a number of previously unknown forms of molecular (DNA sequence) variability within *Homo sapiens* have been described and characterized. Some of these novel forms of genetic variation are useful in human paternity testing and forensic identification analyses. A smaller number have been used in paternity testing among nonhuman primates.

Hypervariable Single-Locus Probes

The human genome contains at least several hundred loci that can be called 'hypervariable' because they each exhibit a large number of readily distinguishable alleles, often 15–20 or more [5–9]. These loci share a common molecular structure. They consist of unique, single-copy DNA sequences surrounding a set of end-to-end repeats of a 10- to 30-bp core sequence. This structure is highly susceptible to mutations that alter the number of repeat units in an allele [10, 11]. Consequently, these hypervariable loci accumulate mutations that produce new alleles differing primarily in length, i.e. in the number of repeat units present in each allele. This number can vary from a few dozen to hundreds [7, 8]. These loci have been called variable number of tandem repeat or VNTR loci [5].

To detect differences among individuals at these hypervariable VNTR loci, human genomic DNA is cut with a restriction enzyme that does not cut within the core repeat sequence, then run out in an electrophoretic gel and hybridized with high stringency to a radioactively labelled probe made from a clone that includes the unique, single-copy DNA sequences that flank the repeat units, and in some cases also includes the repeat unit itself. This procedure reveals two bands per individual. Each band represents the DNA from one copy of the VNTR locus, located on one member of a pair of homologous chromosomes. Since these loci have a large number of different alleles segregating in any one population, the chances that the two bands from any particular noninbred individual are identical in length are small. The chances that two unrelated individuals will have the same two bands are much smaller, and consequently these loci are powerful tools for distinguishing among individuals, and well suited for paternity testing and forensic identification work [8]. (Alternatively, these loci can be investigated using polymerase chain reaction (PCR) amplification [12] as opposed to restriction digestion and hybridization, but the results obtained are the same.) DNA clones or probes are available for about one hundred such loci in the human genome [6]. Furthermore, the value of VNTR loci is not limited to paternity and forensic analyses. They are also very useful in linkage analysis [5], population genetics [7] and other applications.

Hypervariable Multilocus Probes

The approach outlined above has been used extensively in forensic analyses and gene-mapping studies using genetic linkage. In these circumstances, the laboratories seek to determine the phenotype of a person at a single, specific locus, often a locus mapped to a particular location in the human genome. Most analyses of paternity have used a somewhat different approach to detect variation at similar hypervariable loci with multiple alleles consisting of tandem repeats. Jeffreys et al. [13] introduced a method that detects several different hypervariable loci in the same analysis. This technique, which has been described as DNA fingerprinting with minisatellite probes, does not include the unique, single-copy DNA that surrounds a VNTR locus in the probe. Rather, only the repeated core sequence itself is used. Since several VNTR loci share the same or very similar core sequences, the approach of Jeffreys et al. allows the investigator to compare two individuals for several different hypervariable loci simultaneously in a single analysis.

The use of these minisatellite or multilocus hypervariable probes has advantages over the single-locus method described above. First, hybridization of the probe to multiple loci means that more bands are detected and hence more variation is observed in a given analysis. Second, the same probes have proven useful in various human populations, whereas a particular single-locus probe may be more or less polymorphic in different geographic populations. Overall, the minisatellite probes have been very successful in distinguishing fathers from nonfathers in human paternity analyses [11].

But the use of minisatellite probes also suffers from certain disadvantages relative to single-locus hypervariable probes. It is usually impossible to identify which bands observed in the multilocus analyses come from which loci. As a result, the data are not useful for other types of research applications such as linkage analysis or studies of population genetic structure and genetic distances between populations. In addition, because several loci, each with a high mutation rate, are examined at once, the chances of observing a new mutation that will make offspring differ from their parents are increased. New mutations will obscure the true biological relationships among individuals, and it is of course the goal of paternity testing to determine these relationships.

Microsatellites

Other similar loci, called hypervariable microsatellite loci, also occur in the human genome. These microsatellites resemble VNTR loci, but differ in the length of the repeat unit. They consist of single-copy DNA surrounding multiple repeats of a 2-bp sequence (CA). Hundreds of such loci occur in the human genome, and they are valuable in human genetic analyses because, like the other hypervariable loci, they have multiple alleles [14, 15]. Analyses of human paternity have not made substantial use of these markers to date, but they could be of value in that area. Furthermore, microsatellites have all the advantages of the single-locus VNTR probes over the multilocus minisatellite method.

Single-Copy Sequences

Several different approaches have been used to detect sequence polymorphism in single-copy nuclear DNA, i.e. DNA that does not contain multiple repeats like the hypervariable loci. One of the most successful has been the investigation of restriction fragment length polymorphisms (RFLPs). This approach uses restriction endonucleases and probes that detect only a single locus to reveal individual differences in DNA sequence [see ref. 16 for a review]. More than 2,000 human DNA polymorphisms are known, and most of them are RFLPs [6]. The discovery and characterization of human RFLPs have progressed rapidly, partly because there are thousands of different human clones available that can be used as probes.

Human DNA clones have been used successfully as probes in studies of RFLP variation in the single-copy DNA of several cercopithecine and hominoid species. Rogers [17] and Rogers and Kidd [18] examined nuclear DNA polymorphism in a wild population of yellow baboons *(Papio hamadryas cynocephalus)* from Mikumi National Park, Tanzania. Four different probes from human single-copy genes were screened for RFLPs with 6 restriction enzymes each. A total of 13 polymorphisms, at least 2 at each locus, were found. Eleven of these polymorphic systems have heterozygosity values above 0.40. Figure 1 illustrates such RFLPs, showing some of the results obtained with a clone of the human gene for apolipoprotein B. Two different restriction sites are polymorphic in this case, which results in a three-allele system. Other RFLPs detected with the same probe provided additional information regarding differences among individuals at this locus.

Using the approach of Ewens et al. [19] Rogers [17] and Rogers and Kidd [18] calculated that the average proportion of nucleotides polymorphic across these 4 loci is between 1 and 1.5%. This indicates that, assuming that these 4 loci are representative of other loci in the baboon genome, approximately 1–1.5% of basepairs in the nuclear DNA are polymorphic in this local population. If this preliminary estimate proves to be correct, then a very large number of RFLPs could be found in baboons by using available human clones as probes.

Other studies have also described RFLPs in nonhuman primates. Hixson et al. [20, 21] found polymorphisms in the APOA1, LDLR and other genes of captive *Papio* baboons. Lu et al. [22] found several different RFLPs in the glycophorin genes of captive

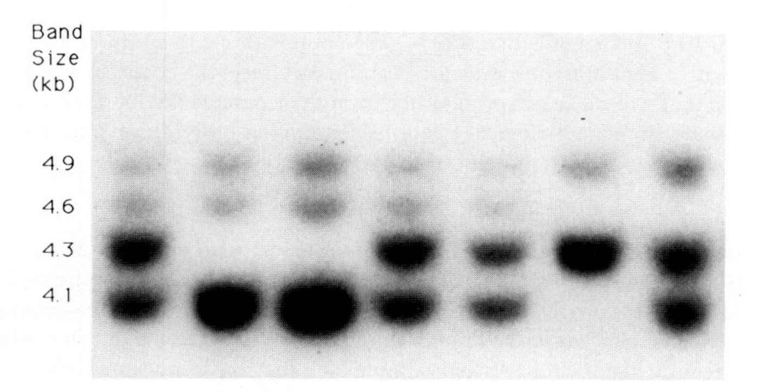

Fig. 1. RFLPs at the apolipoprotein B locus among wild-caught yellow baboons from Mikumi National Park, Tanzania. Total genomic DNA was digested with PvuII and run in a 1% agarose gel. The probe used was a segment of the human clone pABF obtained from J. Scott. Fragment sizes are presented in kilobases (kb) of DNA. These 4 bands result from 2 different polymorphic restriction sites. The presence or absence of one site produces the alternative 4.9- and 4.6-kb bands. The variability in the 4.3- and 4.1-kb bands depends on a different restriction site.

chimpanzees *(Pan troglodytes)*. Rogers et al. [23] describe 9 distinct RFLPs in 5 genes among a sample of 5 captive rhesus monkeys *(Macaca mulatta)*. While it is not possible from these reports to estimate the proportion of basepairs or the proportion of restriction sites that are polymorphic in these species, it is clear that this approach is an effective way to detect new DNA polymorphisms in many primate populations.

Other molecular techniques also have great potential for detecting DNA polymorphisms in primates. One of the most powerful methods for revealing differences among single-copy DNA sequences is denaturing gradient gel electrophoresis (DGE) [24, 25]. This method exploits basic features of the structure and physical chemistry of DNA to detect differences among molecules, and is capable of detecting differences among DNA molecules that RFLP analysis cannot detect. DGE can detect almost any base-pair variation within a small segment of DNA without requiring the investigator to sequence the segment entirely [25].

DGE has been applied to the analysis of variation within and between hominoid species [23, 26]. PCR amplification was used to isolate homologous segments of DNA from the HOX2 gene cluster in humans, chimpanzees, gorillas *(Gorilla gorilla)* and orangutans *(Pongo pygmaeus)*. Analysis using DGE showed that the amplified sequence, which was about 300 bp long, was different in each of the 4 species. No variation was observed among 27 human samples, but intraspecific variability was found in both chimpanzees and gorillas. The 2 alleles found among the 16 chimpanzee chromosomes screened actually differ by 2 bp substitutions. Among the 10 gorillas tested, at least 3 different alleles were observed.

Traditional Studies of Protein Polymorphisms

The tremendous success of molecular genetics in detecting DNA polymorphisms in humans has to some extent overshadowed traditional analyses of protein and blood group polymorphisms. However, numerous serum and red blood cell protein polymorphisms are known in primates, and the value of these genetic markers should not be underrated. Several protein polymorphisms have exhibited high heterozygosity [e.g. 27, 28]. This type of genetic marker can be successfully used for paternity analyses in many primate species [e.g. 2, 29–32]. In some circumstances, such as when (a) the population under study is polymorphic for several known protein systems, (b) the number of potential fathers for each infant is small and (c) all the potential fathers can be genetically tested, then a substantial proportion of fathers can be identified with protein markers alone. The larger the number or degree of heterozygosity of the protein polymorphisms and the smaller the number of potential fathers, the greater the proportion of fathers that will be unambiguously identified. The discovery and characterization of hypervariable loci have clearly revolutionized our ability to detect individual variation among humans. But until the value of these approaches is demonstrated in studies of a variety of nonhuman primates, it would be premature to conclude that analyses of paternity will no longer benefit from the typing of protein systems. Furthermore, the protein polymorphisms can contribute both to the paternity analyses and to other applications such as genetic linkage or population structure analyses.

Application of Molecular Methods to Paternity Testing in Nonhuman Primates

As already noted, a number of researchers have used traditional protein and blood group polymorphisms in paternity exclusion analyses among nonhuman primates [e.g. 2, 29–32]. These studies have been partially successful, determining paternity for a proportion of the infants analyzed. Recently, primatologists have begun to use human hypervariable multi-locus (minisatellite) probes to detect individual variation within nonhuman species [33–36]. Some laboratories have successfully identified fathers by using such probes to exclude all but one possible male [35, 36]. Others have performed the same type of analysis using M13-derived minisatellite probes [37]. But in some of these cases, the number of potential fathers was small, and it is quite possible that analysis using a modest number of single-locus markers such as RFLPs or protein polymorphisms would also have been successful.

Human minisatellite probes clearly represent a very significant advance in the investigation of nonhuman pedigrees. These probes have great power to detect individual variation among humans and thus great power to exclude any specific male as the father of a given infant. But the use of minisatellite probes in studies of nonhuman primates has inherent limitations.

First, not all minisatellite probes are highly polymorphic in all primate populations. Weiss et al. [33] found that two such probes detected substantial variation in *Erythrocebus patas* but much less in *Colobus guereza*. In some cases, the commonly used human minisatellite probes may be no more informative than RFLPs in single-copy sequences or other markers [39]. Preliminary testing of several different probes may be necessary before finding one which is adequately polymorphic in the particular population under study.

Second, minisatellite probes detect loci that will be subject to an unknown but presumably high rate of new mutation. For example, in a study of over 1700 human paternity cases where two commonly used minisatellite probes were applied [11], new mutations which caused an infant to exhibit a band not found in either the mother or the true father occurred in over 25% of families. With one probe, the proportion of new mutant bands in offspring was greater than 0.01. This problem necessitates the testing of additional loci in the human cases, and would lead to the false exclusion of the true father if other data were not available.

Such a new mutation was observed by Ely and Ferrell [37] in a study of 21 paternity cases in captive chimpanzees. Because they were able to obtain DNA from both of the possible fathers, Ely and Ferrell could choose the correct male despite the discrepant band. If this occurred in another study where DNA was not available for all possible fathers, such as a field study, it could lead to false exclusion of the true father. This problem of discrepant bands is addressed using a two-band exclusion criterion by de Ruiter et al. [42], and this may be one way to successfully deal with the problem.

Third, while minisatellites can be used for positive paternity identifications in humans, they can be used only for negative paternity exclusions in nonhuman primates. It is possible in human paternity analyses to calculate the probability that a randomly chosen male that is not the father of an infant will share a given number of bands with that infant. These calculations rely on data describing the proportion of bands shared among unrelated individuals [11]. A large data set is required to calculate these proportions, and it must be obtained for the specific population in which the infant and potential fathers occur. It is very unlikely that this amount of data will be available for any nonhuman population, and thus we must fall back on the method of excluding all potential fathers but one in order to identify the true father in nonhuman analyses. This means that any study in which it is impossible to sample all potential fathers can never achieve complete certainty regarding the paternity of any infant.

Finally, the continuous distribution of band sizes and the complexity of interpreting large numbers of bands per individual necessitate explicit standards for judging whether two bands are the same or different. Laboratories utilizing hypervariable probes in human genetic analyses have adopted explicit decision rules to standardize interpretation [8, 38]. Studies of nonhuman primates should adopt similar rigorous and objective standards.

General Approach to Paternity Testing in Nonhuman Primates

At the present time, human hypervariable multilocus or minisatellite probes are the most powerful tools available for paternity (or maternity) testing in primates. These probes can be used to detect substantial amounts of individual variation in several different species, and therefore provide a high probability of excluding random males as the true father of a particular infant. But the minisatellite approach also has inherent problems, including (a) the low level of variability in some populations, which greatly reduces the likelihood of successful analysis without additional data, (b) the uncertainty caused by high but unquantified rates of mutation, (c) the lack of band frequency data which would allow positive paternity identification rather than negative paternity exclusion procedures, and (d) the inability to apply minisatellite data to other research questions.

On the other hand, other methods are available to detect genetic differences among individuals (e.g. RFLPs in single-copy DNA sequences, protein polymorphisms and DGE). These techniques, though much less powerful than the minisatellite approach, can provide additional information, either to confirm an exclusion based on minisatellite data or to supplement minisatellite data when they are insufficient to resolve paternity [4]. In addition, unlike multilocus minisatellite data, these other single-locus systems provide information useful in other types of analyses, such as studies of population genetic structure, linkage and disease association and the general monitoring of loss of heterozygosity due to inbreeding in captive populations.

Finally, it is not necessary that all paternity analyses result in the unambiguous determination of fathers for all infants. In some cases, the time and resources required to develop hypervariable loci useful in a particular population may not be justified, given that single-locus probe data are either already available or would be very useful in other analyses. The recognition that less precise answers to questions concerning paternity can be adequate alters the relative merits of hypervariable and single-locus approaches. In the future, development of single-locus VNTR or microsatellite polymorphisms in nonhuman primates would provide the power of minisatellite loci for paternity testing and still allow application of the data to problems requiring genotypes for identifiable loci [40, 41].

In summary, we can draw three conclusions concerning paternity analysis in nonhuman primates.

(1) There are several different approaches available to generate genetic data for this purpose.

(2) The most powerful approach at this time for human paternity testing is the use of multilocus minisatellite probes, and this approach holds great promise for studies of nonhuman primates.

(3) The best overall strategy for paternity determinations in primates is a mixed one that combines analysis of hypervariable minisatellites and the most polymorphic of the known RFLPs in single-copy DNA, serum and red cell protein polymorphisms.

Acknowledgment

I wish to thank Dr. R.D. Martin for the invitation to participate in this symposium. I also thank Drs. S. Williams-Blangero, J.L. VandeBerg and especially B. Dyke for helpful discussions and critical comments on an early draft of this manuscript.

References

1 Sprague DS: Mating by non-troop males among the Japanese macaques of Yakushima Island. Folia Primatol 1991;57:156–158.
2 VandeBerg JL, Aivaliotis MJ, Williams LE, et al: Biochemical genetic markers of squirrel monkeys and their use for pedigree validation. Biochem Genet 1990;28: 41–56.
3 Elston RC: Probability and paternity testing. Am J Hum Genet 1986;39:112–122.

4 Smouse PE, Chakraborty R: The use of restriction fragment length polymorphisms in paternity analysis. Am J Hum Genet 1986;38:918–939.
5 Nakamura Y, Leppert M, O'Connell P, et al: Variable number of tandem repeat (VNTR) markers for human gene mapping. Science 1987;235:1616–1622.
6 Kidd KK, Bowcock AM, Schmidtke J, et al: Report of the DNA committee and catalogs of cloned and mapped genes and DNA polymorphisms; Human Gene Mapping 10 (1989): Tenth International Workshop on Human Gene Mapping. Cytogenet Cell Genet 1989;51:622–947.
7 Balazs I, Baird M, Clyne M, et al: Human population genetic studies of five hypervariable DNA loci. Am J Hum Genet 1989;44:182–190.
8 Budowle B, Giusti AM, Waye JS, et al: Fixed-bin analysis for statistical evaluation of continuous distributions of allelic data from VNTR loci, for use in forensic comparisons. Am J Hum Genet 1991;48:841–855.
9 Devlin B, Risch N, Roeder K: Estimation of allele frequencies for VNTR loci. Am J Hum Genet 1991;48:662–676.
10 Jeffreys AJ, Royle NJ, Wilson V, et al: Spontaneous mutation rates to new length alleles at tandem-repetitive hypervariable loci in human DNA. Nature 1988;332: 278–281.
11 Jeffreys AJ, Turner M, Debenham P: The efficiency of multilocus DNA fingerprint probes for individualization and establishment of family relationships, determined from extensive casework. Am J Hum Genet 1991;48:824–840.
12 Boerwinkle E, Xiong W, Fourest E, et al: Rapid typing of tandemly repeated hypervariable loci by the polymerase chain reaction: Application to the apolipoprotein B 3' hypervariable region. Proc Natl Acad Sci USA 1989;86:212–216.
13 Jeffreys AJ, Wilson V, Thein SL: Hypervariable 'minisatellite' regions in human DNA. Nature 1985;314:67–73.
14 Weber JL, May PE: Abundant class of human DNA polymorphisms which can be typed using the polymerase chain reaction. Am J Hum Genet 1989;44:388–396.
15 Litt M, Luty JA: A hypervariable microsatellite revealed by in vitro amplification of a dinucleotide repeat within the cardiac muscle actin gene. Am J Hum Genet 1989; 44:397–401.
16 Williams RC: Restriction fragment length polymorphism (RFLP). Yearb Phys Anthropol 1989;32:159–184.
17 Rogers J: Genetic structure and microevolution in a population of Tanzanian yellow baboons; PhD thesis, Yale University, 1989.
18 Rogers J, Kidd KK: Population genetic structure in Tanzanian yellow baboons as measured by RFLPs in nuclear DNA. Am J Phys Anthropol 1989;78:291–292.
19 Ewens WJ, Spielman RS, Harris H: Estimation of genetic variation at the DNA level from restriction endonuclease data. Proc Natl Acad Sci USA 1981;78:3748–3750.
20 Hixson JE, Borenstein S, Cox LA, et al: The baboon gene for apolipoprotein A-I: characterization and identification of DNA polymorphisms for genetic studies of cholesterol metabolism. Gene 1988;74:483–490.
21 Hixson JE, Kammerer CM, Cox LA, et al: Identification of LDL receptor gene marker associated with altered levels of LDL cholesterol and apolipoprotein B in baboons. Arteriosclerosis 1989;9:829–835.

22 Lu WM, Huang CH, Socha WW, et al: Polymorphism and gross structure of glycophorin genes in common chimpanzees. Biochem Genet 1990;28:399–413.
23 Rogers J, Ruano G, Kidd KK: Variability in nuclear DNA among non-human primates: Application of molecular genetic techniques to intra- and inter-species genetic analyses. Am J Primatol, in press.
24 Lerman LS, Fischer SG, Hurley I: Sequence-determined DNA separations. Annu Rev Biophys Bioeng 1984;13:399–423.
25 Abrams ES, Murdaugh SE, Lerman LS: Comprehensive detection of single base changes in human genomic DNA using denaturing gradient gel electrophoresis and a GC clamp. Genomics 1990;7:463–475.
26 Ruano G, Gray MR, Miki T, et al: Monomorphism in humans and sequence differences among higher primates for a sequence tagged site (STS) in homeo box cluster 2 as assayed by denaturing gradient electrophoresis. Nucleic Acids Res 1990;18: 1314.
27 VandeBerg JL, Cheng ML: The genetics of baboons in biomedical research; in Else JG, Lee PC (eds): Primate Evolution. Cambridge, Cambridge University Press, 1986, pp 317–327.
28 Melnick DJ: The genetic structure of a primate species: Rhesus macaques and other cercopithecine monkeys. Int J Primatol 1988;9:195–231.
29 Curie-Cohen M, Yoshihara D, Luttrell L, et al: The effects of dominance on mating behavior and paternity in a captive troop of rhesus monkeys *(Macaca mulatta).* Am J Primatol 1983;5:127–138.
30 Dyke B, Williams-Blangero S, Dyer TD, et al: Use of isozymes in genetic management of nonhuman primate colonies; in Ogita ZI, Markert CL (eds): Isozymes: Structure, Function and Use in Biology and Medicine. New York, Wiley-Liss, 1990, pp 563–574.
31 Smith DG: Paternity exclusion in six captive groups of rhesus monkeys *(Macaca mulatta).* Am J Phys Anthropol 1980;53:243–249.
32 Smith DG, Smith S: Paternal rank and reproductive success of natal rhesus males. Anim Behav 1988;36:554–562.
33 Weiss ML, Wilson V, Chan C, et al: Application of DNA fingerprinting probes to Old World monkeys. Am J Primatol 1988;16:73–79.
34 Dixson AF, Hastie N, Patel J, et al: DNA 'fingerprinting' of captive family groups of common marmosets *(Callithrix jacchus).* Folia Primatol 1988;51:52–55.
35 Inoue M, Takenaka A, Tanaka S, et al: Paternity discrimination in a Japanese macaque group by DNA fingerprinting. Primates 1990;31:563–570.
36 Collins DA, Casna N, Gergits W, et al: Relatedness and paternity among free-living baboons at Gombe: A comparison of observational data and DNA-based analysis. Am J Primatol 1990;20:182.
37 Ely J, Ferrell RE: DNA fingerprints and paternity ascertainment in chimpanzees *(Pan troglodytes).* Zoo Biol 1990;9:91–98.
38 Lander ES: Research on DNA typing catching up with courtroom application. Am J Hum Genet 1991;48:819–823.
39 Turner TR, Weiss ML, Pereira ME: DNA fingerprinting and paternity assessment in Old World monkeys and ringtailed lemurs; in Martin RD, Dixson AF, Wickings EJ (eds): Paternity in Primates: Genetic Tests and Theories. Basel, Karger, 1992, pp 96–112.

40 Morin PA, Woodruff, DS: Paternity exclusion using multiple hypervariable microsatellite loci amplified from nuclear DNA of hair cells; in Martin RD, Dixson AF, Wickings EJ (eds): Paternity in Primates: Genetic Tests and Theories. Basel, Karger, 1992, pp 63–81.
41 Washio K: Genetic identification of nonhuman primates using tandem-repetitive DNA sequences; in Martin RD, Dixson AF, Wickings EJ (eds): Paternity in Primates: Genetic Tests and Theories. Basel, Karger, 1992, pp 53–62.
42 de Ruiter JR, Scheffrahn W, Trommelen GJJM, et al: Male social rank and reproductive success in wild long-tailed macaques. Paternity exclusions by blood protein analysis and DNA fingerprinting; in Martin RD, Dixson AF, Wickings EJ (eds): Paternity in Primates: Genetic Tests and Theories. Basel, Karger, 1992, pp 175–191.

Dr. Jeffrey Rogers, Department of Genetics, Southwest Foundation for Biomedical Research, PO Box 28147, San Antonio, TX 78228 (USA)

Martin RD, Dixson AF, Wickings EJ (eds): Paternity in Primates:
Genetic Tests and Theories. Basel, Karger, 1992, pp 96–112

DNA Fingerprinting and Paternity Assessment in Old World Monkeys and Ringtailed Lemurs

Trudy R. Turner[a], *Mark L. Weiss*[b], *Michael E. Pereira*[c]

[a] Department of Anthropology, University of Wisconsin-Milwaukee, Milwaukee, Wisc.; [b] National Science Foundation, Anthropology Program, Washington, D.C.; [c] Duke University Primate Center, Durham, N.C., USA

The determination of paternity in nonhuman primates is an essential prerequisite for solving a variety of practical and theoretical problems. Practical problems include the management of captive colonies in zoos and research centers where there is a likelihood of extensive inbreeding. Theoretical problems include the testing of sociobiological hypotheses in free-ranging and semi-naturalistic populations, which may help clarify the evolutionary relationships between genes and behavior.

Several techniques have been used to determine paternity in primates, including examination of blood group systems, allelic differences in blood proteins, HLA markers and restriction fragment length polymorphism (RFLP) analysis. Classical genetic markers such as blood groups, serum proteins and red cell enzymes have limitations. They usually involve proteins and the information they carry is hence removed from the level of the DNA. Allelic variation also tends to underestimate the true amount of variation. For many of the enzymes and proteins there are very low levels of heterozygosity and a reduced ability to differentiate between individuals. (However, see ref. 29, 30 for an alternative view.)

Analysis of RFLP eliminates some of these problems, as it allows for the rapid screening of large numbers of samples at the level of the DNA. Restriction enzymes generally recognize and cut specific sequences with 4 or 6 base pairs. High-molecular-weight DNA is digested with restriction enzymes, generating a tremendous number of fragments of various sizes.

These fragments can be separated according to their size using electrophoresis. Specific fragments are identified by the use of radioactively (or otherwise) labeled probes. Some probes identify specific loci, while others detect families of related sequences. Mutations may create or destroy restriction sites, altering the size of the fragments produced by digestion. The presence or absence of a restriction site can serve as a genetic marker. Because restriction sites can only be present or absent, there are only 2 alleles and the maximum level of heterozygosity is limited to 50%. This limit may be modified if several linked, polymorphic restriction sites are tested simultaneously, but this requires multiple digests.

In 1980, Wyman and White [1] detected a high polymorphic locus now known to be derived from a region of chromosome 14 close to the immunoglobulin genes. Instead of the presence or absence of a particular restriction site, the structural basis for this polymorphism is a variable number of repeats of a short sequence of DNA. This type of polymorphism constitutes a second class of RFLPs that exhibit a greater number of alleles instead of one ubiquitous allele. Jeffreys et al. [2–4] have characterized a series of loci sharing these features in the human genome. Probes used to detect the loci are based on a sequence containing 33 bp originally encountered in the second intron of the human myoglobulin gene. This sequence was known to be repeated in tandem to produce a minisatellite [5]. Embedded in the repeat unit is a core, or conserved consensus sequence 15–16 bp long. Three of the probes prepared from these cloned tandem repeats revealed highly polymorphic conditions. Pedigree analysis indicated that the hybridizing bands are inherited in a Mendelian fashion and that they exhibit somatic and germ-line stability. Jeffreys et al. [6] have estimated that using two probes, 33.6 and 33.15, approximately 60 hypervariable loci can be visualized in human DNA. In any given pattern, roughly half this number of loci will be scorable. In comparing the patterns of two individuals, one is usually looking at overlapping but different arrays of loci. This does not allow for standard population genetic analysis. However, because of the large numbers of loci concurrently screened by Jeffreys' probes and because of the hypervariable nature of at least some of these loci, the likelihood that two individuals will show the same banding pattern is exceptionally low [3]. In humans, using 36 resolvable fragments, the probability that all bands seen in one individual will be seen in a second unrelated individual is 2×10^{-22}.

The germ-line stability and the Mendelian inheritance of minisatellite alleles allow for the identification of parentage in humans. Barring mutations, all bands present in the child can be traced to either the mother or the father. In the case of questioned parentage, usually with respect to the father, fragments common to the offspring and the known parent can be disregarded and the offspring and suspected parent compared for all other bands. The same principle holds for nonhuman primates. At least some of the probes can be hybridized to the DNA of nonhuman primates [7–11]. We are currently engaged in an analysis of DNA fingerprints to determine paternity in an array of primate groups, representing several diverse species living in conditions that range from captivity (zoos) through semi-naturalistic conditions (animals at research centers) to the wild. We present here a summary of our work to date.

Captive Populations

Captive colonies of primates are found primarily in zoos and in biomedical or behavioral research facilities. Some zoos currently function as research-oriented refuges for endangered species. In the last few years, zoo personnel have begun controlled breeding of a limited number of endangered species. Controlled breeding has been aided by the ISIS system, an international breeding record for selected species. Zoo colonies were often initiated with conspecifics of unknown and widely divergent provenance and the genetic constitution of the individuals in any colony was unknown. After a colony was established, there was usually little immigration. Local populations of primates in the wild often have low levels of heterozygosity [12], while widely separated local populations can have highly divergent allele frequencies [13]. Even if a zoo's founding population is constituted so as to maintain heterozygosity, the small size of most zoo colonies and the lack of immigration can soon eliminate any variability and lead to high levels of inbreeding. However, the need to maintain levels of heterozygosity in populations must also be balanced by the need to avoid outbreeding depression, which can result if mating occurs between members of groups that would never mate in the wild.

Biomedical research facilities also maintain colonies of primates. In the past, these animals were obtained from commercial sources. Now, however, these facilities must breed animals and maintain viable breeding colonies. If these populations are housed in sufficient space and are allowed to form normal social groups, they can also be monitored for social interactions. Information on captive colonies housed in this manner can potentially contribute to the answering of theoretical questions about kin selection, but this necessitates a knowledge of paternity. There is, however, an advantage in that the number of potential fathers in a captive situation can be limited and controlled.

We have examined several groups of primates in zoos and in research facilities. These include macaques *(Macaca silenus, Macaca fuscata),* patas monkeys *(Erythrocebus patas)* and colobus monkeys *(Colobus guereza)* at zoos in Detroit and Seattle, guinea baboons *(Papio hamadryas papio)* at Brookfield Zoo in Chicago, Barbary macaques *(Macaca sylvanus)* in France, and ringtailed lemurs *(Lemur catta)* at the Duke University Primate Center.

Zoo Populations – Macaques, Patas and Colobus

DNA fingerprints were generated by hybridizing samples with probes 33.15 and 33.6. The DNA fragments detected show substantial variability and the DNA fingerprint appears to be highly individual-specific. For all species studied, the relationship between the animals is known and, in cases where DNA is available from both parents, all bands in the offspring can be traced to either the mother or the father [7]. The degree of band sharing between unrelated individual patas monkeys, about 37%, is intermediate between the levels in humans (25%) and in dogs and cats (46%) [14]. The two probes do not seem to cross-react as strongly with the *Colobus* samples and samples examined show relatively little variation in DNA fingerprint patterns. This could be a result of inbreeding in this colony. However, an animal which zoo records show to be wild-born shares many bands with a captive-born animal. Broader screening could confirm this lack of variability in this species. However, it was still possible to confirm the paternity of 2 of 5 offspring in this group of *Colobus.*

Brookfield Zoo (Chicago Zoological Park) – Baboons

Brookfield Zoo housed a colony of 50 guinea baboons *(P. h. papio)* in a semi-naturalistic setting. The colony was first established in 1938, when 58 animals were placed on an island. Only limited information is available about the history of this colony over the next 30 years. Additional females were added in 1939 and 1940. Some of these were guinea baboons, but others may have been olive or hamadryas baboons *(P. h. anubis* or *P. h. hamadryas).* As the colony increased in size, some animals, primarily males, were removed. For over 20 years, the colony has been maintained at between 40 and 50 animals, half of them adults. Since 1970, maternity

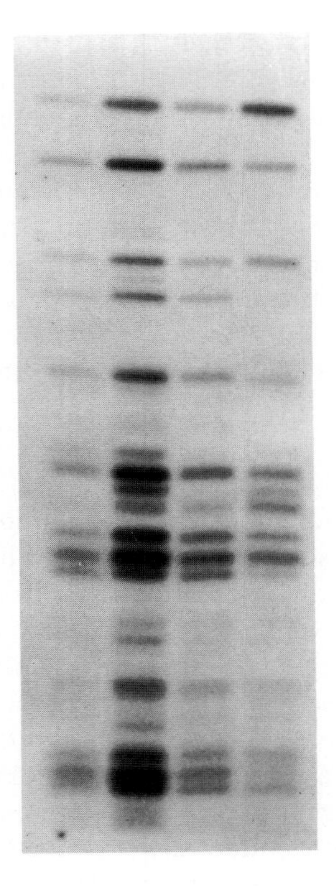

Fig. 1. Four randomly chosen *Papio hamadryas papio* hybridized with probe 33.15. While the probes yield a number of bands, this population is virtually invariant.

records have been kept, but the presence of multiple adult males in the breeding unit has prevented the determination of paternity [15].

An attempt was made to determine the paternity of the baboons in the colony by examining electrophoretic data on blood proteins [15]. Of the 32 loci examined, 31 were invariant. This low level of heterozygosity prohibited determination of paternity using allelic data. We subsequently screened a number of animals from this colony with DNA fingerprints using probes 33.6 and 33.15. Figure 1 shows the effects of generations of inbreeding on 4 members of the colony chosen at random. While the complexity of the pattern is reasonably high, the usually distinctive individual nature of the fingerprints is lacking. DNA samples of wild baboons reflect a more varied gene pool. VandeBerg [30] has suggested, however, that guinea baboons may be less variable than the other subspecies. Thus, the

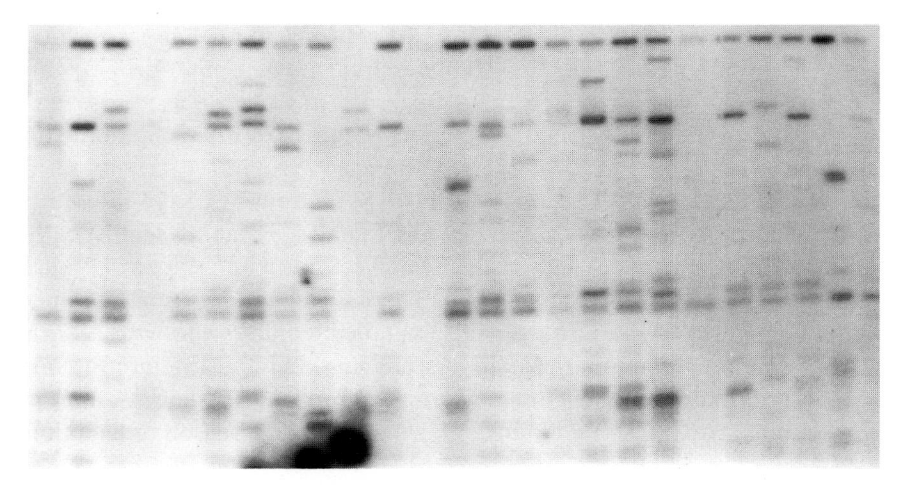

Fig. 2. Macaca sylvanus DNA hybridized with probe 33.6. Five micrograms of DNA was run for each adult animal; samples for infants were often insufficient to provide more than 2–3 μg. Hence, a number of tracks appear empty although weak banding was apparent on the original autoradiographs. A few tentative paternity assignments can be made. Several of the most prominent bands are invariant.

low variability of the zoo population is not characteristic of baboons, nor is it an artifact of hybridization. The animals at Brookfield seem to be extremely inbred, as a result of what amounts to several generations of brother-sister matings.

Rocamadour – Macaques (Macaca sylvanus)

Three colonies of Barbary macaques are maintained in France and Germany. Figure 2 shows a fingerprint for a number of animals in the Rocamadour enclosure. These 'unrelated' individuals share about 45% of their bands. This is approximately the same level of band sharing as that found in a sample of dogs of different breeds and is considerably higher than in humans (25%) [6]. Data from wild populations of Barbary macaques would indicate whether this high level of band sharing is an artifact of captivity or whether it is characteristic of the species. Kuester et al. [34] have found 25–52% band sharing in wild-caught Barbary macaques housed at Affenberg Salem. Even though band sharing is significant, there are enough scorable bands to aid in colony management.

Duke University Primate Center – Ringtailed Lemur (Lemur catta)

The lemur colony at the Duke University Primate Center provides an example of the ways in which DNA fingerprinting can be used to determine paternity and test sociobiological hypotheses in a semi-naturalistic population.

Ringtailed lemurs *(L. catta)* typically live in groups of 8–25 individuals, including several males and several females. Females agonistically dominate males. Females are seasonally polyestrous with periods of sexual receptivity lasting between 4 and 24 h. There is intense male/male competition for estrous females. In the wild, ringtailed lemurs are territorial with strict home range boundaries, where scent marking and intergroup conflicts occur. Males migrate annually between groups, while females appear to be philopatric. Most male transfer occurs during the season of lactation [16].

Duke University maintains a semi-free-ranging study colony of ringtailed lemurs. One group, Lc1, has been under observation since 1977. The following summary is based on data on male/male agonistic relations, on female mating preferences and on female aggression towards males during the birth season and on paternity determinations using DNA fingerprints. It is thus possible to investigate the role of female choice in avoiding inbreeding [16].

In order to demonstrate inbreeding avoidance, it is important to know the degree of genetic relatedness in all possible male-female pairs and the identities of all pairs that reproduce. Female mating preference in ringtailed lemurs is relatively easy to observe because all females agonistically dominate all adult males and because female sexual receptivity is so restricted in time. DNA fingerprinting analyses allowed for the determination of the 17 matings leading to conception and birth that occurred in the Duke colony over a 5-year period [16].

Blood samples were obtained from all animals in the study except 2. DNA from 1 of those individuals was prepared from frozen kidney. However, DNA from the other animal, a male and a potential father of some of the animals, was not available. It was assumed that any offspring sired by this male would be identified by the exclusion of all other potential fathers. In total there were up to 14 potential fathers, 5 mothers and 17 infants. High-molecular-weight DNA was prepared and digested with Alu I using standard techniques. It was electrophoresed and transferred to nitrocellulose filters and then hybridized with probes 33.6 and 33.15 [16, 17].

All the samples yielded informative DNA fingerprints. Figure 3 indi-

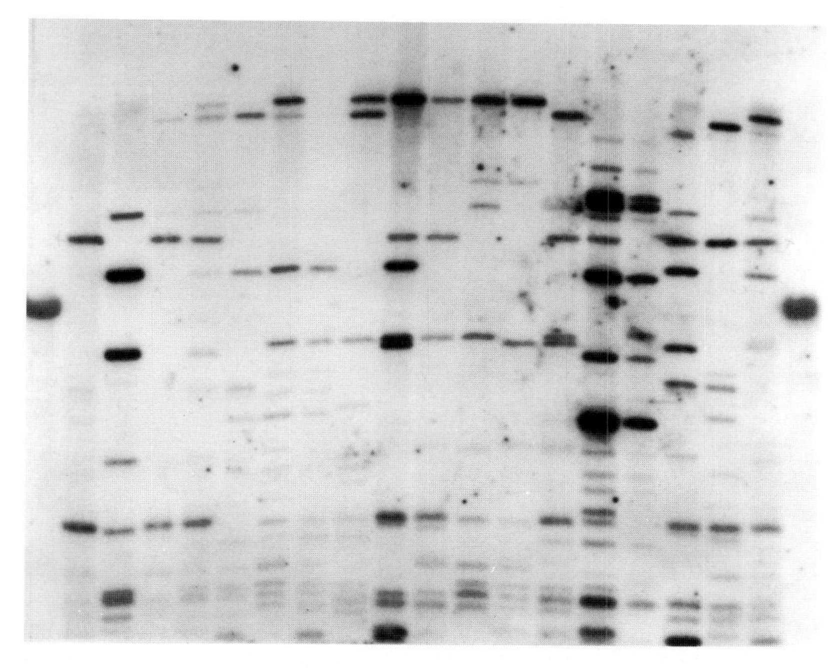

Fig. 3. Lemur catta samples from the Duke colony. Significant amounts of variation are clearly visible in this series of DNAs probed with probe 33.6.

cates the degree of variability in this population. A female, her offspring and four potential fathers are shown in figure 4. In this case, the patterns exclude all males but male 4 from paternity. All but 1 male could be excluded for 15 out of the 17 infants. For the remaining 2 infants, all males for which DNA was available were excluded and paternity was assigned to a single deceased male. This assignment was internally consistent, as the 2 offspring share paternally derived bands. A control for paternity assignment was provided by 2 matings leading to births for which colony management records allowed only 1 possible father. In both, the DNA fingerprints were in agreement with the records.

The paternity information made it possible to document inbreeding avoidance by females. Pooled parentage information from 1982 to 1986 demonstrated that 4 of the 5 reproductive females always mated with males likely to share 25% or fewer of their genes or with the otherwise

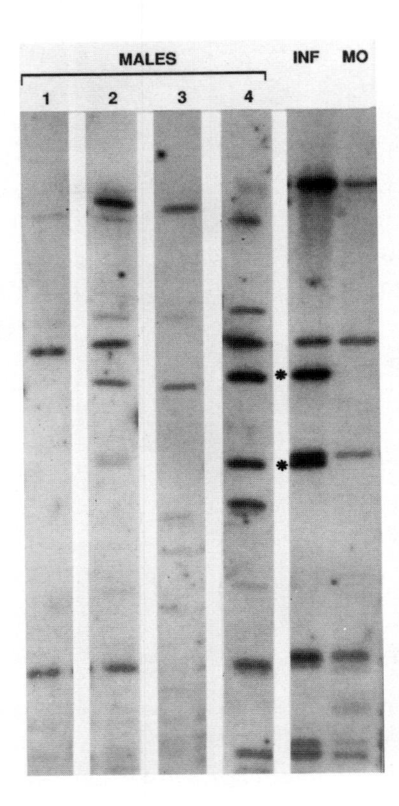

Fig. 4. A mother (MO) and offspring (INF) from the Duke lemur colony. The four potential sires are shown; all but male 4 are excluded. The bands indicated by * indicate male 4's paternity of this infant.

least-related males. There were 15 conceptions where the female had a choice between closely and distantly related males. In 12 of these cases, the female produced offspring fathered by a male which shared 25% or fewer of her genes.

There was no evidence of a female ever reproducing with a son or a matrilineal nephew. Two of the 3 females that had the opportunity to breed with their fathers avoided doing so. One female mated with her father twice in 5 copulations. There were 4 other relatively inbred matings, each involving a matrilineal cousin. In 2 cases, the male was the older of the 2 least-related males available. He was also the only patrilineal half-brother to breed. In sum, females avoided breeding with close matrilineal kin, but insufficient evidence was obtained to test for avoidance of mating with patrilineal kin.

To simulate male migration and diversify female mating opportunities, unrelated males were twice introduced into the colony. The resident males attacked the potential immigrants for several months while at least 5 adult females engaged the males in mutual grooming. DNA fingerprints revealed that immigrant males that remained bottom-ranked in the male hierarchy impregnated a disproportionally large number of females. A conclusion from these data is that female mating preferences and avoidance of inbreeding override the dominance relations among males to influence male mating success.

Pereira and Weiss [16] have proposed a model to explain the lemur mating system and male dispersal: females refuse to mate with their sons or brothers. In a small social group, this limits their reproductive opportunities. Incest avoidance by females is viewed as the trigger for the evolution of male dispersal. Male infanticide has been observed in lemurs. The threat of infanticide is seen as responsible for male/female bonding. In an attempt to deter males from infanticide, lemur mothers target males that could not be the fathers of their infants for months of attack. If a female experiences a reproductive failure in one year, she usually has a reproductive success the following year. A female will choose a male with a history of agonistic prowess in male/male interactions. She is thus choosing a male that may protect her infant. Alternatively, she may choose a male that has a known history of success as a father. Pereira and Weiss [16] thus envision a coevolved reproductive system including female choice, male/male competition and avoidance of infanticide as the basis for male/female interactions.

Wild Populations

Among the questions being asked by primatologists doing field research are: What is the reproductive success of high-ranking males? Does reproductive behavior match reproductive success? How much inbreeding actually occurs? How does female choice influence reproductive sucess? All of these questions require paternity determination. Behavioral data can provide information on the mother of an infant, on priority of access to resources and on the amount of physical contact and social relations between individuals. This information can provide rough estimates of relatedness, but it cannot accurately describe the genetic contribution of any single individual to the group. It does, however, provide a crucial starting point for further genetic analysis. If the mother is known and the potential fathers can be ascertained, this facilitates genetic identification. However, one complication inherent in a wild population is that there may be extra-group copulations during which nonresident males impregnate females. A combination of RFLP analysis and DNA fingerprinting could be the best strategy for screening wild populations, especially when samples are small and resampling is impossible. RFLPs can be used to exclude most of the potential fathers and DNA fingerprinting can then be used to determine paternity. We are currently engaged with E.O. Smith (Emory University) and Jeffrey Rogers (Southwest Foundation for Biomedical Research) in applying this type of analysis to the Tana River baboon population.

The Tana River Primate Reserve is located in Kenya, north of Malindi. The reserve includes a narrow band of riverine forest, brushland and dry plains grassland. There are several species of primates in the reserve, including some that are highly endangered. The Department of Anthropology 4, Yerkes Regional Primate Research Center of Emory University and The National Museums of Kenya are engaged in a long-term collaborative research project to study the behavioral ecology of nonhuman primates in the Tana Reserve. E.O. Smith of Emory University is currently engaged in a behavioral study of male migration of yellow baboons in the area. Baboon troops are being censused, focal observation on adult, subadult and juvenile males is being recorded, and ad libitum data are being collected. These data will provide information on the patterns of male transfer and tactics used by individual males in immigration. Male reproductive success will be determined from an analysis of paternity of infants. Paternity will be determined by polymorphism in single-copy genes in nuclear DNA combined with DNA fingerprinting.

To date, Smith has been collecting behavioral data for over 3 years. He has trapped 57 out of 80 animals of the focal troop. Aliquots of samples were sent from Kenya to both Rogers' and Turner's laboratories. DNA was extracted from all samples. Additional samples are being collected during the summer of 1991. Rogers will initially do an RFLP analysis of the samples to narrow the number of potential fathers. Turner, St. George (University of Wisconsin-Milwaukee) and Weiss will then analyze DNA fingerprints to pinpoint paternity. The total number of potential fathers for this group is not as great as might be expected for a multi-male, multi-female troop. Males outside the focal group that have been observed near the group will also be sampled.

Discussion

DNA fingerprinting is a powerful technique for the determination of paternity in primates. We have demonstrated its applicability in several species of primates. However, in the same way that heterozygosity levels differ in protein polymorphism between different species, the amount of variability and band sharing in fingerprinting differs between groups. Macaque species show considerably greater protein polymorphism, with average heterozygosities of 0.018–0.177, than do baboons, with average heterozygosities of 0.019–0.096. Baboons, in turn, are more variable than vervet monkeys, which have average heterozygosities of 0.021–0.05 [17]. A preliminary examination of DNA fingerprints of wild populations of baboons and macaques indicates that band sharing and the level of variability parallel patterns of protein polymorphism, with macaques being the most variable group [Weiss, unpublished data].

A question of particular importance concerns the difference in band sharing between captive populations of a species and wild populations of the same species. High levels of band sharing were observed in a zoo population of *Papio* and a colony population of *Macaca sylvanus.* Does this apparent lack of variability represent the condition in the wild or is it an artifact of inbreeding in a local group? The amount of inbreeding in wild populations of primates is not clearly known. In most groups, one of the sexes transfers and mates in a group other than the natal group. However, as Cheney and Seyfarth [18] have shown for vervet monkeys *(Cercopithecus* aethiops), male monkeys transfer out of their natal group as subadults but they transfer to a troop where their older brothers have gone. This behavioral study corroborates several genetic studies in nonhuman primates [19–22] which indicate that most of the variability in primate populations is within and not between groups and that local populations are fairly homogeneous. However, the question of the degree of inbreeding in wild populations remains open.

In addition, DNA fingerprinting works better for some species than for others. We have reported that cross-reactivity is not as good in *Colobus* species as in macaques or baboons. There are thus two related problems with the use of DNA fingerprinting: (1) Does the technique work well in a particular species? (2) Is there enough variability in the species to determine paternity realistically? One way out of this dilemma is the use of locus-specific probes for particular species. Locus-specific probes are designed to hybridize under conditions of high stringency with the alleles for a particular locus [23]. Jeffreys has developed several locus-specific probes for humans. Many of these do not cross-react with other species. However, Gray [31] has found that the MS1 probe cross-reacts with macaques and New World monkeys. Scheffrahn [pers. comm.] has found that this probe yields high levels of variability in these groups. Locus-specific probes do have one advantage that fingerprints do not: as they allow for the determination of specific alleles at a specific locus, they can be used for traditional population genetics surveys.

The determination of paternity in wild populations of primates is of particular interest to investigators trying to assess the relationship between behavior and genetics. In order to test hypotheses such as male reproductive success and male protection of their own infants, paternity must be known. However, in the wild, males commonly move in and out of groups, and it may be difficult to find and identify all males that had access to all the females of a group when the females were in estrus.

One of the limitations of DNA fingerprinting is that all the animals involved in a paternity determination should be run on a single gel and in fact should be in close proximity on that gel. It is, of course, theoretically possible to compare between gels if there are enough marker lanes on the gel and if preparation conditions can be standardized. This would not be acceptable in a court of law in the United States, as it would not meet standards of proof mandated by constitutional requirements of due process. Even within a single gel, there may be minor variations between lanes due to the composition of the gel and end osmosis. Between gels, the composition of the buffers used, the length of time the gel is run and the composition of the gel itself may all affect mobility. These differences between gels can be mitigated somewhat by the use of the same standards, such as lambda cut with Hind III, on each gel. However, in reading a DNA fingerprint, one attempts to make fine distinctions between band mobilities. Large mobility differences may be apparent between gels; subtle ones may not. In fact, the experimental error between gels may be greater than the allelic differences. de Ruiter et al. [32] have used a commercial marker every 3–4 lanes to mitigate this problem.

In paternity determinations of multi-male, multi-female groups of primates it is conceivable that each paternity determination would require a separate gel. The question then arises as to how much DNA sample is available for each individual, particularly for the males and the infants. When trapping wild primate populations, there is usually only a single opportunity to trap the animals and there is a maximum amount of blood that can be taken from the animals. A substantial amount of DNA would be required to run multiple gels. In order to determine paternity in this type of case, a combination of techniques may be the most productive, such as the combination of RFLP and DNA fingerprinting analysis for the Tana River baboons. Additional techniques may also prove useful, including polymerase chain reaction (PCR) amplification and single locus probes. PCR amplifies small amounts and small sections of DNA. Probes used to hybridize with these amplified samples are always single-locus probes.

A potential line of research for the future would involve samples collected from wild primate populations in the past. There have been several studies of baboons [24–26] macaques [20, 21] and vervets [19, 27] in which hundreds of samples from wild animals were collected. All of these surveys were conducted on local groups of known provenance. Demographic data are also available for these groups. In some cases [19–21], a considerable body of behavioral data is available. All of these studies were designed to use protein polymorphism to study the actions of gene flow and genetic drift. In all of the surveys involved, serum was separated from red cells and red cells were washed to eliminate white cells. Therefore, until now it was impossible to go back to these samples to obtain material for DNA analysis. However, with a technique known as RAPD or random amplified polymorphic DNA, which is a variant of PCR [28], it may be possible to obtain sufficient DNA for analysis from washed red cells. The probes used in this analysis hybridize to DNA close to unique sequences which are polymorphic in a wide variety of species. There may be very little, but nonetheless sufficient, DNA in washed red cells that can be amplified and examined for these polymorphisms. As large samples of wild-trapped primate populations are rare, it would be of considerable interest to use them for the testing of hypotheses relating to sociobiological models and population genetics.

Thus, even though there are problems with cross-reactivity and amount of variability in DNA fingerprinting, the technique can be used to determine paternity. This is especially true in captive or semi-naturalistic populations where the numbers of potential fathers are known and not very numerous. These types of situations also allow for repeat sampling if necessary. The technique is also useful for wild populations, but not in isolation. The best strategy is somehow to limit the potential fathers in a given study by any means available. Appropriate aids would include behavioral data, protein polymorphism and RFLP analysis. DNA fingerprinting could then be conducted to determine paternity definitively from a selected group of potential fathers. A possible strategy would be to use behavioral data first to establish an appropriate age cohort of males that might be potential fathers. Migrants from outside the group could also be noted. The best method for handling the blood sample would be to centrifuge in the field and separate serum from red cells and white cells. After the blood samples are taken, they could be examined first for any protein polymorphism. While polymorphism may not be extensive in any particular species, it might provide some useful information. In any case, examination of protein polymorphism uses serum and red cells which are plentiful in a blood sample. If protein polymorphism is not to be examined, there are alternative ways of collecting blood samples in specific buffers (0.2 *M* NaCl, 0.1 *M* EDTA, 2% SDS; mixed 1:1 with whole blood) that allow storage without refrigeration prior to shipment to laboratories for analysis. DNA would then be extracted from the samples. Limited amounts of DNA could then be used for RFLP analysis. A small group of potential fathers would then be identified. Remaining DNA could subsequently be used for fingerprinting and determining paternity.

If locus-specific probes are available, population genetics data can also be collected. Additionally, with PCR and RAPD technologies, it would be possible to obtain fragments of DNA from hair samples, thus allowing field primatologists to obtain samples by noninvasive means [33]. There is also the possibility that hundreds of samples collected in the past from wild populations of known provenance and demographic structure may become available for paternity determination using the new RAPD technology.

Primatologists aware of the potential of DNA fingerprinting and its limitations can help answer some of the basic questions that underlie the theoretical and practical concerns of the discipline. Molecular techniques provide a powerful tool for determining the interplay of behavior, genetics and evolution.

Acknowledgments

The authors thank A.J. Jeffreys for use of the polycore probes. He and his associates have provided much assistance and support over the years. The minisatellite probes are subjects of Patent Applications: address commercial inquiries to ICI Diagnostics, Gadbrook Park, Rudheath, Northwich (UK). We also thank the following for providing samples: Robert Lacy (Brookfield Zoo), the staff of Duke University Primate Center, Meredith Small (Cornell University) and Ellen Merz (La Montagne des Singes). Dr. Ann Sodja (Wayne State University) provided much technical support. This research was supported by NSF-BNS 8818405.

References

1 Balazs I, Purrello M, Rubinstein P, et al: Highly polymorphic DNA site D14S1 maps to the region of Burkitt lymphoma translocation and is closely linked to the heavy chain g1 immunoglobin locus. Proc Natl Acad Sci USA 1982;79:7395–7399.
2 Jeffreys AJ, Wilson V, Thein SL: Hypervariable 'minisatellite' regions in human DNA. Nature 1985;314:67–73.
3 Jeffreys AJ, Wilson V, Thein SL: Individual-specific 'fingerprints' of human DNA. Nature 1985;316:76–79.
4 Jeffreys AJ, Neumann R, Wilson V: Repeat unit sequence variation in minisatellites: A novel source of DNA polymorphism for studying variation and mutation by single molecule analysis. Cell 1990;60:473–485.
5 Weller P, Jeffreys AJ, Wilson V, et al: Organization of the human myoglobin gene. EMBO J 1984;3:439–446.
6 Jeffreys AJ, Wilson V, Thein SL, et al: DNA 'fingerprints' and segregation analysis of multiple markers in human pedigrees. Am J Hum Genet 1986;39:11–24.
7 Weiss ML, Wilson V, Chan C, et al: Application of DNA fingerprinting probes to Old World monkeys. Am J Primatol 1988;16:73–79.
8 Dixson AF, Hastie N, Patel I, et al: DNA 'fingerprinting' of captive family groups of common marmosets *(Callithrix jacchus).* Folia Primatol 1988;51:52–55.
9 Ely J, Ferrell RE: DNA 'fingerprints' and paternity ascertainment in chimpanzees *(Pan troglodytes).* Zoo Biol 1990;9:91–98.
10 Tynan KM, Hoar DI: Primate evolution of a human chromosome 1 hypervariable repetitive element. J Mol Evol 1989;28:212–219.
11 Washio K, Misawa S, Ueda S: Individual identification of non-human primates using DNA fingerprinting. Primates 1989;30:217–222.
12 Melnick DJ, Pearl MC: Cercopithecines in multi-male groups: Genetic diversity and population structure; in Smuts B, Cheney RN, Seyfarth RW et al (eds): Primate Societies. Chicago, University of Chicago Press, 1987.
13 Fooden J, Lanyon SM: Blood protein allele frequencies and phylogenetic relationships in *Macaca:* A review. Am J Primatol 1989;17:209–241.
14 Jeffreys AJ, Morton DB: DNA fingerprints of dogs and cats. Anim Genet 1987;18: 1–15.
15 Lacy RC, Foster ML: Determination of pedigrees and taxa of primates by protein electrophoresis. Int Zoo Yb 1988;27:159–168.
16 Pereira ME, Weiss ML: Female mate choice, male migration, and the threat of infanticide in ringtailed lemurs. Behav Ecol Sociobiol 1991;28:141–152.
17 Weiss ML, Turner TR: Hypervariable minisatellites and VNTRs; in Devore E (ed): Molecular Approaches to Physical Anthropology. Cambridge, Cambridge University Press, in press.
18 Cheney DL, Seyfarth RN: Nonrandom dispersal in free-ranging vervet monkeys: Social and genetic consequences. Am Nat 1983;122:393–412.
19 Dracopoli NC, Brett FL, Turner TR, et al: Patterns of genetic variability in the serum proteins of the Kenyan vervet monkey *(Cercopithecus aethiops).* Am J Phys Anthropol 1983;61:39–49.
20 Melnick DJ, Jolly CJ, Kidd KK: The genetics of a wild population of rhesus monkeys *(Macaca mulatta).* I. Genetic variability within and between social groups. Am J Phys Anthropol 1984;63:341–360.

21 Melnick DJ, Pearl MC, Richard AF: Male migration and inbreeding avoidance in wild rhesus monkeys. Am J Primatol 1984;7:229–243.

22 Ober C, Olivier TJ, Sade DS, et al: Demographic components of gene frequency change in free-ranging macaques on Cayo Santiago. Am J Phys Anthropol 1984;64: 223–231.

23 Wong Z, Wilson V, Jeffreys AJ, et al: Cloning a selected fragment from a human DNA 'fingerprint': Isolation of an extremely polymorphic minisatellite. Nucleic Acids Res 1986;14:4605–4616.

24 Olivier TJ, Buettner-Janusch J, Buettner-Janusch V: Carbonic anhydrase isoenzymes in nine troops of Kenya baboons, *Papio cynocephalus* (Linnaeus, 1766). Am J Phys Anthropol 1974;41:175–190.

25 Phillips-Conroy JE, Jolly CJ: Changes in the structure of the baboon hybrid zone in the Awash National Park, Ethiopia. Am J Phys Anthropol 1986;71:337–349.

26 Hamilton III WJ, Bulger JB: Natal male baboon rank rises and successful challenges to resident alpha males. Behav Ecol Sociobiol 1990;26:357–362.

27 Turner TR: Blood protein variation in a population of Ethiopian vervet monkeys *(Cercopithecus aethiops aethiops).* Am J Phys Antropol 1981;55:225–232.

28 Williams JG, Kubelik AR, Livak KJ, et al: DNA polymorphisms amplified by arbitrary primers are useful as genetic markers. Nucleic Acids Res 1991;18:6531–6535.

29 Smith DG, Rolfs B, Lorenz J: A comparison of the success of electrophoretic methods and DNA fingerprinting for paternity testing in captive groups of rhesus macaques; in Martin RD, Dixson AF, Wickings EJ (eds): Paternity in Primates: Genetic Tests and Theories. Basel, Karger, 1992, pp 32–52.

30 VandeBerg JL: Biochemical markers and restriction fragment length polymorphisms in baboons: Their power for paternity exclusion; in Martin RD, Dixson AF, Wickings EJ (eds): Paternity in Primates: Genetic Tests and Theories. Basel, Karger, 1992, pp 18–31.

31 Gray JC, Jeffreys AJ: Evolutionary transience of hypervariable minisatellites in man and the primates. Proc R Soc Lond [Biol] 1991;243:241–253.

32 de Ruiter JR, Scheffrahn W, Trommelen GJJM, Uitterlinden AG, Martin RD, van Hooff JARAM: Male social rank and reproductive success in wild long-tailed macaques. Paternity exclusions by blood protein analysis and DNA fingerprinting; in Martin RD, Dixson AF, Wickings EJ (eds): Paternity in Primates: Genetic Tests and Theories. Basel, Karger, 1992, pp 175–191.

33 Morin PA, Woodruff DS: Paternity exclusion using multiple hypervariable microsatellite loci amplified from nuclear DNA of hair cells; in Martin RD, Dixson AF, Wickings EJ (eds): Paternity in Primates: Genetic Tests and Theories. Basel, Karger, 1992, pp 63–81.

34 Kuester J, Paul A, Arnemann J: Paternity determination by oligonucleotide DNA fingerprinting in barbary macaques *(Macaca sylvanus);* in Martin RD, Dixson AF, Wickings EJ (eds): Paternity in Primates: Genetic Tests and Theories. Basel, Karger, 1992, pp 141–154.

Dr. Trudy R. Turner, Department of Anthropology,
University of Wisconsin-Milwaukee, Milwaukee, WI 53201 (USA)

Martin RD, Dixson AF, Wickings EJ (eds): Paternity in Primates:
Genetic Tests and Theories. Basel, Karger, 1992, pp 113–130

Application of DNA Fingerprinting to Familial Studies of Gabonese Primates

E. Jean Wickings, A.F. Dixson

Centre de Primatologie, Centre International de Recherches Médicales de Franceville, Gabon

Gabon's vast natural rainforests contain a wealth of primates, including two species of apes *(Gorilla gorilla, Pan troglodytes),* monkeys ranging from the large colourful mandrill *(Mandrillus sphinx)* to the diminutive talapoin *(Miopithecus talapoin),* and five species of prosimians, including one of the smallest, *Galago demidovii.* Several of these primates are the subjects of research by scientists at the Centre International de Recherches Médicales de Franceville (CIRMF), and are either held at the Primate Centre in Franceville or studied in their natural rainforest habitat. Our own studies concentrate on primate reproductive behaviour and determination of mating systems, with the mandrill as the focus of several current studies. Pivotal to these investigations is the need to determine parentage with a high degree of certainty. Because of the large maternal investment in each infant, maternity can be established and infants identified before they are weaned, but we have no equivalent indicator of paternity. Hence, over the last 2 years we have been attempting to establish at CIRMF the technique of DNA fingerprinting, which will have applicability to a wide range of species.

The human myoglobin minisatellite core sequence [1] has been shown to be highly conserved throughout the living world [2, 3]. The 2 multi-locus probes 33.15 and 33.6 derived from this sequence, which formed the basis of the original description of the DNA fingerprint [4], have so far been shown to be almost universally applicable in their ability to recognize individual-specific polymorphisms in these minisatellite tandem repeat sequences. Subsequently, simpler synthetic oligonucleotides based on the myoglobin core sequence [5] and indeed oligonucleotides rich in guanidine and cytidine [6] have also been shown to be capable of producing fingerprints as individual in character as those of the larger double-stranded DNA probes. We have adopted the strategy of using oligonucleotide probes, and demonstrate their applicability to several Gabonese species of primates.

Methodology

Extraction of DNA

Blood samples for the extraction of DNA were collected into vacutainers containing sodium citrate during routine captures for health checks. Blood was immediately diluted 1:2 with 1 × SSC [7] and then stored frozen at –60 °C.

Thawed samples (equivalent to 4 ml blood) were lysed for 30 min at 4 °C in 15 ml lysis buffer (0.32 *M* sucrose; 10 m*M* Tris-HCl, pH 7.5; 1% Triton X-100), the cell pellet washed with 10 ml lysis buffer and then digested with proteinase K (1 mg/ml in 10 m*M* Tris-HCl, pH 8.0, 0.1 *M* NaCl, 10 m*M* EDTA, 40 m*M* DTT, 2% SDS) overnight at 37 °C. The aqueous protein digest was extracted twice with an equal volume of salt-saturated phenol-chloroform-isoamylalcohol (25:24:1 vol/vol) and once with the same volume of chloroform-isoamylalcohol (24:1 vol/vol). DNA was precipitated by the addition of 1/10th volume of ammonium acetate (7.5 *M*) and 2 vol of ice-cold ethanol. The DNA pellet was washed twice with cold 80% ethanol, air-dried and taken up in 50 μl TE buffer. The DNA concentration in each sample was determined spectrophotometrically.

Restriction Enzyme Digestion

All enzymes (Boehringer-Mannheim) were used with the manufacturer's recommended buffer system at a concentration of 1 U/μg DNA in a volume of 40 μl. Ten micrograms DNA were digested for 4 h at 37 °C in the presence of spermidine trichloride (4 m*M*). The following enzymes were tested on pools of DNA generated for each species: Alu I, Bam HI, Hae III, Hind III, Hinf I. However, this approach was of limited value in determining the most suitable enzyme for differentiating the degree of band heterogeneity in a given population. Initial results showed that Alu I and Hinf I yielded suitably divergent banding patterns for all four species, and further work concentrated on differences between the patterns generated with these two enzymes and the various oligonucleotide probes. All samples from study groups were digested and run on the same gels where possible, up to 19 samples plus control pools at a time.

Electrophoresis and Vacuum Blotting

DNA digests were loaded onto agarose gel (0.7% in 1 × TAE buffer) and electrophoresed for approximately 40 h at 20–25 V (1 V/cm) in 1 × TAE buffer, with buffer circulation. Ethidium bromide staining of the molecular weight markers run on each gel revealed the extent of migration of DNA (generally the 2.03 kb marker was allowed to migrate approximately 16 cm from the origin). Under 40 mbar vacuum, the gel was depurinated and denatured before the DNA was transferred onto a Nylon membrane (Hybond N-Plus, Amersham; presoaked in 2 × SSC for 2 min) under alkaline conditions

Core Sequence(17bp)

GGAGGTGGGCAGGA A/G G

Myoglobin Minisatellite Probes

33.6 (33bp) $((AGGGCTGGAGG)_3)_{18}$

33.15 (16bp) $(AGAGGTGGGCAGGTGG)_{29}$

Oligonucleotides

o33.6 (37b)
TGGAGGAAGGGCTGGAGGAGGGCTCCGGAGGAAGGGC

o33.15 (16b) GAGGTGGGCAGGTGGA

$(GTG)_5$ (15b) GTGGTGGTGGTGGTG

Fig. 1. Sequences of the three oligonucleotide probes o33.15, o33.6 and $(GTG)_5$ used in fingerprinting gorillas, chimpanzees, guenons and mandrills, highlighting their similarity to the two double-stranded myoglobin minisatellite probes 33.15 and 33.6, and to the myoglobin 'core' sequence [1].

(0.25 *M* NaOH in 1.5 *M* NaCl for 75 min), and finally fixed onto the membrane by treatment with 0.4 *M* NaOH for 20 min and briefly rinsed with 5 × SSC before being wrapped in Saranwrap and stored at 4 °C until hybridization.

Hybridization

Figure 1 shows the base sequences of the three synthetic oligonucleotide probes used for hybridization and their relationship to the core sequence of the parent double-stranded minisatellite probes. It also highlights the similarity between a short 5-base sequence GGTGG in o33.15 and $(GTG)_5$.

Oligonucleotides (30 ng per reaction) were 5'-end-labelled with γ-$AT^{32}P$ (Amersham; 100 µCi per reaction) using polynucleotide kinase (Boehringer-Mannheim, 1 U per reaction; at 37 °C for 40 min). Membranes were briefly wetted with 2 × SSC and prehybridized in a quick oligonucleotide hybridization mix (QOHM; 0.05% BSA, 0.05% PVP, 0.05% Ficoll 400, 0.1% SDS, 0.1% sodium pyrophosphate in 5 × SSC) at 48 °C for 60 min in a shaking water bath. Hybridization was carried out in the same solution, following the addition of the radiolabelled probe (0.5 ng/ml QOHM) to the hybridization sachet, at 48 °C for 5–6 h in a shaking water bath. The membranes were then washed at low stringency (4 × 5 min in 4 × SSC containing 0.1% SDS and 0.1% sodium pyrophosphate at 50–55 °C) before being wrapped in clingfilm and exposed to XAR-5 film, with 2 intensifying screens.

For additional probing of the same membrane, the previous probe was stripped during a 2-hour wash in buffer (1.0 m*M* Tris-HCl, pH 8.0, 1.0 m*M* EDTA, pH 8.0, 0.1 × Denhardt's reagent) at 75 °C in a shaking water bath, followed by a brief rinse in 2 × SSC. The membrane was exposed to XAR-5 film overnight to check that all radioactivity had been stripped from the membrane.

Scoring of Fingerprints

All bands visualized from approximately 3.5–23.13 kb were scored. At lower molecular weights, most bands were common to all DNA digests, as these smaller alleles have a higher frequency in the population and hence a lower genetic variability. For each species, the degree of individuality in banding patterns between (assumed) non-related subjects was calculated from the number of characteristic bands (shared between 2 individuals A and B) divided by total number of bands generated for the two individuals. A band-sharing index to denote the degree of relatedness (D) was calculated for each parent-offspring dyad from the ratio of parentally inherited bands to the total number of bands scored, according to the formula:

$$D = \frac{2N_{AB}}{N_A + N_B}$$

D = 0 would indicate no bands shared, or non-relatedness, while at the other extreme D = 1 (all bands shared) occurs only in homozygous twins. Typically $D < 0.25$ would characterize non-related individuals, and an index of $D \cong 0.5$ would indicate first-degree relationships [8]. However, not all band-sharing ratios reached the expected values, or two males occasionally produced similar D values. In this case, at least two unique bands from one male's fingerprint should be readily identifiable in that of the offspring [9]. This rationale formed the basis of paternity allocation for the following examples.

Primate Studies

Guenon: Cercopithecus solatus Harrison 1988

Guenons are widely distributed throughout the continent of Africa, and many species are indigenous to the tropical forests of the Zaire basin [10]. This Gabonese guenon was first described in 1988 [11], originating from a small isolated area in the central region. CIRMF holds the only breeding colony of this species; the group comprises 6 monkeys: 2 adult males (Mbaya and Bilembi), 1 adult female (Lawagni) and her 2 offspring (Okanda, born 1988 and Achouka, born 1990), and a younger female (Okoumé); at the time of the first birth, only Mbaya was considered to be fully mature. All monkeys were included in the analysis.

The two probes o33.15 and o33.6 bound only weakly to Hinf I digests of *C. solatus* DNA, despite prolonged autoradiography exposures, and results were uninterpretable. The $(GTG)_5$ probe produced multiple bands with molecular weights between 3 and 10 kb, with an average number of approximately 11 (SD ± 3; range 8–15) bands per individual. Out of the total of 65 bands scored, none was shared between all animals, but 29 of 65 (44.6%) were found in ≥2 animals and only 22 of 65 (33.8%) were unique

Table 1. Matrix of band-sharing indices for Hinf I DNA digests probed with $(GTG)_5$ from the 6 *C. solatus,* used to infer relatedness between individuals

	Mba	Law	Ach	Oka	Bil	Okou
Mbaya	–	0.16	0.17[1]	0.43[1]	0.23	0.21
Lawagni	–	–	0.67[2]	0.14[2]	0.29	0.04
Achouka	–	–	–	0.25	0.32	0.19
Okanda	–	–	–	–	0.32	0.10
Bilembi	–	–	–	–	–	0.17

[1] Father-infant dyads (inferred).
[2] Mother-infant dyads.

to any individual. The band-sharing index between assumed non-related individuals was 0.17 ± 0.08 (range 0.04–0.32) (table 1). For the 2 mother-infant dyads the index was lower (0.14; Okanda) or higher (0.67; Achouka) than predicted for Mendelian inheritance of bands. Hence $D \neq 0.5$ for both known mother-infant relationships. Similarly, the band inheritance between the inferred father (Mbaya) and the 2 infants was skewed from that expected, with $D = 0.43$ (Okanda) and $D = 0.17$ (Achouka). However, the 2 bands inherited by Okanda from Mbaya were unique to Mbaya; Bilembi could be excluded from paternity as two of the three bands common with Okanda were also found in Mbaya's fingerprint. Bilembi was excluded as a possible sire of Achouka because all common bands were also present in the mother's fingerprint; bands shared between Achouka and Mbaya were otherwise unique to Mbaya.

Hence, by using both inclusion and exclusion criteria, we can generate familial relationships for this small but unique group of guenons. Nevertheless, this study highlights some of the problems of inbreeding between small numbers of non-related individuals that were originally from a restricted location, and probably a restricted population. The question arises as to the minimum degree of commonality required between fingerprints for a family relationship to be confirmed. In this study, at least, all possible candidates were available for analysis, and hence exclusion criteria could also be used in the computation. When more material is available, a second series of fingerprints using Alu I digests will be generated, in order to verify the initial results. However, on the basis of these results any further breeding within the group at CIRMF will be restricted in the next generation to non-related individuals to reduce inbreeding and maintain the genetic fitness of the group.

Gorilla: Gorilla gorilla

The CIRMF gorilla colony comprises 10 animals; a young adult male (13 years), one parous female (19 years) with 2 daughters (one aged 6 years and the second 3 years), 1 other adult female (12 years), 3 subadult males (7–11 years), and 2 juvenile females (3 and 5 years). The 2 daughters were born at CIRMF to the same parents, but unfortunately the father died shortly after the second birth and no tissue was available for DNA analysis. The remaining 7 animals are presumably not related. All gorillas in the group were included in the analysis.

All three probes were investigated on Hinf I digests of DNA from all 10 gorillas run on the same membrane (results for 6 individuals shown in fig. 2). The fingerprints generated by o33.15 and $(GTG)_5$ were very similar, although more bands were detected with the latter probe and the overall intensity was greater, perhaps due to stronger binding of the probe to the membrane. This similarity in patterns results from the short GGTGG sequence common to both probes (fig. 1). Table 2 compares the distributions of unique and shared bands across the 3 autoradiographs. All three probes gave the expected index of non-relatedness for the 28 pairs of 8 presumably non-related individuals in the group ($D < 0.25$), and a high degree of commonality was found between the banding patterns of the 2 mother-infant dyads. In both instances, the index value $D > 0.5$ indicated a higher-than-expected inheritance of maternal bands; $D > 0.5$ between the two siblings would also indicate a first-degree relationship, i.e. full sibship, with the same mother and the same father. In fact, it was possible to identify bands with all three probes which were common to the 2 offspring, but absent in the mother, and which are therefore assumed to be of paternal origin (fig. 2).

Individual fingerprints generated by this method could serve as a useful 'passport' for members of endangered species; a parental-offspring triad would convincingly identify the captive origin of an animal.

Chimpanzee: Pan troglodytes

Sixty-one chimpanzees are held at CIRMF in 9 social and family groups. Infants remain with their mothers throughout infancy and adolescence and, until recently, females had been allowed to reproduce at their natural frequency. Spatial constraints have necessitated the curtailing of

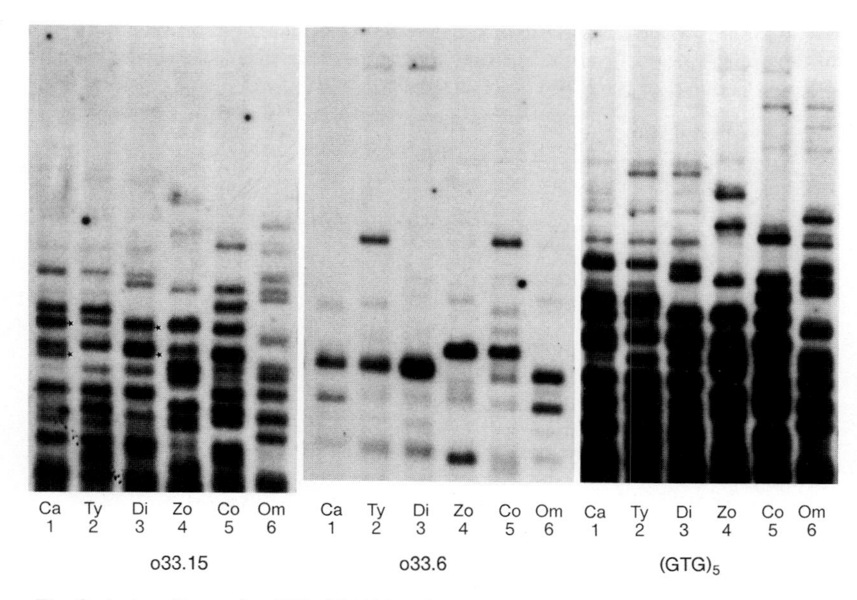

Fig. 2. Autoradiograph of Hinf I DNA digests from 6 gorillas probed with o33.15, o33.6 and $(GTG)_5$. Lanes 1–2 and 2–3 show 2 mother-offspring dyads; lanes 4–6 show 3 other non-related gorillas in the group. Stars on the fingerprints for probe o33.15 indicate paternally derived bands.

Table 2. Comparison of indices of relatedness and non-relatedness in 10 gorillas, as generated from Hinf I DNA digests probed in sequence with o33.15, o33.6 and $(GTG)_5$

	o33.15	o33.6	$(GTG)_5$
Bands scored, n	120	62	157
Extent of band sharing, %			
Unique bands	18.3	17.7	18.5
Shared by ≥ 2 unrelated individuals	55.0	59.7	73.8
Relatedness (D)			
Non-related subjects	0.22 ± 0.10[1]	0.23 ± 0.13	0.21 ± 0.08
Mother-daughter (1)	0.67	0.55	0.70
Mother-daughter (2)	0.56	0.77	0.69
Full siblings	0.78	0.60	0.69

[1] Mean ± 1 SD; n = 28.

Table 3. Age, sex and social status of 4 maternal subgroups of chimpanzees and the fathers of the 8 offspring, as determined from Hinf I DNA digests labelled separately with two oligonucleotide probes, o33.6 and $(GTG)_5$

Name	Sex	Status	Age years	Father	
				o33.6	$(GTG)_5$
Berthe (1)	F	mother	22		
Bitam (2)	M	son-Berthe	10	Boueni	Boueni
Bertrand (3)	M	son-Berthe	5	?	?
Ntebe (5)	F	mother	16		
Nuria (6)	F	daughter-Ntebe	5	Boueni?	Mvadi
Ntoum (7)	F	daughter-Ntebe	2	Mvadi	?
Gemini (9)	F	mother	25		
Gérard (10)	M	son-Gemini	8	Boueni?	Boueni?
Géraldine (11)	F	daughter-Gemini	2	Mvadi	?
Chiquita (14)	F	mother	14		
Claire (15)	F	daughter-Chiquita	5	Mpassa?	?
Chico (8)	M	son-Chiquita	1	Bitam	Bitam/Mvadi
Mpassa	M	son-Dodo	13	Boueni	Boueni

Number in parentheses refers to lane numbers in figure 3.

unplanned pregnancies, but to maintain the full breeding potential of the colony all breeding females have been fitted with intrauterine devices. Any future breeding policy will take into account existing paternal relationships, as revealed by DNA fingerprinting.

In the group studied, there are 4 parous females, each with 2 offspring, one non-parous female and 1 adult male (1 of the putative fathers). However, 1 other male in an adjacent cage has access to receptive females and at least 1 male offspring is now sufficiently mature to be considered a potential father; 1 further adult male had been removed from the group previously and was also included in the analysis. All 4 putative fathers have been included in the analysis along with 2 additional, non-related males without access as controls (total n = 17; table 3).

Hinf I and Alu I DNA-digests were probed successively with the three oligonucleotides o33.6, o33.15 and $(GTG)_5$. Fewer fragments resulted

Table 4. Indices of relatedness in 17 chimpanzees, as computed from autoradiographs of Hinf I DNA digests probed in sequence with o33.6 and $(GTG)_5$

	o33.6	$(GTG)_5$
Bands scored, n	151	182
Relatedness index (D)		
Unrelated (n = 29)	0.18±0.11[1]	0.28±0.13
	(0–0.45)[2]	(0.08–0.52)
Mother-offspring (n = 8)	0.51±0.15	0.55±0.09
	(0.24–0.70)	(0.45–0.70)
Siblings (n = 4)	0.36±0.11	0.48±0.12
	(0.25–0.48)	(0.32–0.57)
Father-offspring (n = 5)	0.40±0.07	0.32±0.07
	(0.29–0.47)	(0.26–0.43)
Paternal half-sibs (n = 4)	0.29±0.04	0.44±0.19
	(0.24–0.35)	(0.21–0.64)

[1] Mean ± 1 SD.
[2] Range.

from Alu I digestion for all three probes, and these were not as informative as those produced by Hinf I digestion (results not shown); hence all further work on chimpanzee DNA was carried out using Hinf I. The banding pattern generated with o33.15 resembled that of a single-locus probe, with only 2 distinct bands located, and this approach was also abandoned. A higher number of bands within the molecular weight range 3.5–23 kb was generated with $(GTG)_5$, on average 10.7, as compared with 8.9 per individual with o33.6. The index of relatedness, D, for both probes distinguished clearly between non-related animals and mother-offspring dyads, but allocation of paternity also relied on band inclusion for the highest-rated male; an index value of D = 0.5 was not achieved in any of the paternal pairings (table 4). For all other degrees of relatedness, the $(GTG)_5$ probe gave the values closest to the expected; only in the category of paternal allocation was the discrimination of o33.6 better (fig. 3). With $(GTG)_5$, however, it was possible to confirm in an indirect fashion the allocation of paternity, by recognizing the paternal half-sibships between the following 4 pairs of chimpanzees – Bitam:Mpassa, Bitam:Gérard, Gérard:Mpassa (father = Boueni) and Ntoum:Géraldine (father = Mvadi) (table 3). Overall, we were successful in allocating paternity in 9 of the 18 cases (including

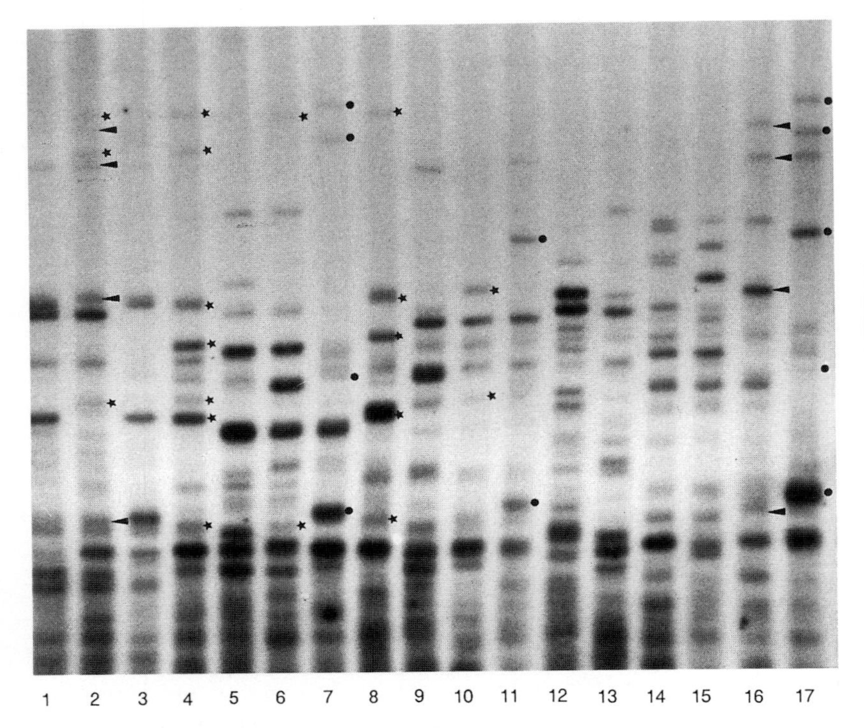

Fig. 3. Autoradiograph of Hinf I DNA digests from 17 chimpanzees probed with o33.6, showing paternity allocation. Lanes 2 (Bitam), 4 (Boueni), 8 (Mpassa) and 17 (Mvadi) show potential fathers, lanes 1–3, 5–7, 9–11 and 14–16 show the 4 maternal subgroups (see table 3); lanes 12 and 13 show 2 unrelated adult male chimpanzees. Matched symbols denote band sharing between father-offspring dyads.

Mpassa, revealed as Boueni's son); equivocal results were produced in 4 cases, and 5 attempts were completely unsuccessful.

Mandrill: Mandrillus sphinx

Observations on the social organization of the mandrill in its rainforest habitat have been few in number, and there is still a controversy as to whether the one-male group of approximately 15–50 monkeys represents the core unit [12], or whether the huge hordes which result when these smaller units coalesce [up to 450 individuals with 11 adult males; C.E.G. Tutin, CIRMF, Gabon, pers. commun.] form the basis of a multi-male society. The question of paternity, or how many males in a mandrill troop contribute to the gene pool of the next generation, is pivotal to defining the mating system of this species. The CIRMF group was initially established with 7 young males of approximately the same ages, 6 of which remain in the enclosure and are now adult (9–11 years) but of differing physical stature and status within the group. There have been several changes of dominance within the male hierarchy over the 8-year history of the group, the latest occurring in 1989, when male No. 7 was deposed by male No. 14. Observations of the group throughout the 1990 mating season provided data on socio-sexual behaviour and the hierarchy between the adult males. Hence, the overall aim is to identify the pattern of paternity since the foundation of the group, but the immediate goal is to identify the fathers of the infants resulting from sexual activities documented during the 1990 mating season.

Fifty-six mandrills are held in a 6-ha natural rain forest enclosure, and this group provides an excellent opportunity to study all aspects of their socio-sexual behaviour. There are 6 adult males, 7 matriarchal lines (with a total of 13 breeding females), 3 subadult males, 22 juveniles and 12 infants; the overall male:female ratio is 31:25 (table 5). Infants are usually identified and tagged according to their maternal line before weaning. However, in some instances infants were already independent at the time of capture (e.g. those born in 1989 and 1990); but at least the analysis is simplified as it is known which of the females gave birth in these years. All mandrills will eventually be included in the analysis, and 38 of the 56 have so far been investigated, including 20 from 4 matriarchies.

Initial work concentrated on Hinf I DNA digests probed with o33.6, which had yielded valuable results for the apes; but this system was not as successful in allowing the unequivocal assignment of identity in all mandrills. Maternal lines could be distinguished, but banding patterns were insufficiently different between the 7 adult males to allow individual identification. Figure 4 shows the DNA fingerprints of 4 of the 5 infants born in 1989, with those of all the adult females ($n = 9$) in the colony. As it was known which 5 females had given birth, it was possible to allocate offspring in all 4 cases where fingerprints were generated. However, Alu I DNA digests probed with $(GTG)_5$ eventually proved to be the most informative system, whereby all banding patterns were sufficiently polymorphic to allow assignment of parentage, and specifically paternity.

Table 6 summarizes the indices of relatedness for two of the series analyzed. The 6 adult males showed D-values that did not differ from those for non-related individuals; in each case, the highest D value (0.43)

Table 5. Composition of familial groupings of the CIRMF semi-free ranging mandrill colony

Total number of mandrills	56	Overall sex ratio	31 M/25 F
Adult males	6 (> 10 years)	Subadult males	3 (6–8 years)

Constitution of 7 matriarchal groups:

Founders	2	5	6	10	12	16	17
Offspring			6A		12A	16A	17A
	2B	5B	6B		12A1	16B	17A/91
	2C	5C	6/91	10C	12A/91	16/91	17B
	2C1	5D		10D	12C		(17/91)
	2C/91	5E		10/91	12C/91		
	2D	5/91			12D		
	2D/91				12D/91		
	2/91				12E		
					12F		
					12/91		

Unidentified juveniles: n = 8.
Two potential fathers were previously removed from enclosure: No. 3-9/1988, No. 2A-12/1989.

was for the pair No. 7/No. 14. To date, paternity has been determined for 10 of the 12 surviving infants and 5 show a high degree of relatedness with male No. 14, 3 are potentially No. 14's progeny and 2 infants were sired by No. 7 (table 7). As yet, there is no evidence that either the 3rd-ranking male (No. 15) or the 3 solitary males (No. 9, 13 and 18) have fathered progeny, despite having access to females during 'sneaky' copulations.

The shift in dominance between No. 7 and No. 14 occurred during 1989, and it remains to be determined whether all offspring born in the colony before No. 7 was deposed are his progeny. Figure 5 demonstrates this shift in paternity for 1 female line (No. 12, with her 2 previous daughters No. 12C and No. 12D, each now with their own infants No. 12C/91 and No. 12D/91, and No. 12's own baby No. 12/91). Both of these older daughters were fathered by No. 7, the dominant male in the colony in 1986/7, whereas their own offspring result from matings with the current highest-ranking male, No. 14, as does No. 12's latest infant. Current analysis results indicate that both No. 7 and No.14 fathered offspring during the 1989 and 1990 mating seasons. So far, 7 of the 13 infants have been

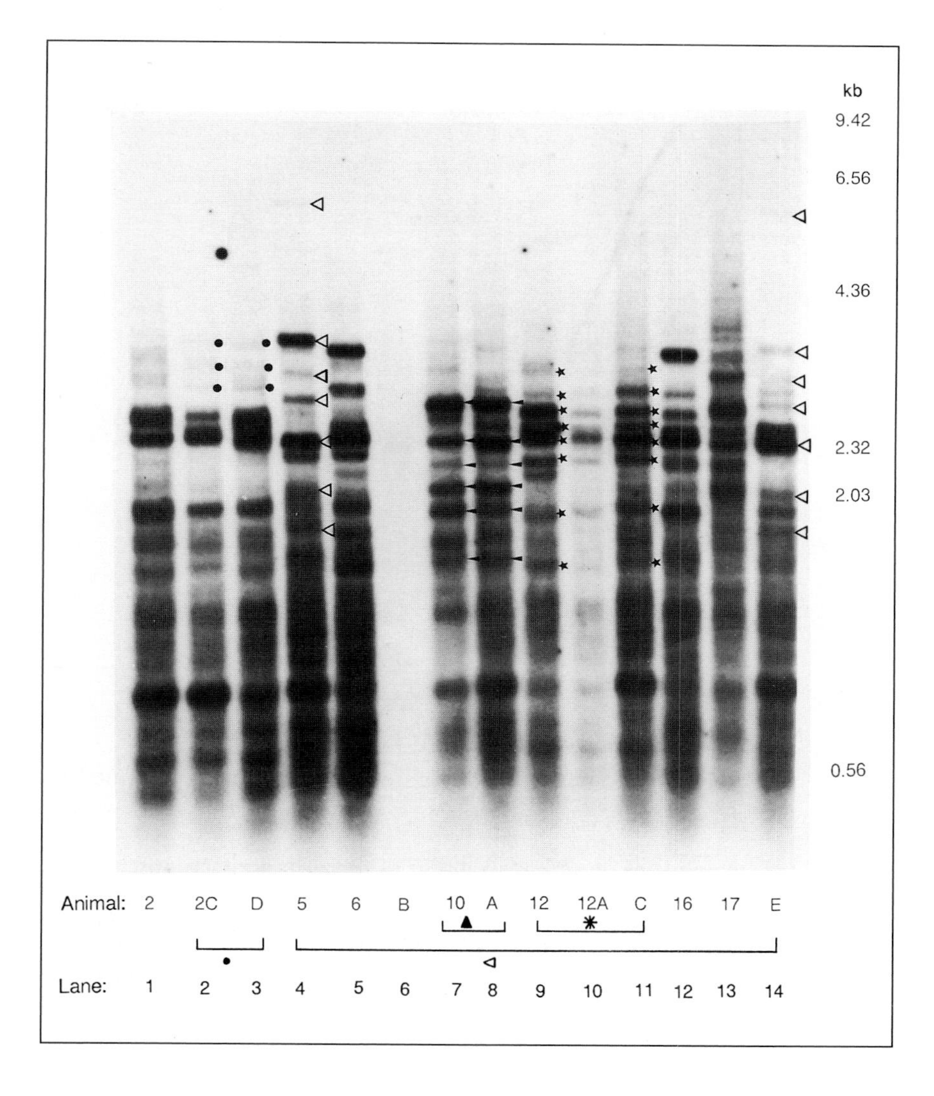

Fig. 4. Autoradiograph of Hinf I DNA digests from 14 mandrills probed with o33.6. Lanes 1, 2, 4, 5, 7, 9, 10, 12 and 13 show banding patterns of all 9 adult females; lanes 3, 8, 11 and 14 show fingerprints of 4 of the 5 infants born in 1989, for which maternity was unknown. Single letters denote offspring. Matched symbols denote band sharing between mother-offspring dyads. Lane 6 (infant B) was incorrectly loaded.

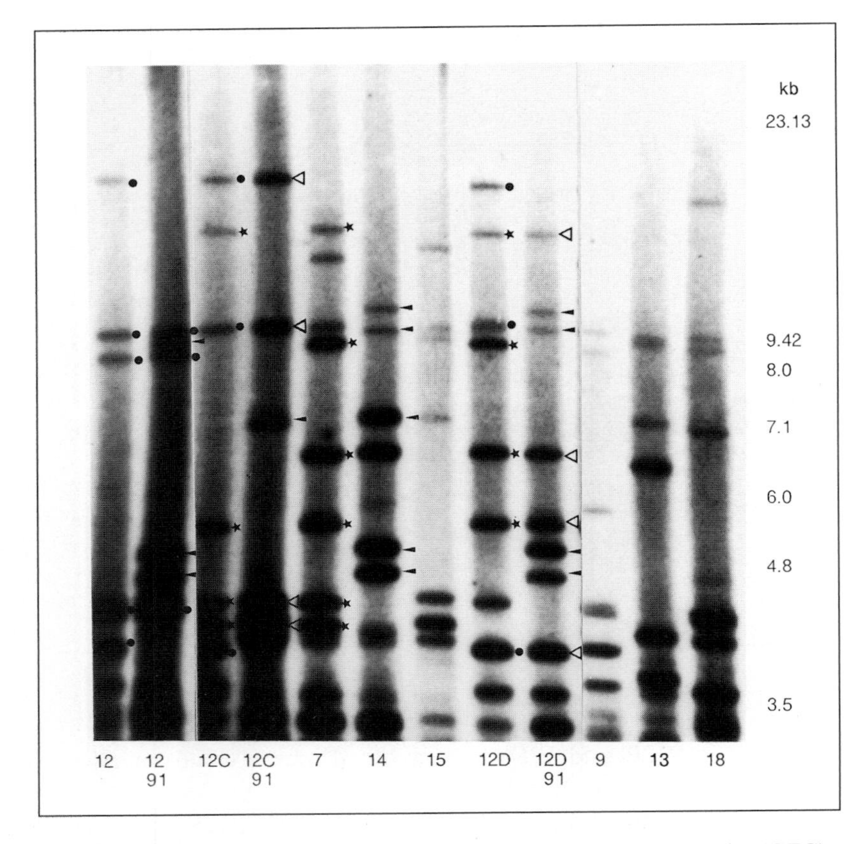

Fig. 5. Autoradiograph of Alu I DNA digests of 12 mandrills probed with $(GTG)_5$, showing fingerprints of all 6 adult males No. 7, 14, 15, 9, 13 and 18, and the matriarchal line of No. 12, her 2 daughters by male 7 (12C and 12D) and their 3 offspring by male 14 born in 1991 (12/91, 12C/91 and 12D/91). ● = Bands inherited from female 12; ◀ = bands inherited from male 14; ★ = bands inherited from male 7; ◁ = bands inherited by 12C or 12D from female 12 or male 7 and passed on to their respective offspring.

analyzed; 5 show a paternal D-index with No. 7 (0.45 ± 0.06) and 2 are more closely related to No. 14 than No. 7. Interestingly, in the maternal lineage of No. 12 it is possible to follow the inheritance of bands from the grandsire (No. 7) through the mothers (No. 12C and 12D) to their offspring (fig. 5). From paternal allocations so far achieved in the colony, it has emerged that those four now-adult females which were fathered by

Table 6. Indices of relatedness in mandrills, as computed from autoradiographs of Alu I DNA digests probed with $(GTG)_5$

	n	Gel 1	n	Gel 2
Bands, n	18	113	14	119
Relatedness (D)				
Non-related mandrills	45	0.17±0.10[1] (0–0.43)[2]	21	0.14±0.10 (0–0.37)
Adult males	15	0.23±0.09 (0–0.43)	15	0.13±0.11 (0–0.37)
Mother/offspring	8	0.50±0.12 (0.31–0.63)	6	0.53±0.09 (0.37–0.63)
Sire/offspring	8	0.45±0.08 (0.33–0.57)	6	0.40±0.11 (0.32–0.56)
Full siblings	3	0.39±0.22 (0.18–0.62)	3	0.49±0.20 (0.33–0.71)

Gel 1 compared the 6 adult males with 5 females and their 8 offspring; Gel 2 compared the same 6 males with 4 females and their 6 offspring.
[1] Mean ± 1 SD.
[2] Range.

No. 7 (2C, 2D, 12C and 12D) have avoided incest, as all their first offspring were fathered by No. 14 (table 7).

The final analysis of patterns of paternity within the mandrill colony will allow a definitive statement on the mating system of this species. Known changes in hierarchy can therefore be correlated with shifts in paternity. The preliminary data presented here indicate that the dominant male is responsible for a high percentage (80%) of progeny. However, he does not enjoy sole access to receptive females (or perhaps one could invoke the expression of female choice here), as the previously dominant male of the group still fathered infants despite his secondary role in the hierarchy. This latest change in dominance among the males fortuitously occurred as the offspring of the previously dominant male matured, so that their offspring would be more likely to have a different sire, and thus avoid incest.

The close relatedness of the second generation of the mandrill colony with both potential sires has led to difficulties in differentiating between sire and grandsire. A high degree of maternal transmission of paren-

Table 7. Individual indices of relatedness for 10 mandrill offspring born following the 1990 mating season, determined for the 10 mother-offspring dyads, the 10 father-offspring dyads and the 10 dyads between the infants and the other putative father (only male No. 7 and male No. 14 were included in the calculations)

	Mother	Male No. 14	Male No. 7
Progeny of No. 14			
2/91	0.67	0.67	0.40
2C/91	0.64	0.59	0.25
2D/91	0.53	0.53	0.27
10/91	0.33	0.40	0
12D/91	0.56	0.56	0.32
Potential progeny of No. 14			
12/91	0.59	0.32	0.10
12C/91	0.37	0.35	0.22
17A/91	0.63	0.33	0.21
Progeny of No. 7			
5/91	0.50	0.31	0.46
6/91	0.59	0	0.43

tally inherited alleles, and a higher-than-expected index of relatedness (0.43, on 2 of 3 analyses; table 6) between these 2 males results in ambiguous results in some second-generation offspring. Again the question arises as to how many characteristic bands are required for a confident allocation of paternity, when all possible males have been included in the analysis. Males No. 7 and 14 show several very distinctive bands in their fingerprints and hence we were able to use band inclusion in allocating paternity.

Conclusions

We have characterized DNA fingerprinting methods for gorillas, chimpanzees, guenons and mandrills, demonstrating the versatility of this technique as well as some of the problems. A variety of methods for DNA fingerprinting of nonhuman primates has been described in the literature using widely different combinations of restriction enzymes and probes, and clearly the nature of minisatellite DNA does not restrict the bench worker to any specific combination [13–15]. Hinf I DNA digests appear to provide the most suitable combination with o33.6 for hominoids and Alu I digests with o33.15 or $(GTG)_5$ for monkeys, although it remains to be seen whether Alu I is as suitable for *C. solatus.*

In all species investigated we were most successful in detecting first-degree relationships (parent-offspring or full sibships); identifying second-degree, half-sibs proved very difficult, especially where there was a higher-than-expected degree of band-sharing between non-related individuals. In the case of the mandrill colony, we will be dependent on identifying first-degree relationships each time, against exclusion of known individuals, and hence the task of establishing paternity in this group, whilst tedious, appears feasible. In all cases, we have assumed non-relatedness between the founder chimpanzees, gorillas and mandrills which came to CIRMF from other colonies in Gabon when the Primate Centre was established; perhaps this assumption will be shown to be unrealistic.

Acknowledgments

The oligonucleotide probes used in this study were synthesized by Dr. N. Hastie, MRC Human Genetics Unit, Western General Hospital, Edinburgh, UK (o33.15 and o33.6) and Dr. P.T.K. Saunders, MRC Reproductive Biology Unit, Edinburgh, UK ($(GTG)_5$); we gratefully acknowledge their generosity and their help with setting up the technique in Gabon. Invaluable help was provided by the veterinary staff of the Primate Centre (CIRMF). We thank Prof. A.J. Jeffreys for his support and advice throughout this work.

References

1 Jeffreys AJ, Wilson V, Thein SL: Hypervariable 'minisatellite' regions in human DNA. Nature 1985;314:67–73.
2 Hill WG: DNA fingerprints applied to animal and bird populations. Nature 1987; 327:98–99.
3 Dallas JF: Detection of DNA 'fingerprints' of cultivated rice by hybridisation with a human minisatellite DNA probe. Proc Natl Acad Sci USA 1988;85:6831–6835.
4 Jeffreys AJ, Wilson V, Thein SL: Individual-specific 'fingerprints' of human DNA. Nature 1985;316:76–79.
5 Wickings EJ, Dixson AF: DNA fingerprinting of a semi-free ranging colony of mandrills *(Mandrillus sphinx)* in Gabon (abstract). First Int Conf on DNA Fingerprinting, Berne, October, 1990.
6 Schäfer R, Zischler H, Birsner U, Becker A, Epplen JT: Optimized oligonucleotide probes for DNA fingerprinting. Electrophoresis 1988;9:369–374.

7 Burke T, Bruford MW: DNA fingerprinting in birds. Nature 1987;327:149–152.
8 Wetton JH, Carter RE, Parkin DT, Walters D: Demographic study of a wild house sparrow population by DNA fingerprinting. Nature 1987;327:147–149.
9 Ely J, Alford P, Ferrell RE: DNA 'fingerprinting' and the genetic management of a captive chimpanzee population *(Pan troglodytes).* Am J Primatol 1991;24:39–54.
10 Lernould J-M: Classification and geographical distribution of guenons: A review; in Gautier-Hion A, Bourlière F, Gautier J-P (eds): A Primate Radiation: Evolutionary Biology of the African Guenons. Cambridge, Cambridge University Press, 1988, pp 54–78.
11 Harrison MJS: A new species of guenon (genus *Cercopithecus*) from Gabon. J Zool, Lond 1988;215:561–575.
12 Jouventin P: Observations sur la socio-écologie du mandrill. Terre et Vie 1975;29: 493–532.
13 Washio K, Misawa S, Ueda S: Individual identification of non-human primates using DNA fingerprinting. Primates 1989;30:217–221.
14 Weiss ML, Wilson V, Chan C, Turner T, Jeffreys AJ: Application of DNA fingerprinting probes to Old World monkeys. Am J Primatol 1988;16:73–79.
15 Dixson AF, Hastie N, Patel I, Jeffreys AJ: DNA 'fingerprinting' of captive family groups of common marmosets *(Callithrix jacchus).* Folia Primatol 1988;51:52–55.

Dr. E. Jean Wickings, CIRMF, B.P. 769, Franceville, Gabon (Africa)

Martin RD, Dixson AF, Wickings EJ (eds): Paternity in Primates:
Genetic Tests and Theories. Basel, Karger, 1992, pp 131–140

Paternity Testing in Captive Japanese Macaques *(Macaca fuscata)* using DNA Fingerprinting

Miho Inoue, Fusako Mitsunaga, Hideyuki Ohsawa, Akiko Takenaka, Yukimaru Sugiyama, Aly Gaspard Soumah, Osamu Takenaka

Primate Research Institute, Kyoto University, Inuyama, Japan

It has been impossible to determine paternity in Japanese macaque groups from behavioral observations because of their multimale group structure and promiscuous mating patterns. Electrophoretic analyses of blood proteins and immunoserological analysis have also failed to establish paternity because there is little detectable polymorphism. Paternity discrimination in Japanese macaques requires novel genetic markers. Recently, Jeffreys and others have explored hypervariable regions of DNA called 'minisatellites' [1–6]. The present report describes the results of paternity discrimination in enclosed Japanese macaque groups employing hypervariable genetic markers for DNA generated by synthesized minisatellite probes.

During the mating season, we observed all mating behaviors of one captive group of Japanese macaques in the daytime to calculate the share of mating cost, indicated by the number of copulations, performed by each male with different female partners [7]. We were able to identify paternity for all 8 infants born in the following spring. These results permitted accurate comparisons of the reproductive success of each male with that expected on the basis of observations of his mating activity.

Materials and Methods

Two captive Japanese macaque groups at the Primate Research Institute, Kyoto University, were the subjects of this study. The first (Wakasa group) was introduced from western Japan in 1974 and kept in an open enclosure with an area of 500 m². In 1988, the

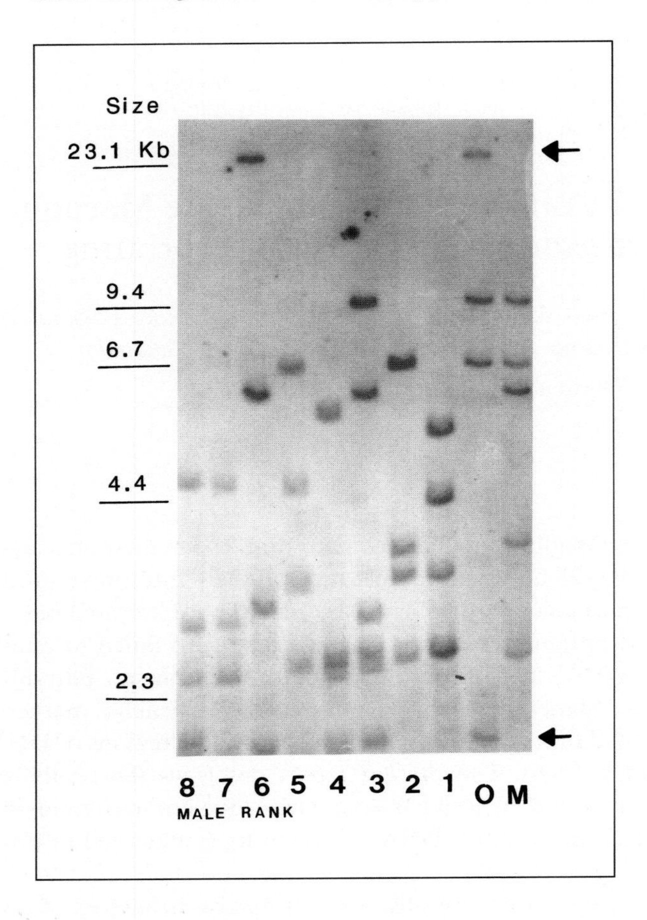

Fig. 1. An example of paternity discrimination in Japanese macaques. High-molecular DNA was prepared from peripheral blood samples. Four micrograms of each DNA sample was digested with the restriction enzyme, Hae III. The digests were separated by electrophoresis through agarose gel. The DNA was denatured and transferred to a Nylon membrane filter (Hybond-N). The filter was hybridized with ^{32}P-labeled single-stranded minisatellite probe [1]. The consensus sequence of the probe was 5′-TGGAGTCTGGGTGTCGTGCGTCAGAGT-3′. We also used another enzyme and another 3 probes for discrimination [2–4]. The band patterns of the mother (M), her offspring (O), and 8 adult male candidates for paternity are shown. Two bands indicated by ← are present in the offspring but not in its mother. One of the 8 male candidates, No. 6, possesses both of these bands. This data set indicates that the two nonmaternal bands of the offspring were inherited from its father (No. 6).

group consisted of 23 males and 34 females of different ages. Each individual was identified by tattooing, and the complete maternal family tree is available. The second (Arashiyama group) was introduced from Kyoto city in 1980 and kept in an enclosure near the Wakasa group. We discriminated paternity for 48 offspring born to the Wakasa group in 1975–1988 and for 22 offspring born to the Arashiyama group in 1982–1987. Figure 1 shows an example of paternity discrimination.

The authors continuously observed the mating behaviors of all individuals in the Wakasa group from October 1987 to February 1988, with daily observations covering the period from 1 h before sunrise to 1 h after sunset. Total observation time was about 1,500 h. In this group, males were observed to ejaculate after reaching 4 years of age, while females were found to conceive at ≧ 4 years. We analyzed behavioral data of all macaques in the group aged ≧ 4 years (11 males and 18 females) The socionomic sex ratio in the group was hence 0.61 adult males for every adult female.

Results and Discussion

Paternal Lineages

Male Rank and Number of Offspring. Figure 2 shows the numbers of offspring fathered by individual adult males in the Wakasa and Arashiyama groups in each year. In the Wakasa group, for a number of years after the establishment of the group, only male No. 1 fathered offspring. Subsequently, however, various other males joined in the breeding of the group. Male No. 3, born in 1976, first impregnated females in the 1981 mating season at the age of about 5.5 years, although male Japanese macaques generally begin to mount and ejaculate at 4.5 years. There was no change in dominance rank between the 2 adult males (No. 1 and No. 2) in the period 1974–1979, or among the 3 adult males (No. 1, No. 2 and No. 3) in the period 1980–1988. It is often speculated from observations of mating behavior that the higher the dominance rank of a male the more offspring he can father. In total, 41 offspring were born in the period after 1981 when all 3 males in the Wakasa group had reached sexual maturity. In none of the years from 1981 to 1988 did the number of offspring of the three males consistently reflect their rank order.

Preliminary results from work in progress on wild troops of Japanese macaques (data not shown here) have provided some valuable additional information. In wild troops, males can migrate between troops. The choice of mating partners in wild troops is wider than in captive groups. Therefore, lower-ranking males have more opportunities to mate compared with captive groups, because they can hide or run away from higher-ranking males when attacked or disturbed. Inter-troop copulations are also ob-

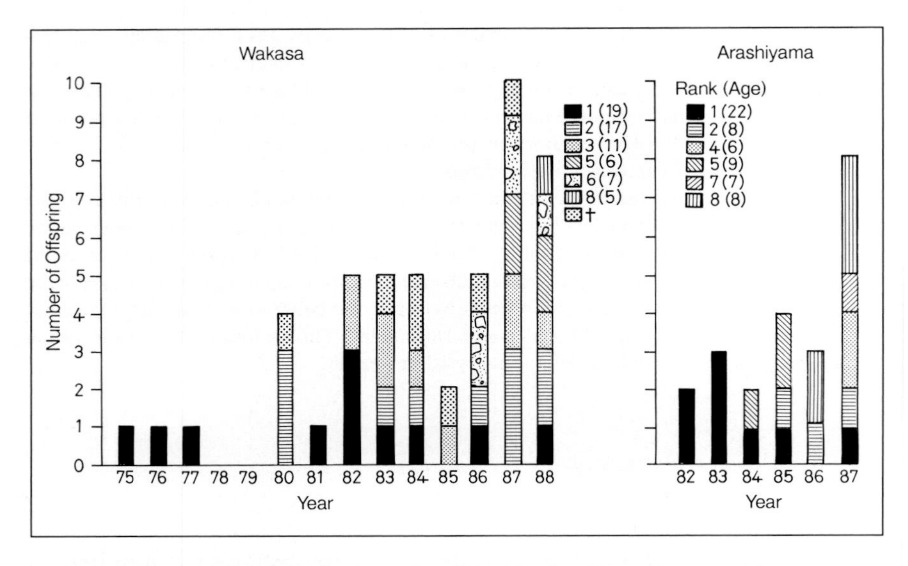

Fig. 2. The number of offspring of each male in the Wakasa and Arashiyama groups in each year. 'Rank (age)' indicates the rank (age) of each male in 1987. The cross (†) indicates that the father was thought to be one of the males that subsequently died.

served in the wild. We have conducted analyses to permit paternity discrimination for two wild, partially provisioned troops living at Koshima, south Japan. Capture and release fieldwork was carried out in 1990. We took blood samples from 85 macaques in the main troop and 15 in the branch troop, out of a total of 101 macaques (1 animal remaining untrapped). We were able to narrow down the possibility of paternity for 41 out of 44 offspring born in 1979–1989. Although two or more candidates for paternity remained for most offspring, every male was excluded from paternity for at least 85% of the offspring, with the exception of the 4th-ranking male in the main troop and the 5th-ranking male in the branch troop. Both of these males were also excluded from paternity in more than 70% of cases. It is therefore obvious that no male in a troop was able to monopolize the paternity of offspring, in agreement with the results obtained for the captive groups.

Inbreeding Avoidance. In the two groups studied which were kept in enclosures, males were unable to emigrate from their natal groups. In spite of this, inbreeding was avoided in any given maternal lineage. Figure 3a

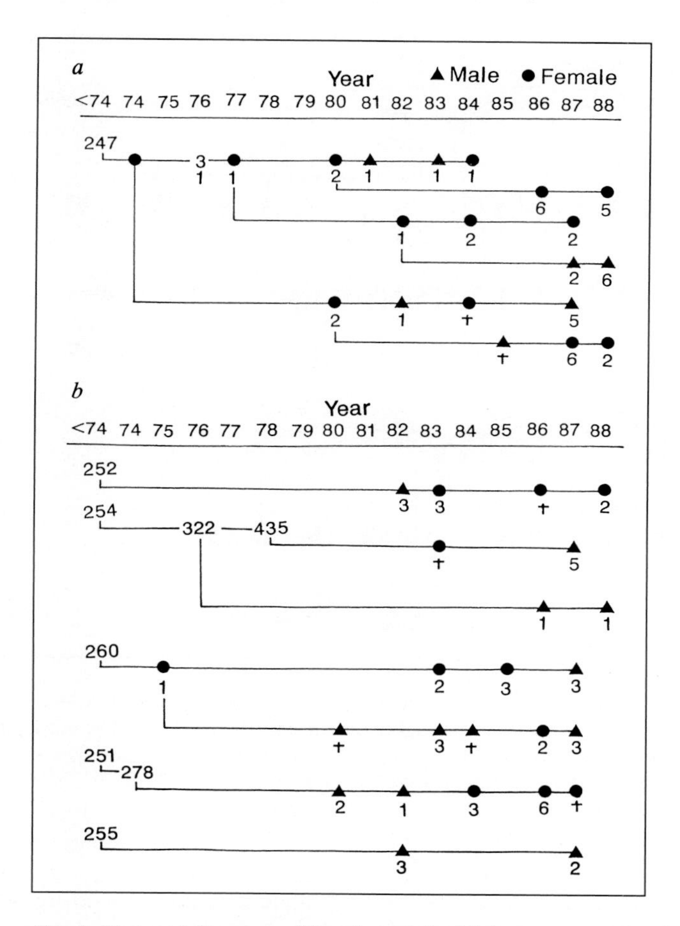

Fig. 3. Maternal lineages of females of the Wakasa group. The lineages of female No. 247 (*a*) and No. 252, 254, 260, 251, 255 (*b*) are shown. Male No. 3 was the offspring of female No. 247 and is indicated by this number, rather than by a symbol as for other offspring. The number below each symbol indicates the identifying number of the father. † = Infant of one of the males which died.

shows the lineage of female No. 247 in the Wakasa group; male No. 3 was one of her sons. Among the 41 offspring born since 1981, at which time No. 3 became sexually mature, 16 were members of his matrilineal family and 25 were members of other families. He fathered no offspring in his own family, but was the father of 9 in the other families. (This difference is

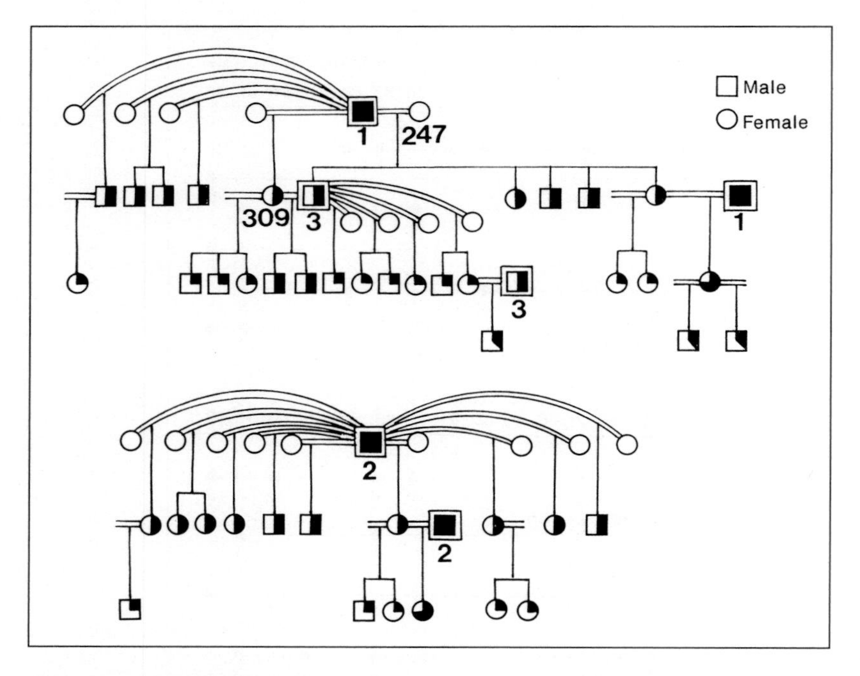

Fig. 4. Paternal lineage tree. The upper part of the figure describes the lineage of males No. 1 and No. 3, while the lower part gives the lineage of male No. 2. The black part of each mark represents the proportion of genes inherited from No. 1 (upper diagram) or No. 2 (lower diagram).

significant; binominal test: $p < 0.05$.) This suggests that inbreeding was avoided in the same matrilineal family, although incestuous copulations among maternal kin were in fact observed.

By contrast, inbreeding was not avoided in father-daughter relationships, nor between paternal half-siblings. For example, female No. 247 gave birth to a daughter in 1977 whose father was No. 1 (fig. 4). The daughter gave birth to a grandchild of No. 247 in 1982, whose father was also No. 1. As seen in figure 4, male No. 3 and female No. 309 were half-siblings, and they had two offspring together.

Fathers of Offspring Born from the Same Mother. Figures 3a, b show maternal family trees of females that gave birth to 2 or more offspring. Except for 1 female (No. 322), all females gave birth to offspring of at least

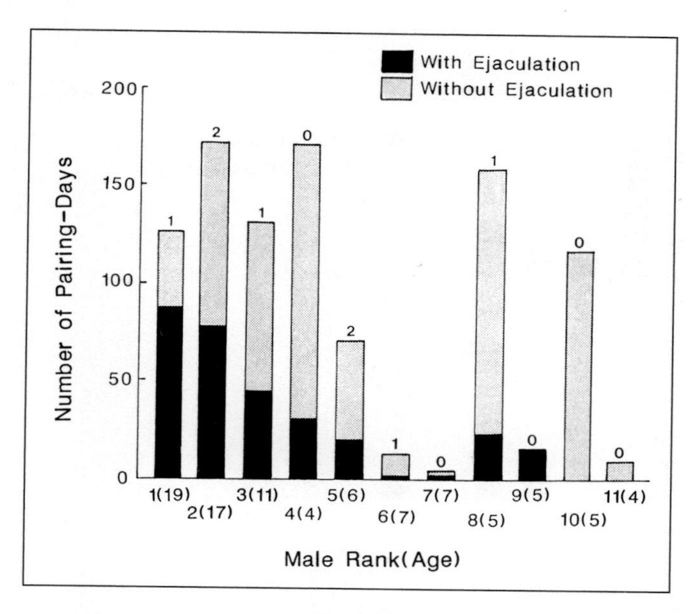

Fig. 5. Male dominance rank order (ages in parentheses) and number of pairing days in the 1987–1988 mating season. The number above each column indicates the number of offspring fathered in 1988 by each male.

2 different males. Thus, female Japanese macaques generally do not tend to produce offspring fathered by the same males year after year, indicating there are few mating pairs that are stable over the years.

From the results of paternity discrimination in the Arashiyama group, almost the same tendencies were inferred as for the Wakasa group.

Mating Behavior and Paternity

Male Rank and Number of Copulations. Figure 5 shows the number of observed copulations in relation to male rank. The Japanese macaque has a multimount system of copulation. Copulation is terminated by ejaculation, by disturbance through a high-ranking male, or by some other factor. The number of pairing days refers to the total number of days in which copulation between a male and any individual female was observed. For example, if a male mated with three different females in a day, his number of pairing days is counted as 3. The sum of pairing days across all males was 1,355. The 11 males $\geqq 4$ years were divided into 3 classes for conve-

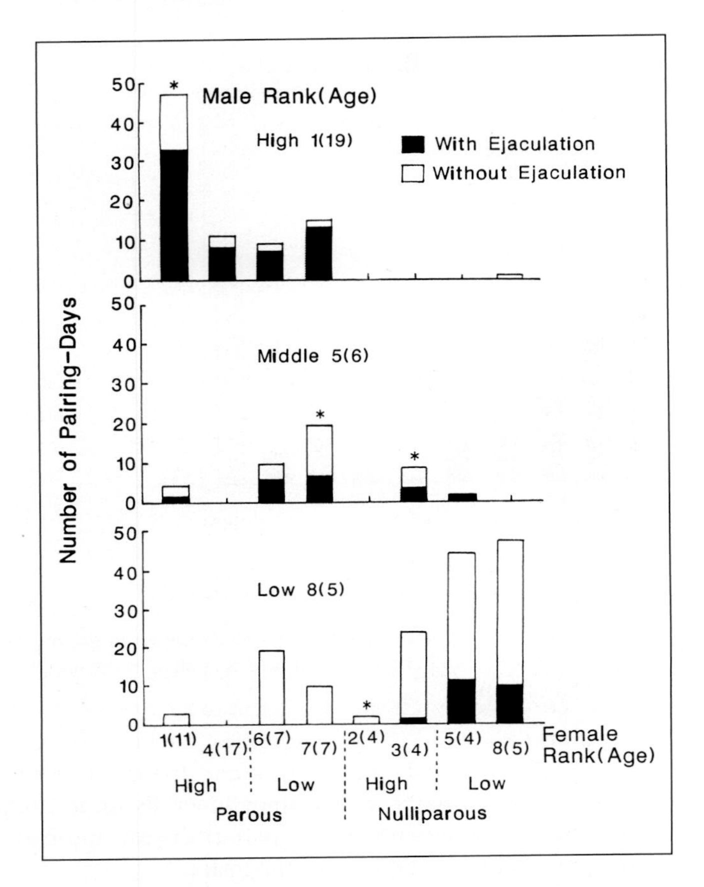

Fig. 6. Male rank and pairing day frequency for matings with individual females. Pairing day frequencies are given for 3 males in each rank class (high, middle and low) for matings with 8 females that gave birth in 1988. Females are divided into 4 categories; high-ranking and low-ranking parous, high-ranking and low-ranking nulliparous. Females are numbered in the order of their dominance rank (numbers in parentheses indicate their ages). Females marked with * gave birth to offspring of a given male in that class.

nience, the first 4 males as 'high-ranking', the next 3 as 'middle-ranking', and the last 4 as 'low-ranking'. Male dominance rank order in this group generally followed age order, with the exception of the 4th-ranking 4-year-old male. He was given agonistic support by his high-ranking mother.

Ejaculation frequency reflects male dominance rank ($r_s = 0.84$, $p \leq 0.01$), but not the number of offspring ($r_s = 0.44$, $p \geq 0.05$). In other words, observed mating behavior alone is a poor indicator of paternity. The middle- or low-ranking males must have counter-strategies enabling them to father offspring despite having far less frequent copulations than high-ranking males.

Selection of the Partner. Figure 6 illustrates the selection of copulation partners by a male in each of the 3 rank classes. Eight females that gave birth in the following spring are ordered according to parity and rank. The number of copulations with each partner changed in accordance with male rank and age. Males in the same rank class showed similar patterns. A high-ranking male tended to copulate with high-ranking adult females that gave birth to his offspring next spring. One middle-ranking male fathered two offspring in spite of his small number of pairing days. One low-ranking 5-year-old male copulated frequently with young, low-ranking females; but he copulated on only a few occasions with a female that subsequently gave birth to his offspring. The percentage of pairing days involving a male and a female that gave birth to an offspring varied greatly.

These results indicate that males of each rank class followed different reproductive tactics. Each high-ranking male fathered one or two offspring following frequent copulations with the mother(s). Middle- and low-ranking males apparently copulated with as many females as possible while avoiding attacks by dominant males. The high-ranking males seem to follow a 'secure' strategy. By constrast, the middle- and low-ranking males may follow a sneaky 'chance' strategy, although it was difficult to predict any positive choice of females for copulation (e.g. male-initiated approach) from observations of mating behavior.

We need more data about the condition of female estrus for more precise discussion of the relationships involved.

Conclusions

(1) The number of offspring of the three fully mature males did not consistently reflect their rank order.

(2) Inbreeding avoidance existed within maternal lineages.

(3) Inbreeding was not avoided in father-daughter relationships nor between paternal half-siblings.

(4) Females did not produce offspring fathered by the same males throughout their lives.

(5) Males can father offspring from the age of 5 years.

(6) Male dominance rank was positively correlated with the number of copulations accompanied by ejaculation.

(7) The number of copulations with ejaculation observed for any given male was not correlated with the number of offspring fathered. Although low-ranking males had only limited opportunities for copulation, they were still able to inseminate females.

(8) The selection of copulation partners by males can be divided into 3 categories according to their age and social rank.

Acknowledgment

We wish to express our thanks to the staff of the Laboratory Primate Center of the Primate Research Institute, Kyoto University, for their assistance in collecting blood samples and for providing us with demographic data for the macaques.

References

1 Nakamura Y, Leppert M, O'Conell P, et al: Variable number of tandem repeat (VNTR) markers for human gene mapping. Science 1987;235:1616–1622.

2 Inoue M, Takenaka A, Tanaka S, et al: Paternity discrimination in a Japanese macaque troop by DNA fingerprinting. Primates 1990;31:563–570.

3 Jeffreys AJ, Wilson V, Thein SL: Hypervariable 'minisatellite' regions in human DNA. Nature 1985;314:67–73.

4 Kominami R, Mitani K, Muramatsu M: Nucleotide sequence of a mouse minisatellite DNA. Nucleic Acids Res 1988;16:1197.

5 Weiss ML, Wilson V, Chan C, et al: Application of DNA fingerprinting probes to Old World monkeys. Am J Primatol 1988;16:73–79.

6 Wetton JH, Carter RE, Parkin DT, et al: Demographic study of a wild house sparrow population by DNA fingerprinting. Nature 1987;327:147–149.

7 Inoue M, Mitsunaga F, Ohsawa H, et al: Male mating behaviour and paternity discrimination by DNA fingerprinting in a Japanese macaque troop. Folia Primatol 1991;56:202–210.

Miho Inoue, Primate Research Institute, Kyoto University,
Inuyama, Aichi, 484 (Japan)

Martin RD, Dixson AF, Wickings EJ (eds): Paternity in Primates: Genetic Tests and Theories. Basel, Karger, 1992, pp 141–154

Paternity Determination by Oligonucleotide DNA Fingerprinting in Barbary Macaques *(Macaca sylvanus)*

Jutta Kuester[a], *Andreas Paul*[a], *Joachim Arnemann*[b]

[a] Institut für Anthropologie, Universität Göttingen, BRD;
[b] National Institute for Medical Research, London, UK

Although it is of utmost importance for countless questions concerning primate sociobiology, a key parameter, *male reproductive success,* is extremely elusive, as methodological problems have hindered routine application of 'classical' paternity determination. The discovery of hypervariable DNA sequences in the genome of eukaryotes [1–4] led to a wave of optimism that a powerful tool had been found for solving this and other problems concerning identity and relatedness. However, the degree of variability of DNA fingerprint patterns shows species-typical differences among primates [5, 6]. So far, probes producing individual-specific patterns, or 'true' fingerprints, have not become available for nonhuman primates and they are known only for a few other species [7, but see ref. 28]. Furthermore, sample sizes have been small in most studies of nonhuman primates [6, 8–10], and paternity analyses often demonstrated only the inheritance of bands from both parents in individuals with known genetic relationships. The first results from a fairly large sample of Japanese macaques became available only recently [11], but the general applicability of this method to large samples, including complete social groups with many potential fathers, is still unknown. An attempt to tackle this problem was made in the present study with individuals from a semifree-ranging population of Barbary macaques *(Macaca sylvanus)*. This species shows a seasonal, polygynandric mating system and a female-bonded social grouping pattern with overlapping generations. Background information on demographic development, social structure, mating system and life histories was available, covering more than a decade. An earlier attempt at paternity identification with serological markers failed, because of insufficient polymorphism [W. Scheffrahn, University of Zürich, pers. commun.]. Mating data cannot be used even to speculate about paternity, as female Barbary macaques usually mate with several males at all stages of estrus [12–14].

The study was conducted with synthetic simple-repeat oligonucleotide probes. A pilot study on rhesus monkeys had previously revealed variable banding patterns with such probes and had indicated that paternity discrimination is possible in principle [8; unpubl. data; for details on this 'family' of probes, see e.g. ref. 1, 15–18].

Materials and Methods

Study Sample

The study was conducted with the semifree-ranging Barbary macaque population of Affenberg Salem, a 14.5-ha outdoor enclosure in Southwest Germany. The park was founded in 1976 with a complete social group of 144 individuals and a small party of 20 immature animals, translocated from two similar enclosures in France. More than 200 individuals had originally been captured in the late sixties and early seventies in the Middle Atlas for these enclosures [19]. At least 60 of the original wild-caught individuals were included in the founding generation of the new park in Salem. About half of the captive-born animals transferred to Salem arrived there without one or both of their parents.

The population grew rapidly and group fissions occurred [20]. Several groups were removed during subsequent years, e.g. in 1986 for a release project in Morocco. Male transfer between groups was common, although some males from large social groups (> 100 individuals) spent varying periods of time in their natal groups after becoming sexually mature [for details on management, group histories, see ref. 20, 21]. Demographic development of the whole population has been completely documented since 1977. Detailed behavioral studies had been conducted on three groups (B, C and F). Matrilineal relationships of all group members born in Salem and many of those born in France were also known.

The study was designed to identify the fathers of 273 individuals. These were all infants born to females of groups B, C and F during 1977–1988 for which blood samples were available, including some infants sired before group establishment (table 1). The infants were descendants of 27 mainly wild-caught matriarchs. Matrilineally, most of them belonged to the first (35%) or second (53%) captive-born generation. The total sample size, including the mothers and all available potential fathers was 396 individuals.

On the basis of more than 10,000 recorded matings, it was concluded that infants were principally sired by males of their own group, and all sexually mature males (>4 years of age), living in a group during a mating season were regarded as potential fathers of all infants sired during that season.

In total, 140 males had to be considered as potential fathers. Among them were all 19 wild-caught males and 44 captive-born males that were matrilineally unrelated with

Table 1. Study sample

Year of birth	Group	Infants	Potential fathers[1]	
			adult	subadult
1977	(B/C)[2]	20	19	6 (2)
1978	B	10	8	7 (1)
	(C)[2]	5	14	5 (1)
1979	B	2	5	12 (2)
	C	3	3	2
1980	B	10	9 (1)	21 (2)
	C	4	4	5 (1)
1981	B	5	11 (1)	21 (2)
	C	7	6	10 (1)
1982	B	6	14 (1)	24 (2)
	C	13	9 (1)	12
1983	B	10	26 (2)	12 (1)
	C	9	10 (1)	14 (1)
1984	B	16	28 (1)	16 (6)
	C	17	12	9
1985	B	22	22 (1)	14 (5)
	C	18	8	8
1986	B	13	24 (1)	20 (11)
	C	24	10	13
1987	B/F	12	15	13 (2)
	C	20	11	22
1988	B	8	12 (1)	10
	C	13	17	16
	F	6	9	7
Total		273		

( ) = Number of those males for which DNA was unavailable.
[1] Adult males: age ⩾ 7 years; subadult males: age 4–6 years.
[2] Infants of females of groups B and C, sired before group establishment. Groups B and C fissioned from the translocated group in 1977 and 1978, respectively. Group F fissioned from group B in 1987.

the 273 study subjects. Due to the long study period and male migration between study groups, most infants of the older cohorts became mothers and potential fathers of the infants from the younger cohorts. Closely related males (e.g. maternal half-brothers) were present among potential fathers in many cohorts. Infants had between 5 and 44 potential fathers ($\overline{X} = 25$; table 1), and 7,461 dyads of potential fathers and infants existed, of which all but the 273 'true' father-infant dyads had to be excluded.

Blood samples were taken from the population in winter 1977/78 and in autumn 1988. Only animals born after 1977 could, therefore, be analyzed if they were still present in 1988. This condition was met for 80% of all infants born during the study period to females of group C and for 39% born to females which were or had been members of group B and/or F. The lower proportion for groups B and F was mainly attributable to the release project in 1986. Largely for the same reason, blood samples of 21 males, mostly subadults (table 1), were unavailable, such that 13.3% of the dyads in group B, and 1.9% of the dyads in group C could not be analyzed. However, most of these 'missing' males were born in Salem, and for 16 males, banding patterns of the mother, other maternal relatives and the potential fathers could be investigated, so that information about their possible patterns was not completely lacking.

Biochemical Analysis

In autumn 1988, 0.5–4 ml blood per individual (n = 317) was collected with EDTA syringes and frozen. In addition, 79 samples from 1978 were available for analysis. DNA was isolated using standard procedures with proteinase K and phenol/chloroform extractions. From each sample, 5 µg DNA was digested with the restriction enzymes Alu I, Hae III, and Hinf I, according to the suppliers' recommendations. Restriction fragments were separated on a 0.8% agarose gel in a vertical gel chamber. Gels were run with 1.3 V/cm for approximately 14 h. Gels were dried on a vacuum gel dryer for 1 h at room temperature and for 0.5 h at 60 °C prior to storage at room temperature. Twenty samples, usually of maternal relatives, were run together on each gel. In every case, DNA of one family member was separated on another gel to serve as an internal standard.

The study was initiated with 3 synthetic simple-repeat oligonucleotides probes – $(GACA)_4$, $(GATA)_5$ and $(GTG)_5$ – and subsequently expanded to include the probes $(GA)_8$, $(GT)_8$ and $(GGAT)_4$ (names of probes indicate the nucleotide sequence and the number of repetitions). Oligonucleotides were end-labeled with ^{32}P-τ-ATP using T4 polynucleotide kinase. Prior to hybridization, gels were denaturated for 0.5 h and then neutralized for 0.5 h. Hybridization was carried out overnight in a solution consisting of 5 × SSPE, 10 × Denhardt's reagent, 0.1% SDS, 10 µg/ml salmon sperm-DNA or t-RNA as competitor, and 1.5–2.0 × 10^6 cpm/ml of the radiolabeled probe. Hybridization temperature was 45 °C for $(GTG)_5$ and $(GATA)_5$ and 43 °C for the other 4 probes. Gels were washed twice for 1 h each in 6 × SSC on ice or at room temperature, and 1 min in 6 × SSC at the hybridization temperature. The solution was reused on the following day. For autoradiography, gels were exposed to X-ray films with one intensifying screen for 2–8 days at 4 °C. Reproducibility of the results was tested with a control sample by repeating the experimental procedure with approximately 15% of the individuals. All gels were denaturated and neutralized again and successively hybridized with all probes without loss in quality of the autoradiographs over the course of nearly 2 years [for further methodological details, see ref. 1, 8].

Analysis of Banding Patterns

The limited gel capacity and the high number of potential father-mother-infant trios made a comparison of banding patterns on different autoradiographs inevitable. Homologous bands were identified with the aid of invariable bands, the patterns of closely related individuals and/or the control sample. All variable bands found in the complete sample were determined for any given enzyme/probe combination, and presence/absence of these bands was then recorded for every individual. These transformed one-zero patterns were used in (computer-aided) analysis. The transformation led to a certain loss of information, as 2 narrowly located bands in a pattern had to be treated as a single band, if correct identification on *all* autoradiographs was impossible. Analysis of faint bands was limited because the intensity of autoradiographs, despite great efforts to achieve standardization, varied. Such bands were analyzed only visually, with special consideration given to overall intensity.

Visual comparisons were principally necessary when inconsistencies occurred (e.g. exclusion of all potential fathers in cohorts without missing males) and were also usually conducted when the majority of males had been excluded with several probes and only a few males remained as potential fathers. If a father was identified, the infant remained in the sample and all other patterns were also analyzed to control the result.

Results

Characteristics of the Banding Patterns

Variable banding patterns were obtained with all 6 probes (see fig. 1, for an illustration of the 3 probes used first), but none of them produced individual-specific patterns or true DNA fingerprints. Between 7 and 22 variable bands that were suitable for computer analysis were found in the complete study sample with each of the 15 enzyme/probe combinations analyzed (table 2). Most bands were located in the higher-molecular-weight region (> 4.4–6 kb), and 1–2 invariable bands were also present in this region in most patterns. Invariable bands occurred more frequently in the lower-molecular-weight region, where analysis of variable bands was limited because most of them appeared blurred and were of low intensity, making correct identification on separate autoradiographs problematic.

All bands were reproducible in approximately 95% of the individuals of the control sample. Patterns obtained with $(GT)_8$ were somewhat exceptional, as the intensity of homologous bands varied considerably, which made analysis difficult and affected reproducibility on different autoradiographs. This probe was, therefore, of only limited value. All other unreproducible bands were weak and occurred mainly in the lower-molecular-weight region. Since repeated hybridization resulted in identical patterns,

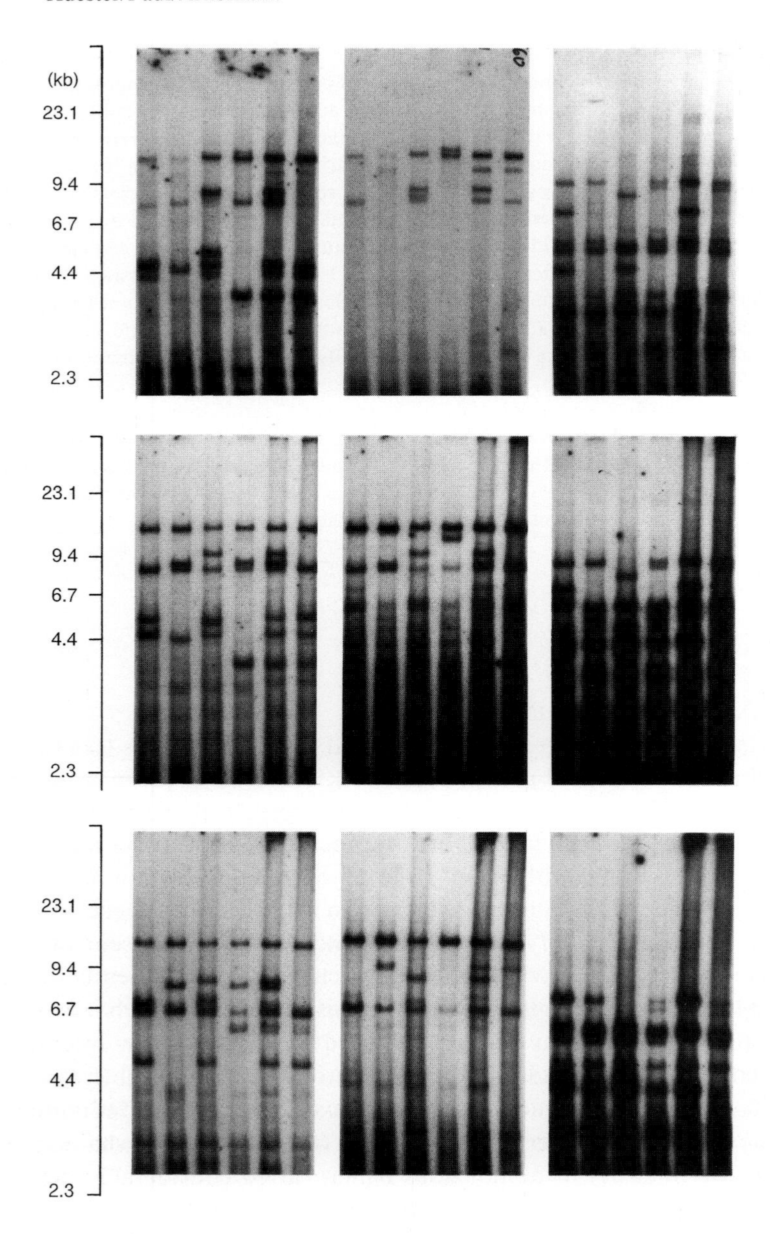

Fig. 1. Banding patterns of 6 unrelated individuals of the first captive-born generation. From left to right: probes $(GACA)_4$, $(GATA)_5$, $(GTG)_5$. From top to bottom: Enzymes Alu I, Hae III, Hinf I. Approximate fragment length is indicated (kilobases).

Table 2. Numbers of variable bands (vb) in the complete study sample and rate of band sharing (bs; variable bands only) among the wild-caught individuals (n = 39–40 for Alu I and Hinf I, n = 42 for Hae III)

		Alu I	Hae III	Hinf I
$(GACA)_4$	vb	11	11	9
	bs Mean	0.38	0.37	0.37
	bs Range	0.10–0.93	0.02–0.93	0.04–0.93
$(GATA)_5$	vb	8	7	8
	bs Mean	0.39	0.34	0.32
	bs Range	0.06–0.93	0.07–0.93	0.05–0.93
$(GTG)_5$	vb	11	8	10
	bs Mean	0.51	0.46	0.52
	bs Range	0.05–0.89	0.05–0.71	0.05–0.90
$(GA)_8$	vb	ND	21	22
	bs Mean	–	0.23	0.32
	bs Range	–	0.02–0.86	0.04–0.90
$(GT)_8$	vb	ND	13	20
	bs Mean	–	0.42	0.56
	bs Range	–	0.04–0.95	0.08–0.96
$(GGAT)_4$	vb	11	15	ND
	bs Mean	0.39	0.36	–
	bs Range	0.05–0.76	0.05–0.70	–

ND = Not determined.

incomplete digestion of DNA, rather than variability in processes during hybridization and washing, was regarded as responsible for these unreproducible bands. Mutations were the probable cause of 2 other inconsistencies. One infant possessed a reproducible band in the Hinf I/$(GTG)_5$ pattern which was absent in all other individuals. In another infant, no mature male in its own or any other study group possessed all (reproducible) paternal-specific bands found in the 3 $(GATA)_5$ patterns, and all males were again excluded when each $(GATA)_5$ pattern and the paternal-specific bands of the other probes were considered.

Because of the relatively low number of variable bands, only a limited number of different patterns were possible and, consequently, considerable band sharing occurred even among the wild-caught individuals (table 2).

Identical patterns in a given enzyme/probe-combination were found not only in some mother-infant pairs but also in more distantly related or even unrelated individuals. The patterns of mother and infant could also differ only in that the infant possessed fewer bands than the mother. As paternal-specific bands were lacking in such cases, none of the potential fathers could be excluded. The frequency of band sharing, however, varied enormously. Some bands were found in only 1 or 2 animals of the founding generation and their descendants, while others were present in almost all individuals. A band could, therefore, be inherited from the father and the mother, but a distinction between the homo- and heterozygous state (e.g. with help of band intensity) was not possible.

Patterns produced with a given probe and the 3 alternative enzymes did not lead to the expected gain in information, as several bands of one pattern corresponded completely with specific bands obtained with both of the other enzymes, i.e. they were either all present or all absent in an individual. Apparently, these bands represented homologous loci for the repetitive sequence. Fragment lengths differed only due to different cleavage sites in the nonrepetitive flanking sequence. Due to this redundancy, not all possible enzyme/probe combinations were analyzed during the second part of the study. A similar linkage between bands within a pattern was not detected, and linkages of bands obtained with one enzyme and different probes were restricted to single bands in $(GACA)_4$/$(GATA)_5$ and $(GGAT)_4$/$(GATA)_5$, respectively.

For comparisons of variability in the banding patterns, the probability of occurrence of identical patterns in 2 randomly chosen individuals was calculated according to Arnemann et al. [8]. Analysis was based on the transformed one-zero patterns of 10 'unrelated' individuals which had no common ancestors in the parent and grandparent generation. Probability values varied considerably and were mainly related to the number of variable bands (table 3). Consequently, probability values decreased with every new probe used. $(GA)_8$ produced the most variable patterns and, in general, patterns obtained with Alu I and Hae III were more informative than Hinf I patterns. The probability for complete identity of all 15 patterns analyzed could not be calculated as this violated the basic assumption of independent loci for all bands.

Five other simple repeats were tested. $(GAG)_5$ and $(GGA)_5$ produced no bands at all, $(GAT)_6$ and $(GTA)_6$ only faint invariable bands, and the patterns of $(GATA)_2GACA(GATA)_2$ were almost identical to the $(GATA)_5$ patterns.

Table 3. Probability of identical banding patterns in 2 randomly chosen individuals (details see text)

	$(GACA)_4$	$(GATA)_5$	$(GTG)_5$	3 probes	$(GA)_8$	$(GGAT)_4$	$(GT)_8$
Alu I	0.019	0.143	0.004	1.2×10^{-5}	ND	0.014	ND
Hae III	0.020	0.229	0.010	4.6×10^{-5}	0.004	0.003	0.013
Hinf I	0.019	0.206	0.019	7.2×10^{-5}	0.002	ND	0.003

ND = Not determined.

Paternity Analysis

An exclusion of all but 1 male with a single probe was only possible in a few cases in which the male possessed and passed on unique bands. The use of several probes, however, increased the exclusion rate considerably. Moreover, although paternal-specific bands were sometimes absent in patterns obtained with one probe, this was never the case with all probes combined.

With the first 3 probes, males were excluded from paternity in 5,086 potential father-infant dyads, i.e. 71% of all dyads which had to be excluded (all but 1 for each infant; n = 7,188; because of missing males, 626 dyads could not be analyzed). For half of the infants, the number of potential fathers was thereby reduced to 4 or less, but for only 43 infants (16%) was it possible to exclude all males but 1, which was then regarded as the father. This proportion increased considerably – to about 50% – with $(GA)_8$, which was used as the 4th probe (with Hae III). Additional exclusions were possible with $(GGAT)_4$ and to some extent with $(GT)_8$. By the end of the study, all analyzed males except 1 had been excluded from paternity for 228 infants. For 21 infants 2 males and for 4 infants 3 males remained as potential fathers. For 10 of these infants, the remaining males were closely related (father/son, half-brothers), although a distinction between such males was possible in many other cases. Close genetic relationships among the remaining males were possible, but unknown, in several other cases.

For 13 infants *all* analyzed males were excluded from paternity, and 'missing' males were among the potential fathers of all these infants. Six infants were assigned either to the only missing male of their respective birth cohort or to the only missing male which could have inherited certain paternal-specific bands. The 234 infants for which paternity could be determined were sired by 72 different males.

As indicated above, post hoc reconstruction of genetic relationships was, however, rarely possible in this study. Only 2 other infants, in which all analyzed males were excluded, possessed paternal-specific bands which could be traced back to a single male of the founder generation. Other bands were absent in his patterns (and in those of 2 of his sons). This male had been the potential father of several missing males and, apparently, was the infants' paternal grandfather. With the help of paternal-specific bands not present in his patterns but in those of some females whose sons were among the missing males, the number of potential sires for these 2 infants could be reduced from 6 to 2 males (which had the same mother) in 1 case, and from 12 to 5 (including the same 2 males) in the other. Such a 'detour' via the grandparent-generation was not possible in other cases, including 2 infants, in which DNA of the mothers was not available. A few infants remained for which analysis was limited for other reasons. Probably, their fathers were also missing from the sample.

Discussion

Paternity analyses using DNA fingerprinting are in principle possible in large social groups living under seminatural conditions with a great degree of 'self-organization' (e.g. group fissioning, male transfer between groups). For the majority of the study subjects, it proved possible to exclude all but one potential father from paternity. However, positive identification of a father, which is sometimes regarded as characteristic for this method, was an exception. Variability of single banding patterns was usually insufficient to produce individual-specific, true DNA fingerprints.

It seems unlikely that the low variability seen was mainly connected with the probes used. A lower degree of variability in nonhuman primates than in humans, and a high frequency of band sharing among unrelated individuals has been found with other probes, too [5, 6, 9]. While most authors have speculated about an unknown history of inbreeding among the ancestors of their study subjects, it seems more likely that there are, in fact, species differences in the degree of variability of a given probe. One probe used in this study – $(GTG)_5$ – produces individual-specific patterns in humans [18], while the other 5 probes either produce patterns that were much less variable or invariable, or yield no patterns at all [pers. obs.].

More variable sequences may exist in the genome of Barbary macaques and of other nonhuman primates, but they are as yet unknown.

Because of the lack of comparative data, it is unknown whether Barbary macaques in general, or just those of the study population, were especially 'difficult' study subjects. It was not expected at the beginning of the study that inbreeding would have a major impact on the results, as the founder population was large and effectively closed for only a relatively short space of time. Male transfer between groups was common, although it could be delayed or sometimes even lacking in large social groups, as they develop under conditions of provisioning. But exceptions to male migration were coupled with an absence of mating between maternal relatives [22]. All males in Salem born in small groups that were comparable in size to wild groups emigrated when reaching sexual maturity [20]. Recent data from an Algerian population has confirmed that male migration is common in this species [23]. Contrary to earlier speculations [14], behavioral traits which are regarded as effective mechanisms to prevent inbreeding in primates are also present in Barbary macaques.

A quantitative comparison of the variability of the patterns with that found in other studies, using statistical analyses, was not possible, because other authors conducted only visual comparisons of patterns on a single autoradiograph, while here the lower effective variability of the one-zero patterns was used. Moreover, some authors [1, 8] did not take into account the redundancy in information obtained with different enzymes. As such calculations often result in seemingly impressive values, special attention should be paid in general to *what* was calculated. The probability of occurrence of identical patterns in randomly chosen individuals is, for example, much lower than the probability that identical patterns can occur in a population.

Because this study was conducted on a closed population in which all individuals (although not the patterns of all individuals) were definitely known, it was not necessary to use such calculations, e.g. for estimating the probability of false exclusion of a male. As all such calculations give valid results only in the case of unrelated individuals, or if specific relations are considered they could not be used to estimate probabilities of paternity chances for missing males, since most of them were sons of the oldest males but precise genetic relationships were unknown. But even in other cases it is not so much the probability of occurrence of an identical pattern in another male that is of interest, but the probability with which the paternal-specific bands in question occur.

Another factor influencing variability of banding patterns is the mating system of a species, more specifically the mobility of prospective mates. In this respect, most nonhuman primates are unsuitable candidates for easy paternity determination. Female macaques, like females of many other Old World monkey species, spend their whole lives in their natal groups. Males typically migrate only over short distances, often into neighboring groups [24]. A male may well migrate into the natal group of his own father, meeting paternal aunts or cousins as prospective mating partners. While this stepwise migration effectively prevents inbreeding depression [25], it clearly negatively affects the degree of variability of banding patterns, as mates are, on average, more closely related than in species with high mobility (e.g. humans).

Close genetic relationships among some potential fathers were not peculiar to the study population. While father/son pairs may be rare under natural conditions, due to a higher rate of mortality and mobility, (half)-brothers or cousins may well be present in the same social group [26, 27].

Paternity analyses would have been impossible without complete information about the matrilineal relationships of the study subjects and about male group membership during the whole study period. Discriminatory power was usually insufficient to reconstruct patterns of missing individuals and false exclusions of some males or other errors cannot be ruled out with certainty. Some infants might, for example, have been assigned to a male which was actually the paternal grandfather, if the true father was one of the missing males. On the other hand, if alternatives are not at hand, standards necessary in human paternity identification need not be demanded when a new technique is introduced.

The sometimes uncritical optimism that DNA fingerprinting will solve all problems concerning identity, paternity and other related topics in all nonhuman primate species under all circumstances, or may even replace other forms of data collection, is not supported by this study. For nonhuman primates, the suggestive term fingerprint is still misleading.

Acknowledgments

We are grateful to G. de Turckheim and E. Merz for permission to take blood samples from their monkeys at the end of our behavioral studies. Dr. W. Angst and the staff of Affenberg Salem greatly facilitated our work over all the years. Dr. W. Scheffrahn generously contributed many blood samples, collected in 1978, to this project. The experiments were conducted at the Institutes of Human Genetics and Legal Medicine, and at the Zentrales Isotopenlabor, University of Göttingen. The support and hospitality of Profs. W. Engel and K. Saternus are gratefully acknowledged. Special thanks go to Dr. J.T. Epplen for his introduction to fingerprinting, to all colleagues of the 'Kellerlabor' who made it a pleasant stay after all the years in the 'wilderness', to Prof. R.D. Martin for valuable comments and to Prof. C. Vogel for his constant support and encouragement during all stages of the Barbary Macaque Project. The study was financially supported by the Deutsche Forschungsgemeinschaft (grants Vo. 124/15-1, 124/18-1).

References

1 Ali S, Müller CR, Epplen JT: DNA fingerprinting by oligonucleotide probes specific for simple repeats. Hum Genet 1986;74:239–243.
2 Jeffreys AJ, Wilson V, Thein SL: Hypervariable 'minisatellite' regions in human DNA. Nature 1985;314:67–73.
3 Jeffreys AJ, Wilson V, Thein SL: Individual-specific 'fingerprints' of human DNA. Nature 1985;316:76–79.
4 Wyman AR, White R: A highly polymorphic locus in human DNA. Proc Natl Acad Sci USA 1980;77:6754–6758.
5 Dixson AF, Hastie N, Patel I, Jeffreys AJ: DNA 'fingerprinting' of captive groups of common marmosets *(Callithrix jacchus)*. Folia Primatol 1988;51:52–55.
6 Weiss ML, Wilson V, Chan C, Turner T, Jeffreys AJ: Application of DNA fingerprinting probes to Old World monkeys. Am J Primatol 1988;16:73–79.
7 Lewin R: Limits to DNA fingerprinting. Science 1989;243:1549–1551.
8 Arnemann J, Schmidtke J, Epplen JT, Kuhn HJ, Kaumanns W: DNA fingerprinting for paternity and maternity in 'group O' rhesus monkeys at the German Primate Center. Results from a pilot study. Puerto Rico Health Sci J 1989;8:181–184.
9 Ely J, Ferrell RE: DNA 'fingerprints' and paternity ascertainment in chimpanzees *(Pan troglodytes)*. Zoo Biol 1990;9:91–98.
10 Washio K, Misawa S, Ueda S: Individual identification of nonhuman primates using DNA fingerprinting. Primates 1989;30:217–221.
11 Inoue M, Takenaka A, Tanaka S, Kominami R, Takenaka O: Paternity discrimination in a Japanese macaque group by DNA fingerprinting. Primates 1990;31:563–570.
12 Kuester J, Paul A: Reproductive strategies of subadult Barbary macaque males at Affenberg Salem; in Rasa AE, Vogel C, Voland E (eds): Sociobiology of Reproductive Strategies in Animals and Humans. London, Chapman & Hall, 1989, pp 93–109.
13 Small MF: Promiscuity in Barbary macaques *(Macaca sylvanus)*. Am J Primatol 1990;20:267–282.
14 Taub DM: Female choice and mating strategies among wild Barbary macaques (*Macaca sylvanus* L.); in Lindburg D (ed): The Macaques: Studies in Ecology, Behavior and Evolution. New York, Van Nostrand & Reinhold, 1980, pp 287–344.
15 Bell GI, Selby MJ, Rutter WJ: The highly polymorphic region near the human insulin gene is composed of simple tandemly repeating sequences. Nature 1982;295:31–35.

16 Tautz D: Hypervariability of simple sequences as a general source for polymorphic DNA markers. Nucleic Acids Res 1989;17:6463–6471.

17 Schäfer R, Zischler H, Epplen JT: $(CAC)_5$, a very informative oligonucleotide probe for DNA fingerprinting. Nucleic Acids Res 1988;16:5196.

18 Schäfer R, Zischler H, Birsner U, Becker A, Epplen JT: Optimized oligonucleotide probes for DNA-fingerprinting. Electrophoresis 1988;9:369–374.

19 De Turckheim G, Merz E: Breeding Barbary macaques in outdoor open enclosures; in Fa JE (ed): The Barbary Macaque: A Case Study in Conservation. New York, Plenum Press, 1984, pp 241–261.

20 Paul A, Kuester J: Life history patterns of semifree-ranging Barbary macaques *(Macaca sylvanus)* at Affenberg Salem (FRG); in Fa JE; Southwick CH (eds): Ecology and Behavior of Food-Enhanced Primate Groups. New York, Liss, 1988, pp 199–228.

21 Kaumanns W: Berberaffen *(Macaca sylvana)* im Freigehege Salem. Z Kölner Zoo 1978;21:57–66.

22 Paul A, Kuester J: Intergroup transfer and incest avoidance in semifree-ranging Barbary macaques *(Macaca sylvanus)* at Salem (FRG). Am J Primatol 1985;8:317–322.

23 Ménard N, Hecham R, Vallet D, Chikhi H, Gautier-Hion A: Grouping patterns of a mountain population of *Macaca sylvanus* in Algeria – A fission-fusion system? Folia Primatol 1990;55:166–175.

24 Pusey AE, Packer C: Dispersal and philopatry; in Smuts BB, Cheney DL, Seyfarth RM, Wrangham RW, Struhsaker TT (eds): Primate Societies. Chicago, University of Chicago Press, 1987, pp 250–266.

25 Melnick DJ: The genetic consequences of primate social organization: A review of macaques, baboons and vervet monkeys. Genetica 1987;73:117–135.

26 Altmann J: Age cohorts as paternal sibships. Behav Ecol Sociobiol 1979;6:161–169.

27 Meikle DB, Vessey SH: Nepotism among rhesus monkey brothers. Nature 1981;294:160–161.

28 de Ruiter JR, Scheffrahn W, Trommelen GJJM, Uitterlinden AG, Martin RD, van Hooff JARAM: Male social rank and reproductive success in wild long-tailed macaques. Paternity exclusions by blood protein analysis and DNA fingerprinting; in Martin RD, Dixson AF, Wickings EJ (eds): Paternity in Primates: Genetic Tests and Theories. Basel, Karger, 1992, pp 175–191.

Dr. Jutta Kuester, Zoologisches Institut der Universität, AG Ethologie,
Kirschallee 1, D-W-5300 Bonn 1 (FRG)

Martin RD, Dixson AF, Wickings EJ (eds): Paternity in Primates:
Genetic Tests and Theories. Basel, Karger, 1992, pp 155–174

Application of Blood Protein Electrophoresis and DNA Fingerprinting to the Analysis of Paternity and Social Characteristics of Wild Barbary Macaques

Nelly Ménard[a], *Wolfgang Scheffrahn*[b], *Dominique Vallet*[a], *Charef Zidane*[c], *Christine Reber*[b]

[a] CNRS, UA 373, Station Biologique de Paimpont, Plélan-le-Grand, France;
[b] Anthropologisches Institut der Universität Zürich, Schweiz;
[c] Laboratoire d'Histologie, Embryologie et Génétique, Faculté de Médecine de l'Université d'Alger, Algérie

Ecologists, ethologists and evolutionary biologists are becoming increasingly interested in obtaining genetic information on their species of interest. For example, the ascertainment of genetic relationships (mainly paternity) between animals in a given social group has now become one of the main topics in the field of sociobiology. For nonhuman primates, one question concerns the degree to which frequency and timing of copulation during the mating season are related to paternity, and a second question is whether genetic relatedness and socialization of offspring might be linked. One of the main tasks is to understand whether certain males have preferential access to females and to what degree such a relationship is associated with paternity. Furthermore, hypotheses postulating a relationship between certain social interactions of males (for example, caretaking of an infant by a certain male) and paternity require attention.

The species *Macaca sylvanus* has a wide geographical distribution in Morocco and Algeria, but it is split into a number of isolated populations and has adapted to different habitats (fig. 1). In Algeria, at least 7 populations of different sizes have been reported [1–4], representing a total number of about 5,000 individuals. These are grouped in 'genetic isolates', defined in this paper as reproductive units in the sense of true populations

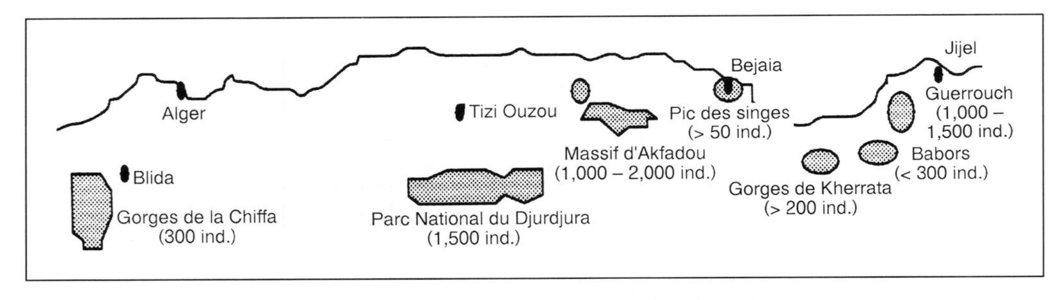

Fig. 1. Outline map showing the geographical distribution of the seven populations of *M. sylvanus* in Algeria [1, 3, 4] and photographs of the habitat at Djurdjura [6].

without any gene flow between them. Demographic subdivisions within these populations are simply regarded as 'groups'.

Barbary macaques live in multi-male multi-female social groups ranging in size from 14 to 88 individuals [5, 37], for which sexual promiscuity has been well documented during the mating season [38]. Furthermore, the Barbary macaque is extremely unusual in showing a close relationship between males and infants [39]. It would be interesting to know whether the degree of genetic relationship influences the type of social relationship between male and offspring.

In order to investigate some of the above-mentioned issues with data on *M. sylvanus,* a long-term research program has been in progress since 1982, involving two Barbary macaque groups, one located in the Akfadou forest and the other in the Djurdjura National Park in Algeria [6] (fig. 1).

Table 1. Size and composition of groups at Akfadou and Djurdjura

		Animals	Adult males	Adult females	Infants
Akfadou	1984	41	10	10	8
	1985	36	7	11	7
	1986	41	7	13	9
Djurdjura	1984	47	9	14	2
	1985	59	10	18	14
	1986	69	16	20	11

With the help of genetic markers such as blood proteins and DNA 'fingerprints' (restriction fragment length polymorphism; RFLP), the genetic relationships between members of a social group (e.g. matrilineages) can be verified. Although they were first detected in 1977 [7], only since 1986 has the use of repetitive DNA sequences as genetic markers begun to play a major role in genetic studies (e.g. of relatedness) on nonhuman primates. In this context, this paper focusses primarily on the ascertainment of paternity, already a well-established topic in primate genetics [8–11]. A description of a three-step procedure is given and first results of paternity testing are presented. Secondly, the relative power of determining blood proteins and DNA separately and in combination has been evaluated for the two groups at Akfadou and Djurdjura. Finally, results of paternity analyses are compared with those for social relationships.

Material and Methods

Field research discussed here has focussed on offspring of cohorts of 1984, 1985 and 1986. The size of the groups under observation at Akfadou and Djurdjura ranged from 36 to 69 animals (7–16 adult males, 10–20 adult females, 2–14 infants), as shown in table 1. For each group, mating activities of sexually mature individuals were studied during one mating season in 1985. Number of copulations and identities of sexual partners were recorded with both focal animal and ad libitum sampling. Only full copulations with ejaculation were considered. Male-infant relationships during birth seasons were analyzed. Time spent by each male caretaking any infant and its identity were recorded with 10-min focal observations of males. 'Care' refers to all close interactions between male and infant involving positive contact (e.g. carrying, holding, grooming). Only infants born at least 8 days before the end of observations were considered. Sexually mature females (> 4 years) and males (> 5 years) were classified as defined by Ménard et al. [6].

Table 2. Paternity cases and exclusion at Akfadou and Djurdjura

Area	Offspring	Potential fathers	Paternity cases	Cases resolved	
				n	%
Akfadou	10	6–8	69	21	30.4
Djurdjura	6	11–14	76	32	42.1
Total	16		145	53	36.6

A paternity case is defined as a mother-offspring-male (potential father) combination. A paternity case is judged as resolved when the male can be excluded from having fathered the offspring on the basis of formal genetics of blood proteins and DNA bands. Males older than 5 years and present in the group over the mating season are considered as potential fathers.

Blood samples were collected from a total of 100 animals at Akfadou (n = 49) and Djurdjura (n = 51). Only cases in which we succeeded in taking blood samples from the infant, its mother and all potential fathers are considered in these preliminary results. All sexually mature males present in a group during the mating season were taken as potential fathers of infants born during the following birth season. We were able to take blood from 10 mother-offspring dyads and 6–8 potential fathers at Akfadou, and 6 mother-child dyads with 10–14 potential fathers in Djurdjura (table 2). A wide selection of enzyme systems was analysed by Arnaud (Toulouse, France). Immunoglobulin systems were typed by J.M. Dugoujon following methods described elsewhere [12]. Various other plasma proteins and the hemoglobin were screened in the laboratory at Zürich. Extensive modifications of conventional methods of isoelectric focussing techniques were necessary to reveal polymorphism in some proteins [13, 14]. Table 3 lists the protein systems tested so far (see Acknowledgments). DNA fingerprinting was conducted in Zürich, using the conventional DNA fingerprinting techniques developed by Jeffreys [15–17], including cleavage of DNA by restriction enzyme Hinf I, fragment electrophoresis, Southern transfer, probe labeling, probe-genomic fragment hybridization (multilocus DNA fingerprint probe 33.15), and print detection by autoradiography.

The blood samples were divided into two classes, due to different degrees of completeness with respect to the genetic systems analysed: (1) sample A with 145 paternity cases and complete protein data, but only partial DNA typing (table 2); (2) sample B with 52 paternity cases and complete protein and DNA typing.

To reduce the theoretically possible number of potential fathers, a three-step procedure has been applied in this study for the paternity testing of males in the Akfadou and Djurdjura groups: (1) preliminary exclusion test through analysis of blood proteins; (2) DNA fingerprint comparison of mother and offspring to reveal nonmaternal bands in the offspring; (3) comparison of nonmaternal bands of the offspring with DNA fingerprints of the remaining potential fathers; evaluation with respect to exclusion/identification of paternity.

Results

Genetic Analysis

Out of 23 potential blood protein markers (9 plasma proteins, hemoglobin, 13 enzymes, table 3), 8 were found to be variable (fig. 2) with an overall rate of polymorphism of about 0.375 for our two groups. However, the relatively highly variable systems of transferrin (TF), amylase (AMY), hemoglobin (HB) and, in particular, DNA could be satisfactorily used in the assessment of genetic relatedness, e.g. paternity. The electrophoretic analysis of TF, AMY and HB revealed significant differences in allele frequencies between the populations of Akfadou and Djurdjura. Furthermore, the Akfadou population showed a lower mean heterozygosity per locus (direct count estimate 0.248 ± SE 0.062) than the Djurdjura population (0.345 ± 0.084). This relatively marked difference in mean heterozygosity between these two populations affects the likelihood of paternity exclusion, as will be shown below.

Table 4 shows the numbers of infants, potential fathers and resolved paternity cases.

Paternity testing was generally initiated by serological and electrophoretic analysis of blood proteins as a first step. With sample A, data for plasma proteins (TF, Gm13, PI, AMY) and hemoglobin (HB) excluded 21% of males from paternity of all offspring born in Akfadou and Djurdjura (analysis of 145 possible paternity cases). We consider this exclusion rate as a relatively standard value for exclusion power by blood proteins in this species and regard this finding as a convincing argument in favor of including proteins as genetic markers in paternity testing. Such protein analysis also serves as a reliable tool to test the power of DNA markers in cases of exclusion or indication of paternity. For sample A, only some DNA fingerprints could be analyzed for complete trios (mother/child/potential father). The additional information from DNA fingerprints increased the overall exclusion rate to 37% (42% for Djurdjura and 30% for Akfadou) (table 2). In sample B from Akfadou and Djurdjura, for which protein findings and DNA fingerprints could be evaluated completely, the rate of exclusion with proteins was 25%. Paternity cases unsolved with blood proteins were analysed with the help of DNA polymorphisms. The rate of exclusion using DNA fingerprints alone was close to 58% for the complete sample. Further, each animal was characterized by a unique banding pattern which, interestingly, would allow for identification in the wild using DNA fingerprinting. A typical DNA autoradiogram for *M. syl-*

Table 3. Blood genetic markers and DNA analysed

		Alleles
Plasma proteins		
ALB	Albumin	1
GC (DBP)	Vitamin D binding protein	1
PI	Protease inhibitor (alpha-1-antitrypsin)	2
TR	Transferrin	2
C3	Complement factor 3	2
BF	Properdin factor B	1
GM	Immunoglobulin GM	2
KM	Immunoglobulin KM	1
BM	Immunoglobulin BM	2
Enzymes		
6-PGD	6-Phosphogluconate-DH, EC 1.1.1.44	1
ADA	Adenosinedesaminase, EC 3.5.4.4	1
AK	Adenylate kinase, EC 2.7.4.3	1
CA1	Carbonic anhydrase 1, EC 4.2.1.1	1
CA2	Carbonic anhydrase 2, EC 4.2.1.1	1
ESD	Esterase D, EC 3.1.1.1	1
ACP	Acid phosphatase, EC 3.1.3.2	1
GLO	Glyoxalase, EC 4.4.1.5	1
PGM	Phosphoglucomutase 1, EC 2.7.5.1	1
PGM	Phosphoglucomutase 2, EC 2.7.5.1	1
AMY	Serum amylases, EC 3.2.1.1	2
DIA1	Diaphorase 1, EC 1.6.2.2	2
DIA2	Diaphorase 2, EC 1.6.4.2	1
Hemoglobin		
HB	Hemoglobin alpha chain	2
DNA RFLP	Restriction fragment length polymorphism of nDNA	

Some plasma proteins have been analysed by agarose gel electrophoresis [14, 33, 34], others, such as immunoglobulins, by serological techniques [12] and isoelectrofocussing. Extensive modifications of conventional methods of isoelectrofocussing techniques were necessary to reveal polymorphism of some proteins [35]. Electrophoresis and staining procedures of the erythrocyte enzymes were mainly done using the methods of Harris and Hopkinson [36]. DNA fingerprinting analysis was done according to the methods of Jeffreys [15, 17].

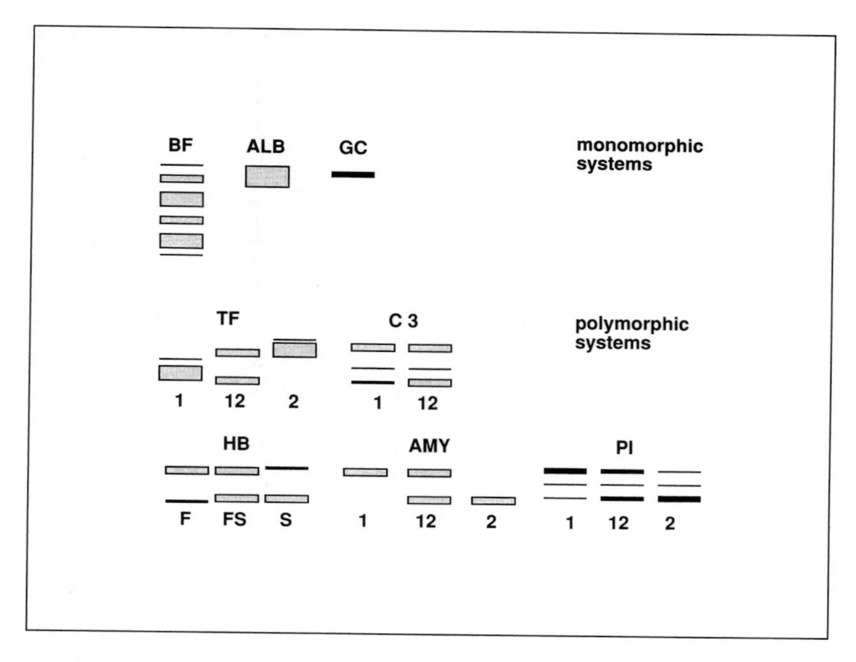

Fig. 2. Phenotypes of plasma proteins and hemoglobin in two populations of *M. sylvanus* in Algeria (Akfadou and Djurdjura).

The marker systems transferrin (TF), complement factor 3 (C3), alpha-1-antitrypsin (protease inhibitor, PI), hemoglobin (HB) and amylase (AMY) yield genetic variation with two alleles, for which the distribution meets the conditions of a genetic polymorphism (the most frequent allele $<99\%$). The plasma proteins properdin factor B (BF), albumin (ALB) and vitamin D binding protein (DBP, GC) are monomorphic in these 2 populations.

vanus is shown in figure 3. We introduced a system of band labeling which was very often helpful in comparing and referencing the bands of different plates. Band comparison was also generally facilitated by reference samples with a well-known band profile.

Comparison of the exclusion values yielded by blood proteins on the one hand and by the DNA on the other shows that the exclusion power of the DNA is markedly greater. Due to the lower degree of genetic variation in Akfadou, fewer cases of exclusion by proteins or DNA bands have been observed than for Djurdjura. The overall exclusion rate (using proteins and DNA combined) for both populations is about 75% (39 exclusions/

Table 4. Genetic markers and paternity exclusion at Akfadou and Djurdjura

Off-spring	Potential fathers									Males excluded
	15 CY	44 NJ	86 CC	1 CR	23 PA	24 RG	35 CI	50 ND	16 BA	
Akfadou										
51 JU	/									0/8
73 NG	/	DNA					DNA		DNA	3/8
57 CQ	/									0/8
56 DO	/									0/8
66 BO	/	/	/	TF AMY	TF AMY	AMY	TF AMY		TF AMY	5/6
75 BB	/	/	/		GM13	HB		HB		3/6
63 CH	/	/	/							0/6
74 CS	/	/	/			HB	DNA	HB		3/6
62 BR	/	/	/			HB DNA	DNA	HB		3/6
67 MI		/	/	PI		DNA	DNA	PI		4/7

	7 MI	14 DG	17 GI	18 NB	20 PP	31 NE	32 SA	33 DI	34 PU	36 CH	37 KH	38 CD	77 BE	29 FA	Males excluded
Djurdjura															
64 LC	/	TF GM13	TF			DNA	HB		/	TF DNA	/	/		DNA	6/10
95 AM	/					DNA			/		/	/			1/10
82 OF	TF				TF	DNA	HB				TF	TF			6/14
100 JU		TF	TF			HB DNA			TF HB	TF HB	TF	HB DNA			7/14
103 FL						DNA		DNA	DNA	DNA		DNA		DNA	6/14
99 NG						DNA		DNA	DNA	DNA		DNA		DNA	6/14

GM13, TF, AMY, P1, HB and DNA fingerprints as genetic markers for exclusion of males (potential fathers, top line) from paternity of offspring (first column). Each individual is designated by its laboratory number and a 2-letter code. / = Paternity ruled out by age or absence from group.

52 cases) in sample B, showing that a combined approach is more effective than the use of DNA fingerprints alone. It should be stressed again here that, as with blood genetic markers (see TF and AMY, fig. 4), DNA analysis in most cases leads primarily to exclusion and only in exceptional cases to positive proof of paternity.

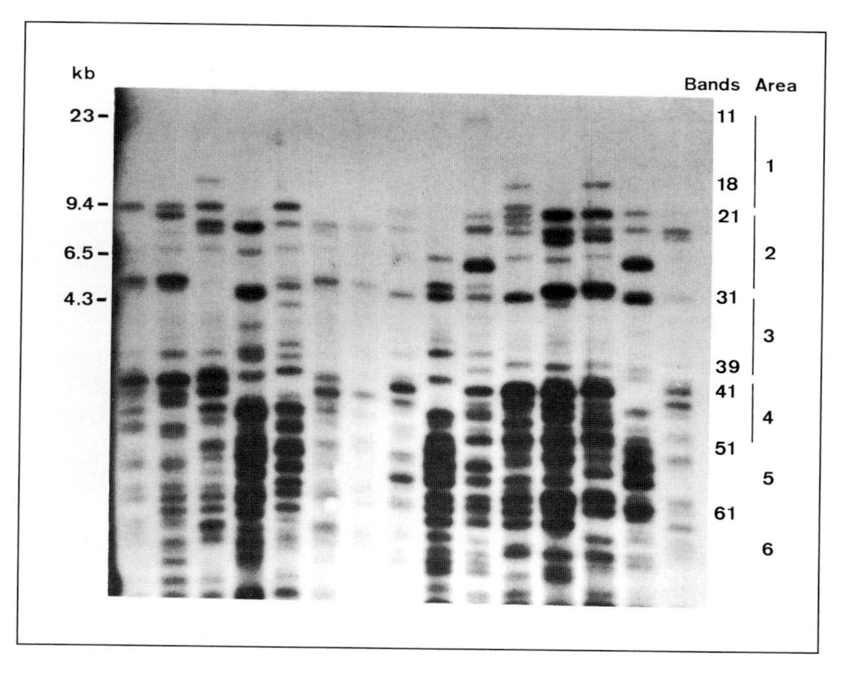

Fig. 3. DNA fingerprints for free-living *M. sylvanus* in Algeria. The DNA autoradiograph (probe 33.15) can be subdivided into 4 banding areas (1–4), at least. We used lambda DNA as marker with Hind III fragments of 23.130, 9.416, 6.559 and 4.361 kb. Each banding area contains several bands. The banding profile begins shortly below the application slot with the large DNA fragments of about 23 kb (labeled as area 1, with bands from 11 to 19). A little further down follow fragments in the size range of about 9–5 kb (labeled as area 2, beginning with band 21) and 4 kb (labeled as area 3 with bands 31 to 39). In most cases, band scoring was limited to bands > 3 kb. In some exceptional cases, clearly distinguishable bands are detectable in the areas of low DNA fragments (labeled as areas 4, 5, 6, 7, 8 and 9).

Sexual Activity of Males and Females

The number of copulations was inequally distributed among males. It varied from 0 to 37 copulations per male in Djurdjura and from 7 to 28 in Akfadou (table 5). The number of sexual partners also varied. One male did not copulate at all, while 5 males in the Djurdjura group and 1 male in the Akfadou group copulated with more than 11 different females.

During the mating period, all females were observed to copulate with more than 1 male and all but 2 with at least half of the available sexually

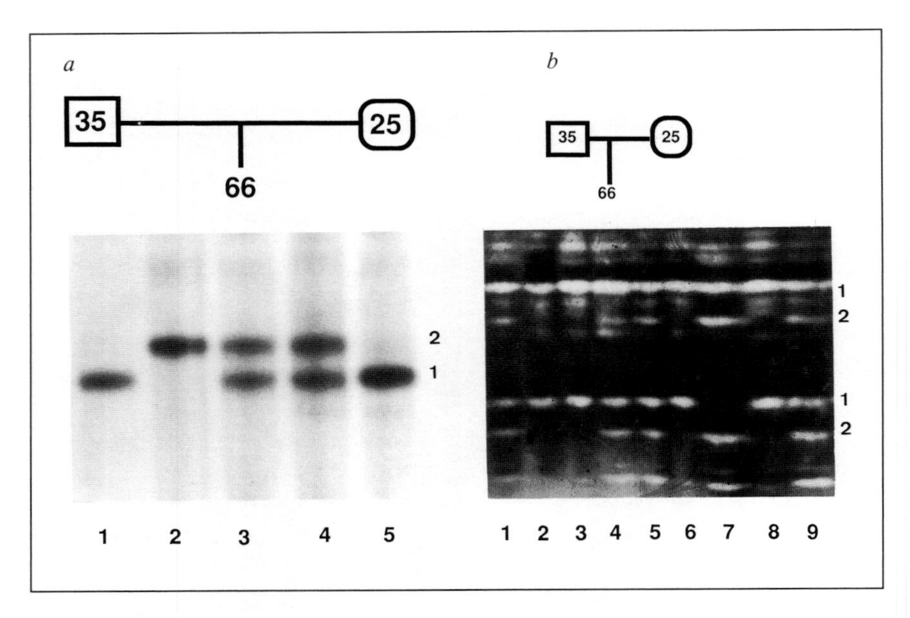

Fig. 4. Phenotypes of transferrin (TF, *a*) and serum amylases (AMY, *b*) in a paternity case

(a)	lane	1	2	3	4	5				
	TF	1	2	12	12	1				
(b)	lane	1	2	3	4	5	6	7	8	9
	AMY	12	1	1	12	12	1	2	1	12

Male 35 is excluded from paternity of offspring 66 (of female 25) with both genetic markers (but not in DNA fingerprints).

mature males (i.e. more than 7 in Djurdjura and more than 4 in Akfadou) (table 6). In most cases, almost equal numbers of matings were performed with all partners.

Distribution of Male Caretaking Behavior among Infants

All but 3 males were involved in caretaking activities (table 7). Most males spent between 10 and 40% of their time in close contact with infants, while a few spent either a lesser or a greater amount of time (up to 70%). Furthermore, the number of infants with which each male interacted

Table 5. Sexual activity and male caretaking behavior in the Akfadou and Djurdjura groups

Males (category)	Mating season 1985			Birth season 1986 time spent caretaking infants %
	sexual partners	frequency of matings observed/100 min (copulations)	total obs. min	
Akfadou				
RG (1)	8	4.7 (26)	554	26.1
PA (1)	9	4.0 (28)	703	20.8
CR (1)	5	3.3 (11)	331	30.7
BA (2)	11	4.4 (28)	635	18.7
CI (3)	5	2.3 (7)	311	32.4
CY (3)	6	1.3 (9)	679	30.9
ND (4)	7	2.9 (14)	476	19.5
Spearman rank $r_s = -0.607$, NS				
Djurdjura				
FA (1)	8	6.9 (11)	159	21.4
SA (1)	14	6.8 (37)	544	22
BE (1)	12	6.2 (28)	450	30.2
DI (1)	11	4.3 (23)	531	24.5
NE (2)	14	6.8 (23)	339	7.1
GI (2)	12	6.5 (35)	538	18.8
PP (2)	8	4.9 (25)	507	14.7
NB (2)	8	4.5 (14)	311	7.9
DG (3)	7	2.9 (15)	512	26
CH (3)	1	2 (2)	100	66.5
PU (3)	2	1.6 (2)	122	49.3
MI (4)	5	2.7 (7)	262	7.6
KH (4)	3	0.9 (3)	328	1
CD (4)	0	0 (0)	17	13.8
Spearman rank $r_s = -0.009$, NS				

Number of adult females: Akfadou: 11; Djurdjura: 18.

varied. Six males took care of only 1 infant, while 21 males took care of 2–3 infants. Finally, 8 males interacted with more than 3 infants (up to 6). Eleven infants out of 39 never received male care, while 22 infants received care from more than 1 male. Only 6 infants had contact with only 1 male.

Table 6. Distribution of female matings among males

Females	Overall matings	Matings/ male (range)	Sexual partners
Akfadou			
CLE	24	1–8	6
ME	17	1–5	7
AR	14	1–4	5
LI	13	1–5	5
CM	12	1–5	5
FR	11	1–2	6
RS	10	1–7	3
Djurdjura			
PA	31	1–6	8
CE	24	1–6	8
AF	14	1–3	10
LO	18	1–5	9
ZO	22	1–3	11
CL	14	1–2	9
UL	23	1–5	8
GR	22	1–6	8
SM	10	1–4	6

Only females that were observed during their maximum swelling period have been considered.
Number of sexually mature males: Akfadou: 7; Djurdjura: 14.

Male Sexual Activity and Caretaking Behavior

For both groups, we found no correlation between male mating frequency and the time they spent caretaking infants born during the following season (table 5). Four categories of males can be distinguished: among males which showed intensive mating activity, some spent a high percentage of time caretaking infants (category 1), whereas others devoted only limited time to infants (category 2). Among males which showed a low level of mating activity, some spent the highest percentage of time caretaking infants (category 3), whereas others spent little time in contact with infants (category 4).

Relationship between Paternity and Caretaking Behavior

If we compare the results of paternity analyses with those of male-infant relationships, different combinations are observed (table 8):

Table 7. Cumulative observations on male-infant relationships for the two groups and the two birth seasons

	Infants per caregiver			
	0	1	2–3	4–6
Caregivers in each case	3	6	21	8
	Percent of time caretaking infants			
	0–10	11–20	21–40	41–70
Caregivers in each case	8	10	16	4
	Caregivers per infant			
	0	1	2–3	4–6
Infants in each case	11	6	9	13

– At least 18 males out of 21 were not seen caretaking 34 infants that they had not sired.
– Eight males (GI, NE, KH, PA, ND, RG, CI, CR) took care of 15 infants that they had not sired. Among them, 6 males (NE, PA, ND, RG, CI, CR) that invested in several infants directed most of their attention to infants they had not sired.
– Male ND did not take care of his own infant BO, but did take care of 2 infants he had not sired.
– For some infants, such as BR, a large amount of care was given both by the 3 potential fathers and by the 3 nonfathers.
– One male RG only took care of 3 infants he had not sired.

Overall, three main points emerge:

– Among males which were excluded from having sired a given infant, 70% did not take care of that infant while 30% did.
– The only male that was identified as the most probable father of 1 infant did not take care of that infant.
– Infants can receive care exclusively from nonfathers.

Table 8. Male caretaking behavior and paternity

Infants	Adult males of the Akfadou group						
	23 PA	16 BA	50 ND	24 RG	35 CI	1 CR	15 CY
75 BB	F*		*	F*		F*	
63 CH	F	I			I		
62 BR	I	F	F*	F*	F*	I	
66 BBO	*	*		I*	*	I*	
74 CS							
67 MI	I*	F	I*	*	I*		F

Infants	Adult males of the Djurdjura group													
	36 CH	32 SA	77 BE	33 DI	20 PP	29 FA	17 GI	31 NE	7 MI	18 NB	14 DG	37 KH	34 PU	38 CD
95 AM	I		I		I			F*						
64 LC	*	*				*	*	*			*			
100 JU	*	I	F	I		I	I*	F*			*	*	*	*
82 OF		*	*		*			*	*		F	F*		*
103 FL	*			*		*		*					*	*
99 NG	*			*		*		*					*	*

F = The infant to which the male directed the majority of his caretaking activity; I = other infants that received less than half of the attention that the male showed to his 'F' infant; * = male excluded from paternity of that infant.

Discussion

On examining our findings with respect to copulation preferences and paternity of offspring, the question arises as to whether the lack of any close relationship between copulation and paternity is linked to the social system in *M. sylvanus.* Some authors state that there is no hierarchical system for males and females in this species; others argue that there is a hierarchical system comparable to those found in other macaque species. Our data, indicating an essentially promiscuous pattern of sexual behavior, support the interpretation that a strong hierarchical social system is lacking in free-ranging populations of *M. sylvanus.* Nevertheless, in a multi-male social group of *M. sylvanus,* most sexually mature males behave so as to maximize their access to estrous females.

M. sylvanus is generally regarded as a macaque species characterized by low genetic variation [10, 18] as far as blood protein variability (genetic polymorphism) is concerned. This reduced genetic variation is considered to be primarily the result of an early phylogenetic bottleneck affecting the wild population. We would argue, however, that the other possible explanation involving bottlenecks or effects of genetic drift associated with the founder principle following migration at a more recent time cannot be ruled out. On the other hand, an assessment of the degree of genetic variation based only on blood proteins seems to be too limited. As outlined in this paper, and as forthcoming publications will demonstrate, there is a higher level of genetic variability in the DNA than in blood proteins in this species [19]. It is one of the main targets of our genetic research on *M. sylvanus* to compare these DNA findings with those on other macaque species to permit a better assessment.

Electrophoretic patterns of blood proteins and DNA fingerprint patterns have been used for clarification of genetic relationships in nonhuman primates on the assumption that they behave as inherited traits, as they do in humans [for DNA fingerprints, see Jeffreys, 20, 21]. However, as long as the inheritance of genetic variants in each species has not been proven through clear-cut formal genetic analysis with the help of samples from undisputed families, some caution is needed. Nevertheless, there is some good evidence that transferring formal genetic findings from one species to another is a justifiable practice.

Although the number of bands in DNA fingerprints seems to show an acceptable level of variability compared to other catarrhine species, we encountered many cases in *M. sylvanus* where only one nonmaternal band in the offspring could be used for the paternity test. While in humans there is a relatively high mutation rate per offspring (band(s) unassignable to either parent) for multilocus probe 33.15 [21], we were obliged to treat exclusions of males based only on one nonparental band as full exclusions if support was provided by the blood protein findings. It seems to be fairly evident that paternity can be positively detected in wild populations of Barbary macaques only when a single male in the group remains unexcluded and when no other male(s) can be considered to have fathered the offspring in question (as indicated by observational data).

The results of our paternity testing in two wild groups of *M. sylvanus* are still incomplete. The remaining set of still-unsolved paternity cases could be further analysed by the application of single-locus probes other than MS 31 and MS 8, which have failed to reveal bands in *M. sylvanus* and in other nonhuman primate species tested in our laboratory to date. Although probe 33.15 has been preferred so far for studies of nonhuman primates, probe 33.6 also delivers a distinctive banding pattern of high individuality for most nonhuman primate species, as has been well documented in recent reports on the use of DNA fingerprints for paternity tests in primates. There are general reports on DNA fingerprint patterns in a variety of primate species [22–24] or in single taxa such as *Callithrix jacchus* [25, 26], *Macaca sylvanus* [27], *Macaca fuscata* [28, 29], *Macaca mulatta* [30] and *Pan troglodytes* [31, 32]. All of these publications agree on the power and usefulness of DNA markers for paternity testing or for the determination of genetic relatedness between group members. The importance of studying these two aspects is becoming increasingly recognized by primatologists. According to kin selection theory [40], we can expect that, in order to maximize his inclusive fitness, a male will behave toward an infant depending on its relatedness, especially on the likelihood that he has sired that infant. Several studies have attempted to assign paternity on the basis of the amount of observed mating activity [41, 42]. However, the validity of paternity assigned on the basis of mating behavior remains doubtful in the absence of paternity exclusion analysis [43, 44].

The results of the present study demonstrate that:

(1) While the rate of mating varied among male Barbary macaques and there is differential access of males to females, all males but 1 and all females actually had access to several sexual partners with which they mated with approximately equal frequencies. In fact, 9 out of 14 adult males accounted for 93.8% (n = 225) of matings in Djurdjura and 5 out of 7 accounted for 87% (n = 123) of matings in Akfadou. Hence mating activities in wild Barbary macaques cannot be recorded as reliable indicators of paternity.

(2) No definite relationships exist between male investment in infants and paternity certainty. Indeed:

- a great proportion of infants receive caretaking from several adult males;
- adult nonfathers regularly took care of infants;
- 1 father totally ignored his infant but took care of 2 others that he had not sired.

(3) Nevertheless, among nonfathers, a higher proportion was seen to ignore infants they had not sired than to take care of them. This could indicate that paternity is one of the factors influencing male investment.

It is important to remember that other types of relationship (e.g. between brothers or cousins) might be as significant as paternities in determining the amount of infant caretaking. This is particularly relevant to stable groups such as those of Barbary macaques, from which males emigrate late in life, such that familiarity with family ties is easy to acquire. Detailed knowledge of these ties might explain why a given male generally invests in several infants and why some infants are more prone to male caretaking than others.

We still need to explain why Barbary macaques developed such a social system in which males invest in infants, whereas the promiscuity resulting from their mating strategies operates to conceal paternities.

Taub [38] has suggested that this promiscuous mating system is the most effective means for females to evoke some care from many males and to obtain cumulative investment in their offspring. However, this attractive hypothesis was not supported by the results of our study, as 11 infants whose mothers copulated with several different males did not receive any male care at all.

Conclusion

The application of genetic markers in paternity cases of two wild populations of *M. sylvanus* from Algeria (Akfadou and Djurdjura) has shown that it is advisable to analyse blood proteins and DNA fingerprints in combination. To render paternity testing reasonably economical by reducing the a priori number of potential fathers, it is helpful to start with electrophoresis of proteins that are already known to show multiallelic systems in the species of interest. In our sample (in cases where both protein and DNA findings were available), with protein systems alone it was possible to exclude 25% of males as fathers. DNA fingerprints alone were more powerful in excluding 58% of males. The combination of both, however, led to the solution of 75% paternity cases. We would stress the fact that, up to now, paternity testing has operated more through exclusion of males than through positive identification of paternity. We were able to identify paternity with a high degree of probability in only 2 cases.

Observations of sexual activities during the mating season confirm the existence of a promiscuous pattern of sexual behavior in *M. sylvanus,* i.e. a large proportion of adult males in a group copulate with a large proportion of the estrous females. Therefore, genetic testing of paternity becomes extremely important in such a social system. First results indicate that there is no strong correlation between paternity and time spent caretaking offspring in *M. sylvanus.*

Acknowledgments

We gratefully acknowledge the inclusion of the results of previous studies of blood proteins kindly provided by Dr. J.M. Dugoujon and J. Arnaud (Hôpital Purpan de Toulouse, France). We thank Dr. A. Gautier-Hion, Mrs. Krisztina Vasarhélyi, Prof. R.D. Martin and Dr. Ch. Pryce for reading the manuscript and for very helpful discussions. Many thanks are due to Mrs. Gerda Greuter for her dedicated photographic assistance. We thank Mr. Gassi (Director of the National Djurdjura Park, Algeria) for permission to carry out the field work. This work received financial support from the CNRS (PICS No. 88) and from the National Geographic Society (grant No. 3766.88).

References

1 Taub JE: Geographic distribution and habitat diversity of the Barbary macaque *Macaca sylvanus* L. Folia Primatol 1977;27:108–133.
2 Taub DM: Aspects of the Biology of the Wild Barbary Macaque (Primates: Cercopithecinae, *Macaca sylvanus* L). Biogeography, the Mating System and Male-Infant Associations, PhD thesis, Davis, 1978.
3 Fa JE: Habitat distribution and habitat preference in Barbary macaques *(Macaca sylvanus).* Int J Primatol 1984;5:273–286.
4 Fa JE, Taub DM, Ménard N, Stewart PJ: The distribution and current status of the Barbary macaque in North Africa; in Fa JE (ed): The Barbary Macaque. A Case Study in Conservation. New York, Plenum Press, 1984, pp 79–111.
5 Ménard N, Hecham R, Vallet D, Chikhi H, Gautier-Hion A: Grouping patterns of a mountain population of *Macaca sylvanus* in Algeria – A fission-fusion system? Folia Primatol 1990;55:166–175.
6 Ménard N, Vallet D, Gautier-Hion A: Démographie et reproduction de *Macaca sylvanus* dans différents habitats en Algérie. Folia Primatol 1985;44:65–81.
7 Gillespie D: Newly evolved repeated DNA sequences in primates. Science 1977;196: 889–891.
8 Scheffrahn W, Glaser D: Polymorphism and formal genetics of transferrins in *Saguinus midas tamarin* LINK, 1795 (Primates, Platyrrhina). J Hum Evol 1977;6: 605–608.
9 Scheffrahn W: Transferrin-Befunde bei *Callithrix jacchus* Linné, 1758 (Primates, Platyrrhina). Anthropol Anz 1978;36:121–126.

10 Schmitt J, Ritter H, Schmidt C, Witt R: Genetic markers in primates: Pedigree patterns of a breeding group of Barbary macaques (*Macaca sylvana* Linnaeus, 1758). Folia Primatol 1981;36:191–200.

11 Smith DG, Small MG, Ahlfors CE, Lorey FW, Stern BR, Rolfs BK: Paternity exclusion analysis and its applications to studies of nonhuman primates; in Hendrickx AG (ed): Advances in Veterinary Science and Comparative Medicine. Orlando, Academic Press, 1984, pp 1–24.

12 Dugoujon JM, Blanc M, Ducos J: New antigens on primate immunoglobulins determined by antiglobulins of normal human sera. Folia Primatol 1981;36:144–149.

13 Ménard N, Scheffrahn W: Ecology, demography and genetic markers of Algerian *Macaca sylvanus* populations, in preparation.

14 Scheffrahn W: Elektrophoretische Methoden; in Knussmann R, Jürgens HW, Schwidetzky I, et al (eds): Anthropologie. Handbuch der vergleichenden Biologie des Menschen. Stuttgart, Fischer, 1991.

15 Jeffreys AJ, Wilson V, Thein SL: Individual-specific 'fingerprints' of human DNA. Nature 1985;316:76–79.

16 Signer EN: DNA-Fingerprints zur Überwachung der Reinerbigkeit von Mäuse-Inzuchtstämmen, Diss, Zürich, 1989.

17 Kirby LT: DNA Fingerprinting. An Introduction. New York, Stockton Press, 1990.

18 Socha WW, Ruffié J: Blood groups of primates: Theory, practice, evolutionary meaning. Monographs in Primatology. New York, Liss, vol 3, 1983.

19 Küster J, Paul A: personal commun.

20 Jeffreys AJ, Wilson V, Thein SL, Weatherall DJ, Ponder B: DNA 'fingerprints' and segregation analysis of multiple markers in human pedigrees. Am J Hum Genet 1986;39:11–24.

21 Jeffreys AJ, Turner M, Debenham P: The efficiency of multilocus DNA fingerprint probes for individualization and establishment of family relationships, determined from extensive casework. Am J Hum Genet 1991;48:824–840.

22 Weiss ML, Wilson V, Chan C, Turner T, Jeffreys AJ: Application of DNA fingerprinting probes to Old World monkeys. Am J Primatol 1988;16:73–79.

23 Gray IC, Jeffreys AJ: Evolutionary transience of hypervariable minisatellites in man and the primates. Proc R Soc Lond [Biol] 1991;243:241–253.

24 Scheffrahn W: Restriction fragment length polymorphism (RFLP) in the Hominoidea. Int J Primatol 1988;8:527.

25 Dixson AF, Hastie N, Patel I, Jeffreys AJ: DNA 'fingerprinting' of captive family groups of common marmosets *(Callithrix jacchus)*. Folia Primatol 1988;51:52–55.

26 Dixson AF, Anzenberger G, Monteiro Da Cruz MAO, Patel I, Jeffreys AJ: DNA fingerprinting of free-ranging groups of common marmosets *(Callithrix jacchus jacchus)* in NE Brasil; in Martin RD, Dixson AF, Wickings EJ (eds): Paternity in Primates: Genetic Tests and Theories. Karger, Basel, 1992, pp 192–202.

27 Küster J, Paul A, Arnemann, J: DNA-fingerprint analysis in a semifree-ranging Barbary macaque population. IPS Proc 1990, Nagoya 1990, p 72.

28 Inoue M, Takenaka A, Takenaka O: Paternity discrimination in the Japanese macaque by DNA fingerprinting. IPS Proc 1990, Nagoya 1990, p 62.

29 Inoue M, Mitsunaga F, Ohsawa H, Takenaka A, Sugiyama Y, Gaspard SA, Takenaka O: Paternity discrimination and male behavior in an enclosed Japanese macaque group by DNA fingerprinting. IPS Congr Proc 1990, Nagoya, 1990.

30 Arnemann J, Schmidtke J, Epplen JT, Kuhn H-J, Kaumanns W: DNA fingerprinting for paternity and maternity in group O Cayo Santiago-derived rhesus monkeys at the German Primate Center: Results of a pilot study. Puerto Rico Health Sci J 1989;8: 181–184.
31 Ely J, Ferrell RE: DNA 'fingerprints' and paternity ascertainment in chimpanzees *(Pan troglodytes).* Zoo Biol 1990;9:91–98.
32 Ely J, Alford P, Ferrell RE: DNA 'fingerprint' and the genetic management of a captive chimpanzee population *(Pan troglodytes).* Am J Primatol 1991;24:39–54.
33 Alper CA, Boenisch T, Watson L: Genetic polymorphism in human glycine-rich beta-glycoprotein. J Exp Med 1972;135:68–80.
34 Teisberg P: High voltage agarose gel electrophoresis in the study of C 3 polymorphism. Vox Sang 1970;19:47–56.
35 Scheffrahn W, Ménard N: Blood proteins in isoelectrofocussing with immobilines and ampholines, in preparation.
36 Harris H, Hopkinson DA: Handbook of Enzyme Electrophoresis in Human Genetics. Amsterdam, North-Holland, 1978.
37 Ménard N, Vallet D: Dynamics of fission in a wild Barbary macaque *(Macaca sylvanus)* group. Int J Primatol, in press.
38 Taub DM: Female choice and mating strategies among wild Barbary macaques (*Macaca sylvanus* L.); in Lindburg DG (ed): The Macaques: Studies in Ecology, Behavior and Evolution. New York, Van Nostrand Reinhold, 1980.
39 Taub DM: Male caretaking behavior among wild Barbary macaques *(Macaca sylvanus);* in Taub DM (ed): Primate Paternalism. New York, Van Nostrand Reinhold, 1984, pp 20–55.
40 Hamilton WD: The genetical theory of social behaviour. I, II. J Theor Biol 1964;7: 1–64.
41 Hausfater G: Dominance and reproduction in baboons *(Papio cynocephalus).* Contrib Primatol, vol 7. Basel, Karger, 1975.
42 Packer C: Inter-troop transfer and inbreeding avoidance in *Papio anubis.* Anim Behav 1979;27:1–36.
43 Curie-Cohen M, Yoshihara D, Luttrell L, Benforado M, MacCluer JW, Stone WH: The effects of dominance on mating behavior and paternity in a captive troop of rhesus monkeys *(Macaca mulatta).* Am J Primatol 1983;5:127–138.
44 Stern BR, Smith DG: Sexual behavior and paternity in three captive groups of rhesus monkeys *(Macaca mulatta).* Anim Behav 1984;32:23–32.

Dr. Nelly Ménard, Station Biologique de Paimpont, Université de Rennes,
F–35380 Plélan-le-Grand (France)

Martin RD, Dixson AF, Wickings EJ (eds): Paternity in Primates:
Genetic Tests and Theories. Basel, Karger, 1992, pp 175–191

Male Social Rank and Reproductive Success in Wild Long-Tailed Macaques

Paternity Exclusions by Blood Protein Analysis and DNA Fingerprinting

Jan R. de Ruiter[a], *Wolfgang Scheffrahn*[b], *Gerjan J.J.M. Trommelen*[c], *André G. Uitterlinden*[c], *Robert D. Martin*[b], *Jan A.R.A.M. van Hooff*[a]

[a] Ethology and Socioecology Group, Utrecht University, Utrecht, The Netherlands;
[b] Anthropologisches Institut, Zürich, Switzerland;
[c] Medscand Ingeny, Leiden, The Netherlands

In captive colonies of macaques dominant males generally, though not always, have priority of access to females at times when they are thought likely to ovulate [1–5]. Surprisingly, in studies that have specifically examined the relationship between rank and reproductive success in cases where such priority of access prevails, no consistent correlation has been found [6–15]. It is possible that unnatural conditions in captivity, for example restrictions on space and inhibition of free migration, account for this lack of correlation [1, 3]. We therefore carried out a combined study of behavior and paternity in long-tailed macaques *(Macaca fascicularis)* living in social groups under natural conditions and found that there is, in fact, a relatively high correlation between rank and reproductive success among males.

Material and Methods

Study Subjects

Long-tailed macaques (*M. fascicularis* Raffles 1821) live in mixed-sex groups in lowland rain forest bordering rivers in the Ketambe Research Area, which is located in the northern part of Sumatra (Indonesia). [For a full description of the study area, see ref. 16].

Fig. 1. Matrilineal family in the early morning in a sleeping tree. Tail shape was an important characteristic for individual identification.

The first author studied the behavior of 3 wild-living groups of these macaques: a large group of variable size containing about 50 individuals and two small groups containing about 18 individuals each. The large group comprised on average about 10 sexually mature males, 12 adult females and their offspring. The small groups comprised 1–4 adult males and 3–5 adult females with their offspring.

Each macaque group has a small number of fixed sleeping trees close to the river and has a travel distance of about 1,500 m per day, visiting trees with ripe fruits such as figs and foraging for arthropods (fig. 1). The members of a group move in a more or less cohesive manner through the forest. Some individuals may wander away from the main party during the day, but they do not leave the exclusive home range (approximately 50 ha) occupied by the group. The behavior of individuals in the main study groups and in a number of adjacent groups in the population has been studied since 1976 [17–19] (fig. 2). In the three main study groups, all group members could be individually recognized by the observer and any changes in group composition, due to birth, death and migration, were recorded at least once a week. Social rank was determined according to the distribution of submissive behavior (preferably on the basis of the bare teeth display) among members of the group [20].

Young males always migrate out of their natal groups to join other, often adjacent groups, between the onset of sexual maturation (at 4.5 years) and about 1.5 years thereafter, before they start copulating (with 2 exceptions, see results). Such males rise in rank in their new groups and may achieve the alpha position shortly after reaching full body size at the age of about 9 years. An alpha male remains at top rank for a limited period (for about 2–3 years in large groups and for up to 5 years in small groups). After his displace-

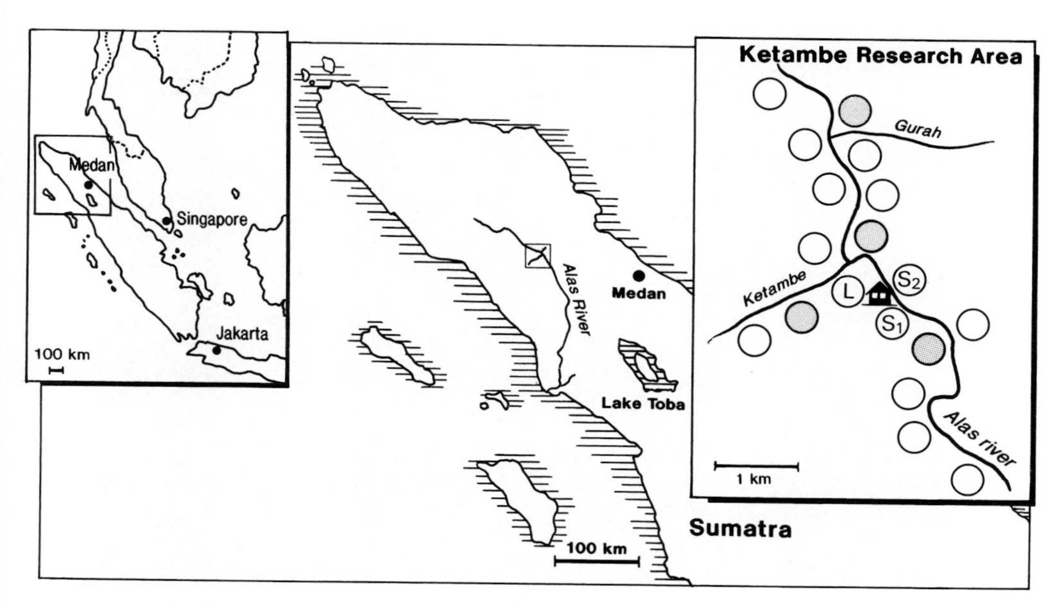

Fig. 2. Study area with the locations of the various study groups. Animals have been captured in groups represented by shaded circles and in the focal study groups: L (the large group), S_1 and S_2 (the two smaller groups). Rivers were no serious obstacle for migrating males.

ment from the alpha rank, he typically remains in the group for another 1–2 years as a beta male. Subsequently, he migrates again to a new group but does not reach top rank again [17; de Ruiter, unpubl. data].

Long-tailed macaques breed seasonally in Sumatra; 90% of the births take place from July to November. A female is attractive to males in that she copulates with males of the group for a period of up to several months. During this period, there are other signs of readiness to male, such as genital swelling and coloration of the sexual skin, which fluctuate roughly in parallel with copulation frequency from week to week, but do not show regular cyclicity [18; de Ruiter, unpubl. data]. Ovulation is concealed, that is, it is impossible – at least for a human observer – to determine precisely when ovulation takes place, and it is an important question whether or not the male macaque can do so. Although there is considerable individual variation in mating behavior between females, they copulate frequently, as often as 3–4 times per hour through the day, often with many different partners. A female may copulate with all group males in a single day in exceptional cases, even in large groups. At periods when females are particularly attractive, high-ranking males generally have better access and may keep other males away. A male-female association, or consortship, may last for a period between less than 1 h and about a week, but is not always exclusive. If a consorting male does not occupy the alpha rank, a higher-ranking male may copulate with the female. Even when the alpha male has formed

Fig. 3. With specially designed traps [21], individuals could be captured and anesthetized conveniently, without much disturbance to the rest of the group.

such a consortship, the female may escape and copulate with another male. In other cases, a female may actively maintain proximity to her consort [18; de Ruiter, unpubl. data].

Genetic Analysis

Paternity can only be established with certainty by means of genetic analysis. We captured 170 individuals from the population, including all individuals of the 3 main study groups, to take blood samples [21] (fig. 3). Blood samples were obtained from all surviving young born into the three study groups during 1984, 1985 and 1986 (n = 28) and, in addition, from 17 young born before and after this period, for which it was also possible to investigate paternity, because all possible fathers in the group had been captured (see below). The samples were analyzed using two methods: electrophoresis of blood proteins [22–24] and DNA fingerprinting [15, 25, 26]. The latter technique is generally thought to have a greater potential, particularly because detectable variability in blood proteins is often too limited in nonhuman primates to permit reliable determination of paternity [27–29]. However, in Sumatran long-tailed macaques there is considerable variability in blood proteins [30, 31], and our study of paternity in the long-tailed macaque was under way before the technique of DNA fingerprinting became available to us. Moreover, DNA fingerprinting with multilocus probes, as applied here, requires larger blood samples than can safely be taken from very young infants. For these reasons, we used both electrophoresis of blood proteins and DNA fingerprinting in our investigation. Blood protein analysis was carried out for all 45 offspring studied. DNA fingerprinting was conducted for 17 offspring for which more than one group male remained as a potential father and for an additional 7 randomly selected offspring from the large group, where it seemed desirable to take extra precautions to confirm the protein results because of the large number of potential fathers in the group.

Blood proteins reflect specific gene loci and hence yields direct information as to the alleles carried by an individual. For a number of the investigated proteins, this has been established for humans. We were able to establish that this also holds for the long-tailed macaque for the proteins we studied, as we knew the mother-infant relationships. Obviously, a male can be ruled out as a possible father of a given infant if, for at least one of the genes investigated, this male lacks an allele that is present in the infant but could not have been inherited from its mother. We investigated the blood samples for a battery of 29 blood proteins. For 7 of these proteins, we found high variability, and the results proved to be reproducible. We analyzed protein variants with two techniques: (1) Isoelectric focussing (IEF): amylase (AMY), protease inhibitor (Pi), group-specific component [Gc or DBP (vitamin-D-binding protein), with immobilines] [22]. (2) Agarose gel electrophoresis (AGE) [22]: isocitrate dehydrogenase 1 (IDH), phosphogluconate dehydrogenase (PGD) [23], properdin factor B (Bf), transferrin (Tf) [24].

For paternity exclusions with DNA fingerprinting, complex and unspecific multilocus banding patterns are compared [32]. Each band probably represents an allele of an unknown polymorphic micro- or minisatellite locus. Therefore, a male can be excluded from paternity if he lacks at least two bands (see Discussion) present in the pattern of the offspring but not in that of the mother [15, 25]. Bands were scored in the size range of 2–12 kb. DNA fingerprinting was carried out essentially as described previously [26, 33], applying probes 33.15 and 33.6 [25] and probe (CAC)n [34].

If, by these means, all but one of the mature males of the group can be excluded as potential fathers, and if it is certain that the female cannot have mated with an extraneous male (see Discussion), paternity is thereby ascertained.

Results

The 21 phenotypes (7 alleles) that we found for the Gc protein are shown in figure 4a. Figure 4b shows variants found at the transferrin locus. Taking this protein together with 6 other variable proteins, we were able to conduct paternity exclusions as illustrated in table 1. One example concerns 2 infants of a particular female; for each of these infants, all males but 1 in the group are excluded as possible fathers. For the other mother-infant pair, 2 males in the group remain candidates.

Figure 5 illustrates the result of DNA fingerprinting with probes 33.15, 33.6 and (CAC)n applied to the first female and her 2 offspring. The DNA patterns confirm paternity in the sense that they exclude all group males except the males that had already been singled out as the only possible fathers within that group on the basis of protein analysis. In the first case, this was the alpha male, in the second case the beta male (No. 3;

Table 1. Paternity exclusion with protein variants

Individual	No.	Proteins						
		IDH	6PGD	Pi	Tf	Gc	Bf	AMY
Mother		1-1	A-A	1-2	2-3	2-2	2-2	1-1
Offspring		1-1	A-F	1-1	1-2	2-2	2-2	1-1
Offspring		1-2	A-A	1-1	1-2	1-2	2-3	1-1
Mother		1-2	A-A	1-2	2-3	1-2	2-2	1-1
Offspring		1-1	A-A	1-1	1-3	1-5	2-2	1-1
Alleles contributed by father								
Offspring	1	1	F	1	1	2	2	1
Offspring	2	2	A	1	1	1	3	1
Offspring	3	1	A	1	1	5	2	1
Potential fathers	1	1-1	A-F	1-1	1-2	2-5	2-2	1-4
	2	1-2	A-A	1-1	2-3	2-5	2-2	1-2
	3	1-2	A-A	1-1	1-3	1-4	2-3	1-2
	4	1-2	A-F	1-1	1-2	3-4	2-2	1-2
	5	1-1	A-F	1-2	2-3	2-2	1-4	1-1
	6	1-1	F-F	1-2	2-2	2-5	2-4	1-2
	7	1-1	A-A	1-1	2-2	2-5	2-2	1-2
	8	1-2	A-A	1-2	1-2	4-5	2-2	1-1
	9	1-1	A-F	2-2	2-3	4-5	2-4	1-2
	10	1-2	A-A	1-2	2-2	2-5	2-4	1-4
	11	2-2	A-A	1-2	2-2	1-4	2-2	1-2
	12	2-2	A-A	1-1	1-2	3-5	2-2	1-1

Rows represent the pairs of alleles for the different proteins (columns) present in each individual. Above: 1 female with 2 offspring and another female with a single offspring. Below: the males in the group arranged according to rank. For the first offspring of the first mother, the alpha male (No. 1) is the only male that carries all alleles that the young has not received from its mother. For the second offspring, this is male No. 3, the beta male at that time (No. 2 had emigrated). For the third offspring, 2 males in the group carry all alleles that the offspring cannot have inherited from its mother.

males are numbered in order of rank at a given time, and No. 2 had emigrated from the group). Figure 6 illustrates how, with three different probes, DNA fingerprinting excludes the alpha male, but not the young male that remained candidate after protein analysis (table 1, offspring 3), for he shows all nonmaternal bands of the offspring in his DNA patterns.

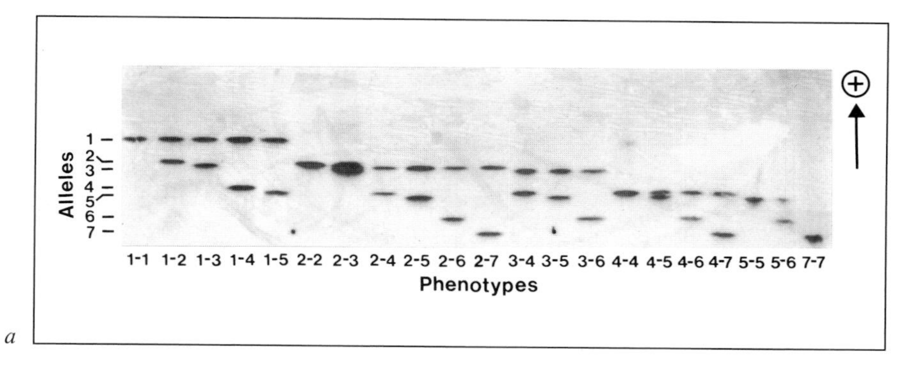

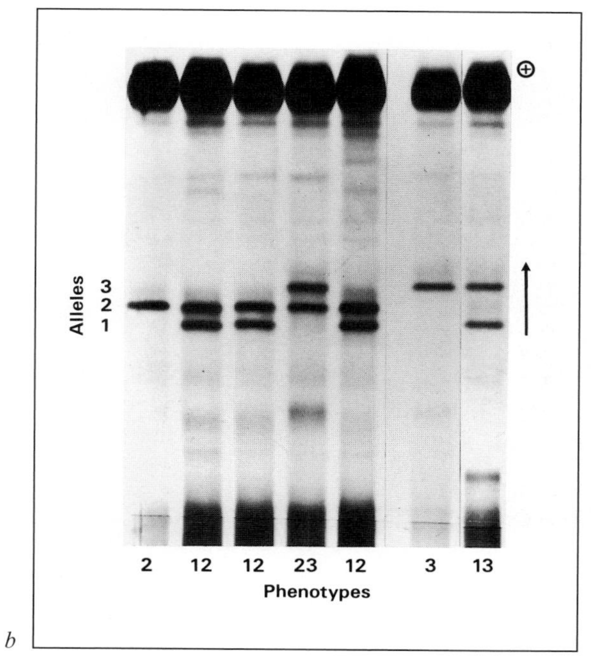

Fig. 4. *a* The different Gc (group-specific component) phenotypes in the population. Individuals with two different alleles (heterozygotes) and two identical alleles (homozygotes) can be distinguished. Proteins vary slightly in electric charge; they were separated by isoelectric focusing on a polyacrylamide gel with a gradient built into it with immobiline. This electrophoretic method enables one to separate molecules with only small differences: the final position of each variant is determined by its electric charge. Small quantities of the protein on the gel can be bound to a specific antibody (anti-Gc) and subsequently stained [22]. *b* Transferrin phenotypes as distinguished with vertical polyacrylamide gel electrophoresis confirmed the results obtained with agarose [22].

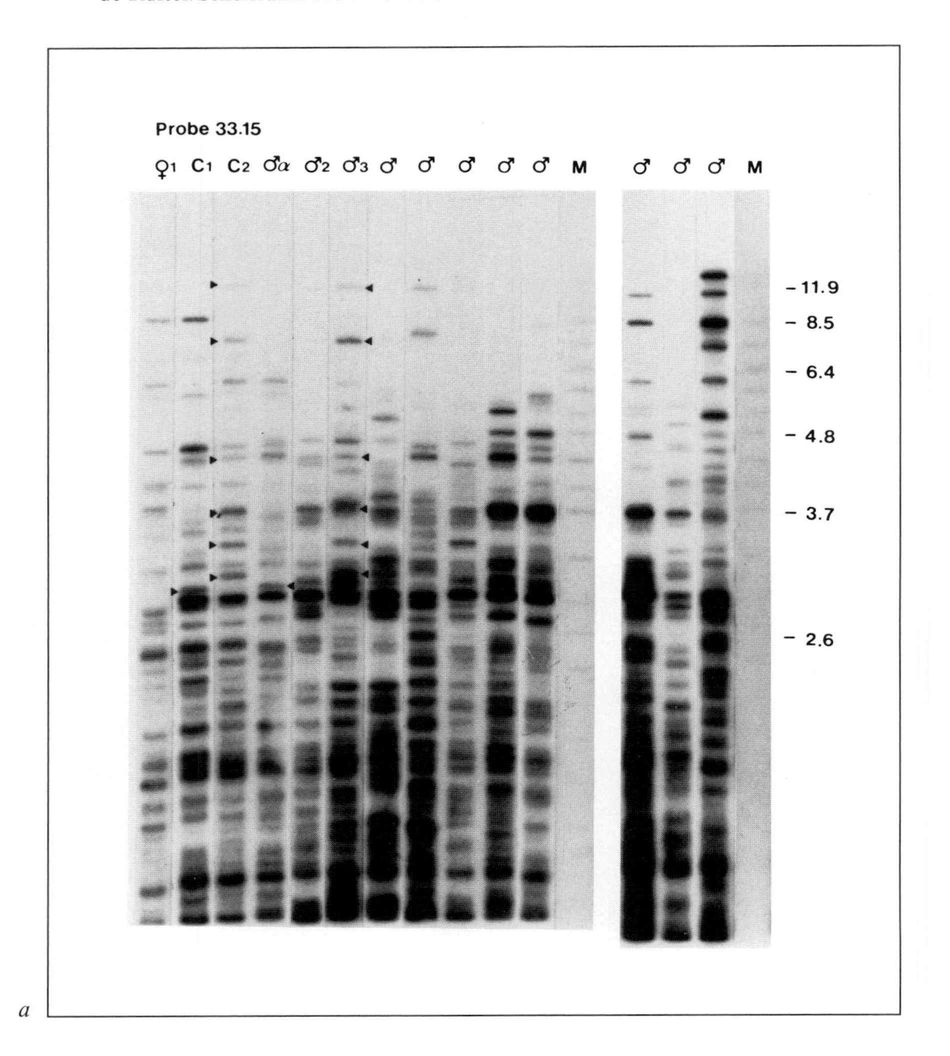

a

Fig. 5. DNA fingerprints of female No. 1 and her 2 offspring (C1 and C2), as in table 1. Next to the female and her offspring are the DNA fingerprints of the males according to rank at a given time. The DNA patterns point to males 1 and 3, respectively, as fathers. Restriction enzyme digestion of genomic DNA was performed using Hae III and hybridization was with probe 33,15 (*a*), 33.6 (*b*) and (CAC)n (*c*). Probe 33.15 produced on average 5 nonmaternal bands in our study population. Probes 33.6 and (CAC)n yielded 4 nonmaternal bands. The figure is composed of autoradiographs with different exposure times to obtain comparable results. On the gels, every 3 or 4

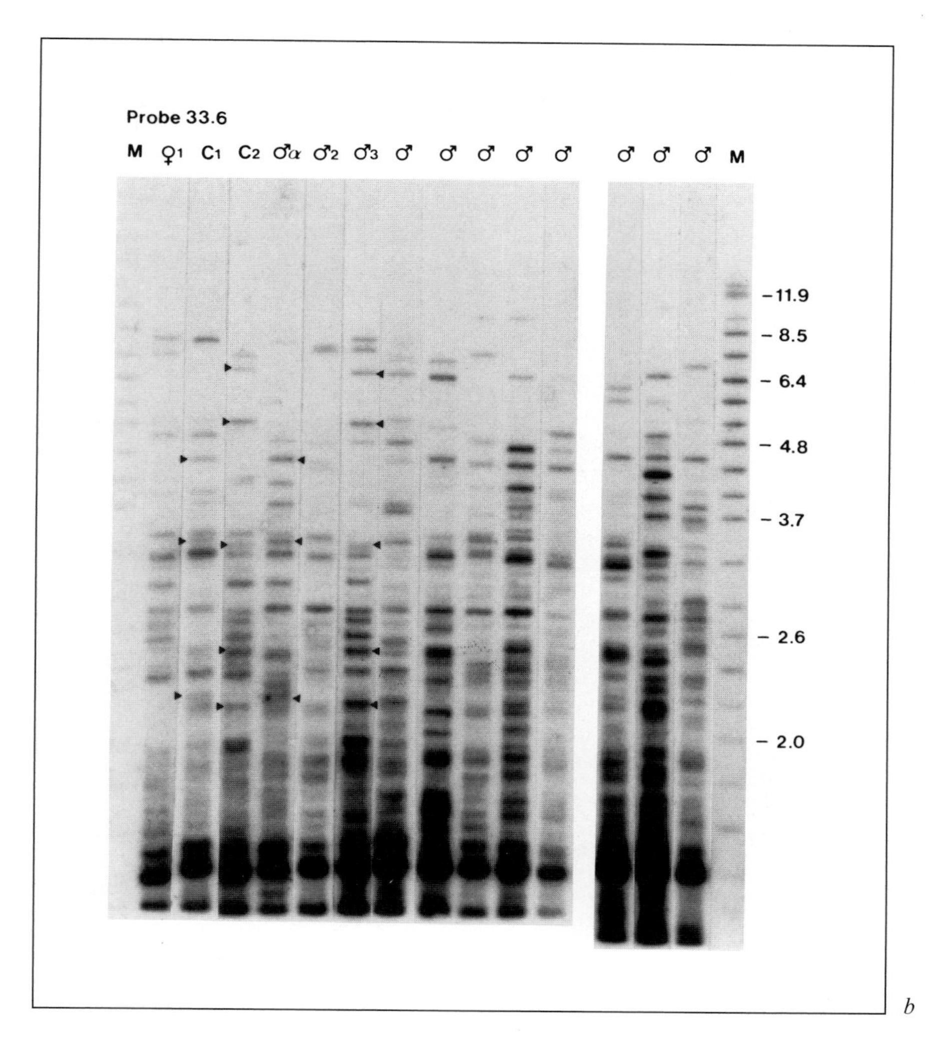

b

lanes a DNA fragment size marker (Promega, USA; sizes are in kilobasepairs; kb) was run. In the composition photo it is represented on the sides (M). This allowed accurate comparison within and across gels. The 3 males at the right are on a different gel. An average of 38 bands in the 2- to 12-kb size range could be discerned per DNA profile (probe 33.15: 21 bands; probe 33.6: 17 bands; probe (CAC)n 18 bands). Band-sharing frequency in the 8 males on the same gel was calculated to be 0.43 (on average for 3 probes). The known average relatedness of these males was 0.10 [de Ruiter et al., in preparation].

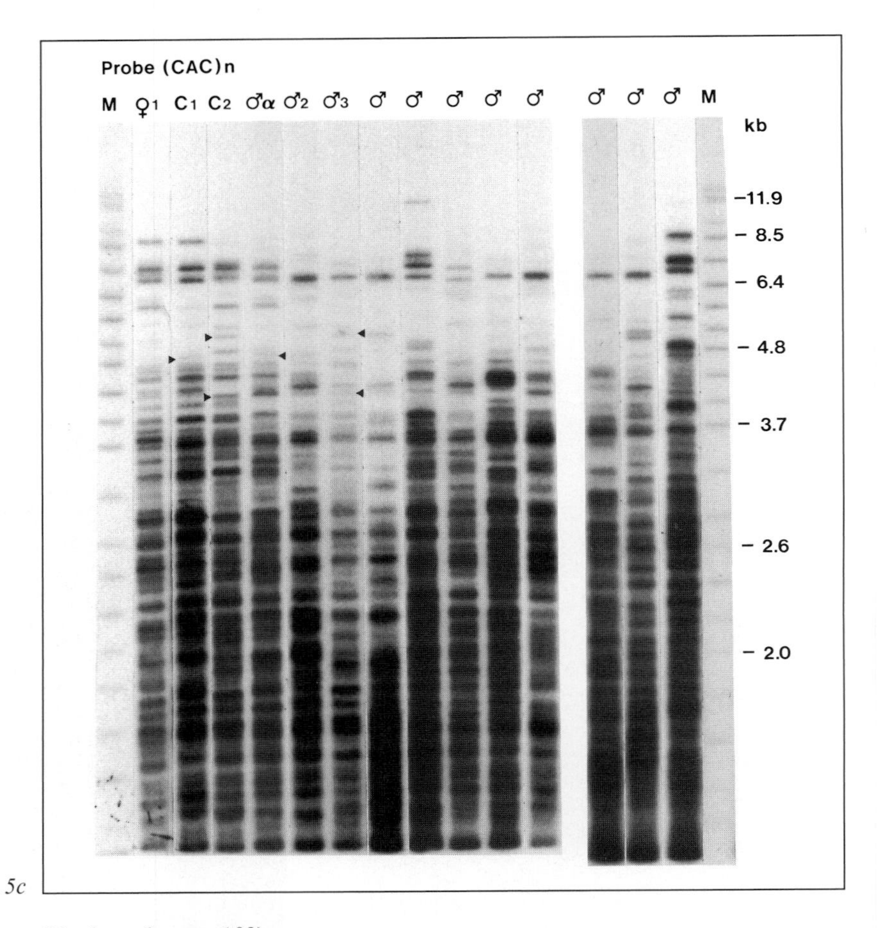

5c

(For legend see p. 182)

It was possible to exclude all but 1 male from the group for 33 out of the total of 45 offspring using blood protein analysis. In 7 randomly taken cases (see above), such inferences of paternity were confirmed by subsequent DNA fingerprinting, thereby demonstrating the validity of the DNA fingerprinting. Of the remaining 12 offspring, for which more than 1 potential father in the group remained after blood protein analysis, their number was reduced to 1 by DNA fingerprinting in 7 cases (in the remaining 5 cases DNA fingerprinting was impossible for technical reasons).

The overall results are summarized in figure 7. In the large group, the alpha male was the only male not excluded from paternity in 52% (11 out

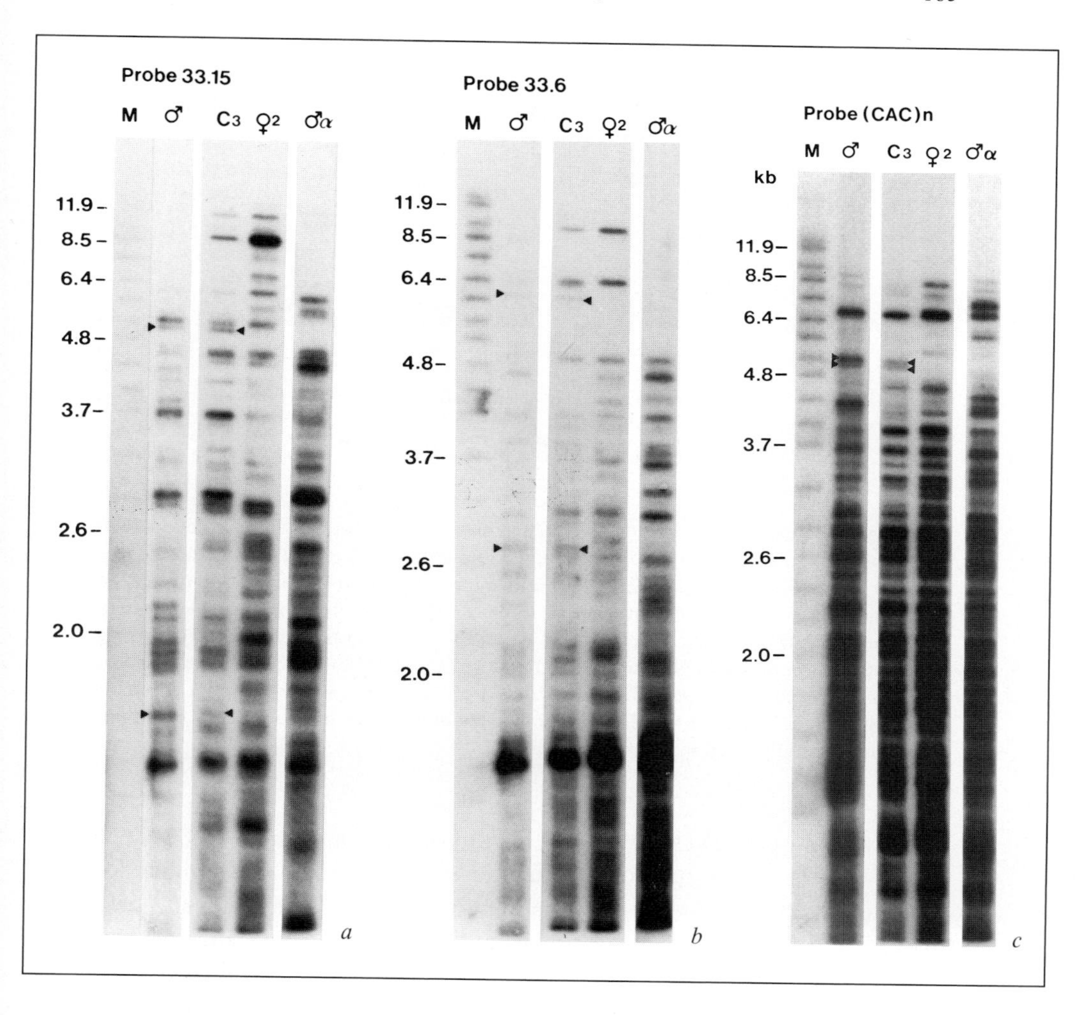

Fig. 6a–c. DNA fingerprints of female 2 with offspring (C_3), as in table 1 with the same 3 probes as in figure 5, together with the DNA fingerprints of the 2 remaining potential fathers in the group on another part of the same gel (the alpha male and male No. 8; the latter is not represented in fig. 5). This is an exceptional case in that a young low-ranking male was not excluded from paternity as the only group male. Reference markers on each autoradiograph (M) allowed for comparison across autoradiographs.

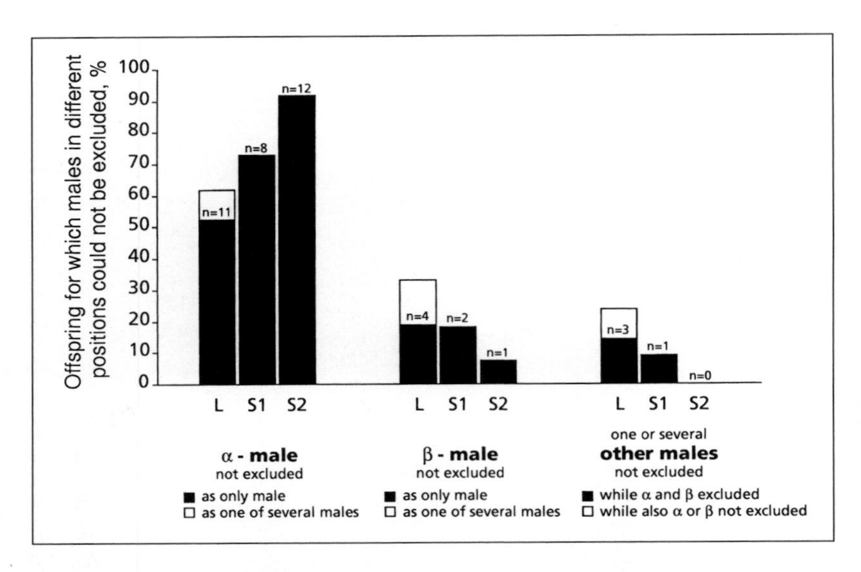

Fig. 7. Percentage offspring for which males in different positions could not be excluded. The data of several individual alpha and beta males are taken together. In the smallest group (S2) the alpha male was the only male in the group at conception of 5 of the offspring (for 1.5 years). The contribution of males other than alpha or beta is very small given the fact that the figures represent a large number of low-ranking males. All surviving offspring born between 1984 and 1986 (L) and 1983 and 1986 (S1, S2) and some additional offspring after this period (up to 1988) are included. L = Large group; S1, S2 = smaller groups; n = number of offspring for which a male with a particular rank could not be excluded as only male in his group.

of 21) of the cases investigated and for the small groups this was 83% (20 out of 24). The beta male of a group was found to be the only male not excluded in 50% (7 out of 14) of the remaining cases. Both alpha and beta males could be excluded for 4 offspring (19% of the cases) and for 2 of these only 1 of the other males in the group could not be excluded (fig. 6). In the large group, the alpha was somewhat less often not excluded for offspring with high-ranking females 73% (8 out of 11) as opposed to low-ranking females 50% (5 out of 10).

Two findings deserve mention with respect to inbreeding. Two males stayed in their natal groups for about 2 years after they had become sexually mature; although they did copulate in their natal groups, they were always excluded from paternity. Father/daughter inbreeding did not occur in a group where the tenure of the alpha was long enough for this to have been possible. In this case, the beta male proved to be the only nonexcluded male in the group.

Discussion

Comparison of Techniques Used

For our natural population of long-tailed macaques, DNA fingerprinting and protein electrophoresis proved to be almost equivalent in terms of results eventually achieved. DNA fingerprint analysis, used in the second instance after protein analysis, was always successful in establishing these unresolved cases. This demonstrates the high discriminating power of DNA fingerprinting in this species. However, each approach has specific advantages and disadvantages. DNA fingerprinting is technically more complicated and, moreover, materials are expensive. Both the time involved and the possibility of procedural mistakes (e.g. confusion of samples) are, however, smaller because DNA fingerprinting involving a single electrophoretic separation of DNA restriction fragments and hybridization with 1–3 probes leads to roughly the same degree of certainty as electrophoresis of a far greater number of proteins. Conversely, although DNA fingerprinting yields a great deal of information, this information is more difficult to interpret. This is partly because corresponding bands do not necessarily have the same intensity and partly because there is a possibility of new bands due to mutations. We therefore used the two-band criterion for exclusion (see 'Genetic Analysis'), although mutation presumably occurred only in a few cases. Moreover, this technique may fail to produce detectable results due to partial decomposition of DNA or as a result of procedural errors (e.g. incomplete digestion). Similar failures in protein electrophoresis can be traced far more easily. New developments in DNA techniques are, however, rapidly taking place and the amount of potentially detectable variation for DNA is much larger than for proteins.

Male Rank and Reproductive Success

As stated above, paternity can only be established by excluding all but 1 of the males that have mated with a female at the time when she conceived. We must bear in mind here that females may possibly mate with a male that does not belong to the group, but is a migrating stranger. However, it seems unlikely that such strangers would be genetically indistinguishable from the alpha or beta male of the female's group in all cases with the techniques that we used. Therefore, we are inclined to conclude from our data that, in this natural population of long-tailed macaques, the alpha males father most offspring, and that the remaining fertilizations are largely attributable to the beta males. This correlation between rank and fertilizations is more marked in the smaller groups. This may be because there are fewer males competing in such groups, and because fewer females are sexually receptive simultaneously.

Although it may seem plausible that high rank would enhance reproductive success, evidence for this from studies of natural populations is scarce. In the red howler monkey, an older male is sometimes tolerated in a harem group, but probably does not copulate, and could be excluded from paternity in most cases [35]. Better evidence was found in two small groups of rhesus macaques, but, according to the author, this is 'not necessarily indicative of a general positive correlation' [36]. On the other hand, many studies on captive populations of various macaques and one lemur species [37] have generally yielded no clear correlation between male rank and reproductive success. In *M. fascicularis,* no correlation was found [14]; in *M. fuscata,* only a weak correlation was discovered [15]; in *M. mulatta* such a correlation was either lacking [6] or transient [7–12]. The positive correlation in *M. sylvanus* reported by Witt et al. [13] has rightly been questioned [3]. The social structure in all of these macaque species is similar in many respects: a linear hierarchy is found among males and no male coalitions occur [with the possible exception of *M. sylvanus;* 4]. Female receptivity is prolonged and breeding is seasonal [3]. Hence, the discrepancy between previous results and our own is probably attributable to the unnatural living conditions experienced by macaques in captivity [1, 3, but also see ref. 38]. In the first place, space restriction prevents males from avoiding one another when conflicts arise and receptive females from separating from the group. Secondly, the pattern of relationships within captive colonies is often unnatural because immature males – which would normally migrate – are not removed, with the possible consequences that (a) such males continue to receive kin support in agonistic conflicts, and (b) they are avoided as mating partners [15]. With respect to this, it is noteworthy that, as the dominant families are the larger ones in most macaques species, this may result in fewer nonkin to copulate with for growing males from dominant families [M. Cords, pers. commun.]. As a general rule, the structure of captive groups may become progressively more abnormal over the years. For instance, an alpha male, which would normally be displaced and migrate again, is compelled to stay in the group with uncertain status. For the colony of *M. fascicularis* kept at the University of Utrecht, it seems as though the alpha male puts so much effort into maintaining his position in such cases that he is unable to also monopolize the fertile females [W.J. Netto, pers. commun.]. In the colony maintained at Zurich University, although the alpha male appears to be dominant, he fails to perform the normally corresponding sexual behavior, probably due to his old age [M. Cords, pers. commun.]. Such effects may explain the inconsistency in observed correlations between male rank and reproductive success.

Finally, it should be noted that in our study, there is a slight trend that offspring likely to have been fathered by the alpha male more often had high-ranking mothers than offspring likely to have been fathered by another male. This is in agreement with what would be expected on the assumption that the alpha male has to choose among simultaneously attractive females, because female receptivity in this species may last for several months and high-ranking females – and probably also their offspring – appear to be at an advantage [1, 19].

The next question, of course, will be in what manner the structure of sexual behavior in wild-living *M. fascicularis* explains the observed correlations between male rank and reproductive success. Some aspects will be dealt with in a subsequent publication by the first author, but a few comments may be included here. On the whole, the alpha male copulates somewhat more than low-ranking males, and, in addition, he tends to restrict his copulations to females likely to be fertile. For the time being, however, with the kind of observational data that the first author collected (which are similar in scope to the data collected in most other field studies), we have found it impossible to predict in specific cases which male will prove to be the father of the offspring born, even though birth dates were known and conception dates could thus be estimated.

Acknowledgments

We thank the Indonesian Institute of Sciences (LIPI) and its Biological Division (Puslitbang Biologi), the Directorate of Forest Protection and Nature Conservation (PHPA) and the Universitas Nasional in Jakarta (UNAS) for sponsoring and permission to carry out the fieldwork. We thank G.P. Baerends, S. Djojosudharmo, M.A. van Noordwijk, C.P. van Schaik, and J. Sugarjito for support and discussion and many others who contributed substantially to the work. We thank Mr. Ery Zulma and Mr. Iskander Zulkarnaen from Singapore Airlines for being especially helpful in assisting us with the blood sample shipments. Thanks are due to Prof. W. Schmid (Zürich) for producing the first DNA fingerprints of our animals in spring 1986. J.G. Robinson and F. Aureli made valuable comments on the manuscript. D. Smit, R. Leito and others prepared the figures. We thank the first author's wife Margo Meijers for continuous support and contributions to all social and scientific activities throughout this study. This work was in part financed by WOTRO (Netherlands Foundation for the Advancement of Research in the Tropics), the H.F. Guggenheim Foundation, Lucie Burger's Stichting voor Vergelijkend Gedragsonderzoek and the A.H. Schultz Stiftung.

References

1 Robinson JG: Intrasexual competition and mate choice in primates. Am J Primatol 1982;1(suppl):131–144.
2 Berenstain L, Wade TD: Intrasexual selection and male mating strategies in baboons and macaques. Int J Primatol 1983;4:201–235.
3 Fedigan LM: Dominance and reproductive success in primates. Yearbook Phys Anthropol 1983;26:91–129.
4 Paul A: Determinants of male mating success in a large group of Barbary macaques *(Macaca sylvanus)* at Affenberg Salem. Primates 1989;30:461–476.
5 Cowlishaw G, Dunbar RIM: Dominance rank and mating success in male primates. Anim Behav 1991;41:1045–1056.
6 Duvall SW, Bernstein IS, Gordon TP: Paternity and status in a rhesus monkey group. J Reprod Fertil 1976;47:25–31.
7 Smith DG: Paternity exclusion in six captive groups of rhesus monkeys *(Macaca mulatta).* Am J Phys Anthropol 1980;53:243–249.
8 Smith DG: The association between rank and reproductive success of male rhesus monkeys. Am J Primatol 1981;1:83–90.
9 Stern BR, Smith DG: Sexual behaviour and paternity in three captive groups of rhesus monkeys *(Macaca mulatta).* Anim Behav 1984;32:23–32.
10 Smith DG, Smith S: Paternal rank and reproductive success of natal rhesus males. Anim Behav 1988;36:554–562.
11 Curie-Cohen M, Yoshihara D, Blystad C, et al: Paternity and mating behavior in a captive troop of rhesus monkeys. Am J Primatol 1981;1:135.
12 Curie-Cohen M, Yoshihara D, Luttrell L, et al: The effects of dominance on mating behavior and paternity in a captive troop of rhesus monkeys. Am J Primatol 1983;5: 127–138.
13 Witt R, Schmidt C, Schmitt J: Social rank and Darwinian fitness in a multimale group of Barbary macaques. Folia Primatol 1981;36:201–211.
14 Shively C, Smith DG: Social status and reproductive success in male *Macaca fascicularis.* Am J Primatol 1985;9:129–135.
15 Inoue M, Takenaka A, Tanaka S, et al: Paternity discrimination in a Japanese macaque troop by DNA fingerprinting. Primates 1990;31:563–570.
16 van Schaik CP, Mirmanto E: Spatial variation in the structure and litterfall of a Sumatran rain forest. Biotropica 1985;17:196–205.
17 van Noordwijk MA, van Schaik CP: Male migration and rank acquisition in wild long-tailed macaques *(Macaca fascicularis).* Anim Behav 1985;33:849–861.
18 van Noorwijk MA: Sexual behaviour of Sumatran long-tailed macaques *(Macaca fascicularis).* Z Tierpsychol 1985;70:277–296.
19 van Noordwijk MA, van Schaik CP: Competition among female long-tailed macaques *(Macaca fascicularis).* Anim Behav 1987;35:577–589.

20 de Waal FBM: The organization of agonistic relations within two captive groups of Java monkeys *(Macaca fascicularis).* Z Tierpsychol 1977;44:225–282.
21 de Ruiter JR: Capturing wild long-tailed macaques. Folia Primatol, in press.
22 Scheffrahn W: Elektrophoretische Methoden; in Knussmann R (ed): Lehrbuch der Anthropologie. Stuttgart, Fischer, in press, vol 1, 2.
23 Blake NM, Saha N, McDermid EM, et al: Additional electrophoretic variants of 6-phosphogluconate dehydrogenase. Humangenetik 1974;21:347–354.
24 Teisberg P: High-voltage agarose gel electrophoresis in the study of C3 polymorphisms. Vox Sang 1970;19:47–56.
25 Jeffreys AJ, Wilson V, Thein SL: Hypervariable 'minisatellite' regions in human DNA. Nature 1985;314:67–73.
26 Jeffreys AJ, Wilson V, Thein SL: Individual-specific 'fingerprints' of human DNA. Nature 1985;316:76–79.
27 Chakraborty R, Ferrell RE, Schull WJ: Paternity exclusion in primates: Two strategies. Am J Phys Anthropol 1979;50:367–372.
28 Smith DG, Small MF, Ahlfors CE, et al: Paternity exclusion and its applications to studies of nonhuman primates. Adv Vet Sci Comp Med 1984;28:1–24.
29 Hayasaka K, Kawamoto Y, Shotake T, et al: Probability of paternity exclusion and the number of loci needed to determine the fathers in a troop of macaques. Primates 1986;27:103–114.
30 Kawamoto Y, Ischak TM: Genetic differentiation of the Indonesian crab-eating macaque *(Macaca fascicularis).* I. Report on blood protein polymorphism. Primates 1981;22:237–252.
31 Kawamoto Y, Ischak TM: Genetic variations within and between groups of crab-eating macaques *(Macaca fascicularis)* on Sumatra, Java, Bali Lombok and Sumbawa, Indonesia. Primates 1984;25:131–159.
32 Jeffreys AJ, Turner M, Debenham P: The efficiency of multilocus DNA fingerprint probes for individualization and establishment of family relationships determined from extensive casework. Am J Hum Genet 1991;48:824–840.
33 Uitterlinden AG, Slagboom PE, Knook DL, Vijg J: Two dimensional DNA fingerprinting of human individuals. Proc Natl Acad Sci USA 1989;86:2742–2746.
34 Ali S, Muller CR, Epplen JT: DNA fingerprinting by oligonucleotide probes specific for simple repeats. Hum Genet 1986;74:239–243.
35 Pope TR: The reproductive consequences of male cooperation in the red howler monkey: Paternity exclusion in multi-male and single-male troops using genetic markers. Behav Ecol Sociobiol 1990;27:469–446.
36 Melnick DJ: The genetic consequences of primate social organization: A review of macaques, baboons and vervet monkeys. Genetica 1987;73:117–135.
37 Pereira ME, Weiss ML: Female choice, male migration, and the threat of infanticide in ring tailed lemurs. Behav Ecol Sociobiol 1991;28:141–152.
38 Inoue M, Mitsunaga F, Ohsawa H, et al: Paternity testing in captive Japanese macaques *(Macaca fuscata)* using DNA fingerprinting; in Martin RD, Dixson AF, Wickings EJ (eds): Paternity in Primates: Genetic Tests and Theories. Basel, Karger, 1992, pp 131–140.

Dr. Jan R. de Ruiter, Ethology and Socioecology Group, Utrecht University, PO Box 80.086, NL–3508 TB Utrecht (The Netherlands)

Martin RD, Dixson AF, Wickings EJ (eds): Paternity in Primates:
Genetic Tests and Theories. Basel, Karger, 1992, pp 192–202

DNA Fingerprinting of Free-Ranging Groups of Common Marmosets *(Callithrix jacchus jacchus)* in NE Brazil

A.F. Dixson[a], *G. Anzenberger*[b], *M.A.O. Monteiro Da Cruz*[c], *I. Patel*[d], *A.J. Jeffreys*[d]

[a]Centre International de Recherches Médicales de Franceville, Gabon;
[b]Department of Psychology, Biomathematical Section, University of Zürich, Switzerland; [c]Universidade Federal Rural de Pernambuco, Recifé, Brazil;
[d]Department of Genetics, University of Leicester, UK

Many authors have attributed the diversity of secondary sexual adornments, genital anatomy and occurrence of 'sperm competition' in primates to effects of sexual selection operating in different types of mating systems [1–6]. A major problem facing such research is the need to define the mating system. In order to do this, information on both maternity and paternity is required. Traditionally, behavioural observations have been used, but it is exceedingly difficult, or impossible, to establish paternity in field studies of primates by using behavioural criteria alone. In some cases, observations of the social composition of primate groups have led to erroneous assumptions about mating systems. Thus 'one-male units' such as occur in many *Cercopithecus* species or in *Erythrocebus* have been equated with 'harem' type or polygynous breeding groups. However, recent field studies show that periodic influxes of additional males can occur in at least some of these species and that females mate with a number of partners [7–9].

Monogamy is believed to be the prevalent mating system in various primates, such as indri *(Indri),* owl monkeys *(Aotus),* titi monkeys *(Callicebus),* marmosets and tamarins (Callitrichidae) and the lesser apes (Hylobatidae) [10]. However, some field studies of marmosets and tamarins indicate that facultative polyandry may occur in some species or populations

[11, 12]. The significance of these observations in terms of paternity within groups is as yet unknown.

Recently, the discovery of hypervariable minisatellite sequences in the human genome has led to the development of DNA fingerprinting techniques which have been used to perform individual identifications and/or to study familial relationships in humans [13, 14], several Old World monkey species [15, 16], and in the common marmoset [17]. This report concerns the first attempted application of DNA fingerprinting to free-ranging marmoset groups.

Materials and Methods

This study was carried out during July and August 1988 at the Tapacura Ecological Station of the Rural University of Pernambuco in NE Brasil, and also in the forest of Dos Irmaos near the city of Recifé. The Tapacura Ecological Station is situated in an isolated block of Atlantic Coastal Forest comprising approximately 392 ha. Details of the reserve and of its ecology have been provided by Hubrecht [18] and Stevenson and Rylands [19]. The forest of Dos Irmaos is also of the Atlantic Coastal type and is situated some 40 km from Tapacura.

Common marmosets were trapped at both sites by means of automatic live animal traps baited with pineapple. Both study areas had been pre-baited for some weeks and animals had been previously trapped at both sites by one of the authors (Monteiro Da Cruz). Some of the marmosets had been marked using neck chains and identity discs, or had been tattooed, as detailed in the results section.

At both sites, animals were transferred to a field laboratory in order to obtain blood samples and to take body measurements. Marmosets were removed from the traps and placed in a restraint device [20]. Blood (1–2 ml) was drawn from the femoral veins using a tuberculin syringe (25-gauge needle) and transferred to EDTA tubes chilled on ice. The following body measurements were made: body weight (g), dental stages and condition (0 = minimal wear; 1 = partially worn; 2 = heavily worn), both testes length and width (mm), and (in the female) pudendal pad width (mm). Additionally, females were palpated in order to determine possible occurrence and stage of pregnancy, and notes were made on any nipple elongation or signs of lactation, as previously described by Scanlon et al. [21].

A photograph was taken of the head of each animal (in frontal view) for individual identification and to show the development of the ear tufts. After measurements had been made, animals not wearing tags or tattooed were marked by cropping portions of ear tufts or hairs of the head, and then released into the forest. Animals trapped in groups were always released together at the trapping site within half a day of the capture. No injuries occurred, and in several cases monkeys were retrapped on subsequent days. Additional blood samples were obtained from retrapped animals, provided that at least 1 week had elapsed following the previous capture.

On the day of collection, the blood samples were transferred to a −80 °C deep freezer of the Federal University of Pernambuco in Recifé. Samples were subsequently air-freighted, packed in dry ice, to the laboratory of A.J. Jeffreys (Leicester, UK). After extraction of DNA, cleavage/digestion with restriction enzymes, electrophoresis, Southern blotting and hybridization using minisatellite probes 33.6 and 33.15, autoradiographs of hybridization patterns were prepared for animals from both areas. Descriptions of the techniques used are described elsewhere (e.g. for humans [13, 14]; for several Old World monkeys [15], and for the common marmoset [17]).

Results

Animals Trapped and Group Composition

We trapped a total of 20 monkeys, 11 at Tapacura (TP) and 9 at Dos Irmaos (DI). Group compositions are shown in tables 1 (TP) and 2 (DI). In the case of the TP group, the study area was divided by transects into 50 × 50 m quadrats. Animals of the TP group were trapped on three separate dates in scattered traps, but always within the same quadrat. On the first trapping, 7 individuals were caught (specimens 1–7). One week later, 3

Table 1. Group composition and body measurements of common marmosets trapped at Tapacura

Specimen No.	Trapping date	Tag No.	Age/sex class	Body weight g	Mean testis vol mm^3	Pudendal pad width mm	Tooth wear	Ear tufts	Nipple length
1	x, y	91	M ad	320	415.3	–	0	3	–
3	x	31	M ad	390	324.0	–	0	3	–
5	x	15	M ad	340	532.3	–	0	3	–
4	x	70	F ad	380	–	11.8	0	3	very small
7	x, y	–	F ad	270	–	9.4	0	3	small
17	y	78	F ad	340	–	12.0	0	3	very small
20	z	–	M sd	200	ND	–	0 (L)	2–3	–
2	x, y	–	F sd	235	–	6.9	0 (L)	ND	ND
19	y	–	M juv	140	53.8	–	0 (L)	1	–
6	x	–	F juv	150	–	5.3	0 (L)	1	0
18	y	–	F juv	140	–	3.8	0 (L)	1	0

M = Male; F = female; ad = adult; sd = subadult; juv = juvenile.
For ear tufts: 1 = little development and no white coloration; 2 = white tufts; 3 = fully grown white tufts.
L = Milk dentition; ND = no data.
x = Captured on Aug 2nd; y = captured on Aug 9th; z = captured singly on Aug 11th.

new individuals were caught (specimens 17 through 19) along with 3 retrappings (specimens 1, 7 and 2). At least one other male individual bearing a tag remained in the vicinity of the group on both occasions but was never trapped. On the basis of previous field data, identity tags, and especially retrappings and resightings of individually-known specimens, it is thought that these monkeys belonged to one social group. On the third trapping date at TP, 3 days later, only one single individual was caught (specimen 20); there were no signs of any other marmosets in the vicinity of the trapping site. In the case of the DI group, all members were caught together when they entered a group of traps on a small platform (2 m × 1.5 m). Again, at least one additional tagged adult male remained in the vicinity of the trapping site until the others were released. From previous field data, from identity tags and first-hand observation, it is thought that these monkeys similarly all belonged to one social group.

On the basis of body weight, genital and dental measurements, the TP group (table 1) consisted of 6 adults (3 males, 3 females), 2 subadults (1 male, 1 female) and 3 juveniles (1 male, 2 females). None of the 3 adult females was pregnant or lactating, although one was clearly heavier than the others (No. 4). The 3 adult males had a minimal degree of tooth wear.

Table 2. Group composition and body measurements of common marmosets trapped at Dos Irmaos

Specimen No.	Tag No.	Age/ sex class	Body weight g	Mean testis vol mm^3	Pudendal pad width mm	Tooth wear	Ear tufts	Nipple length mm
8	15	M ad	310	532.2	–	1	3	–
10	80	M ad	290	497.3	–	0	3	–
13	12	M ad	320	371.5	–	2	3	–
16	11	M ad	290	398.5	–	0	3	–
9	TAT.0	F ad	320	–	11.7	1	3	1.0
15	16	F ad	360*	–	13.3	2	3	1.5
11	–	F sd	200	–	3.4	0 (T)	2	0
12	–	M juv	130	60.4	–	0 (L)	1	–
14	–	F juv	140	–	4.0	0 (L)	1	0

* Female pregnant; TAT.0 = tattooed animal.
L = Milk dentition; T = dentition transitional between milk and permanent condition.
For other information, see footnote to table 1.

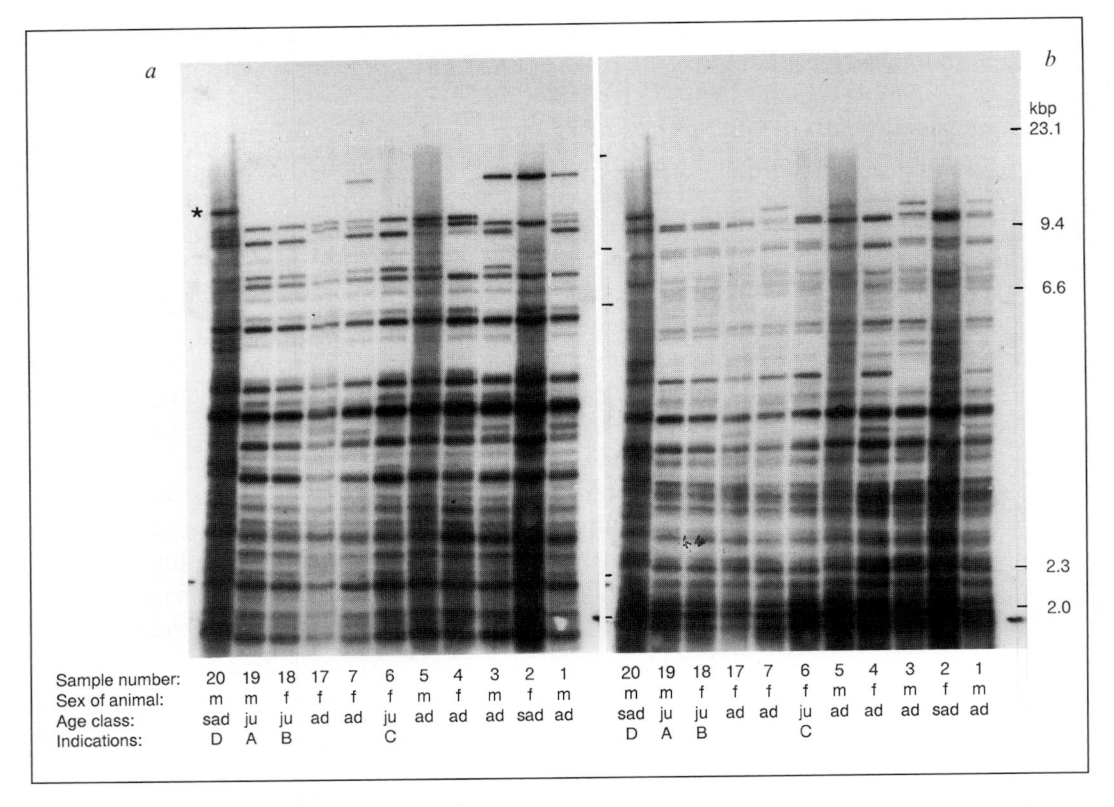

Fig. 1. Autoradiographs showing hybridization patterns in the marmoset group trapped at Tapacura. A + B + C = Triplets; D = specimen showing a unique band (*). m = Male; f = female; ad = adult; sad = subadult; ju = juvenile. *a* Using human minisatellite probe 33.6. *b* Using probe 33.15. kbp = Kilo base pairs.

The three juveniles were strikingly similar in body weight, dental condition and development of the pelage. The DI group (table 2) consisted of 6 adults (4 males, 2 females), 1 subadult female, and 2 juveniles (1 male, 1 female). One of the 2 females was in an advanced stage of pregnancy and her dentition was very worn, with the upper incisors missing. Of the 4 adult males, 1 of them (No. 13) had very worn teeth and may therefore have been older than the others. The two juveniles closely resembled each other in body weight and development of the pelage.

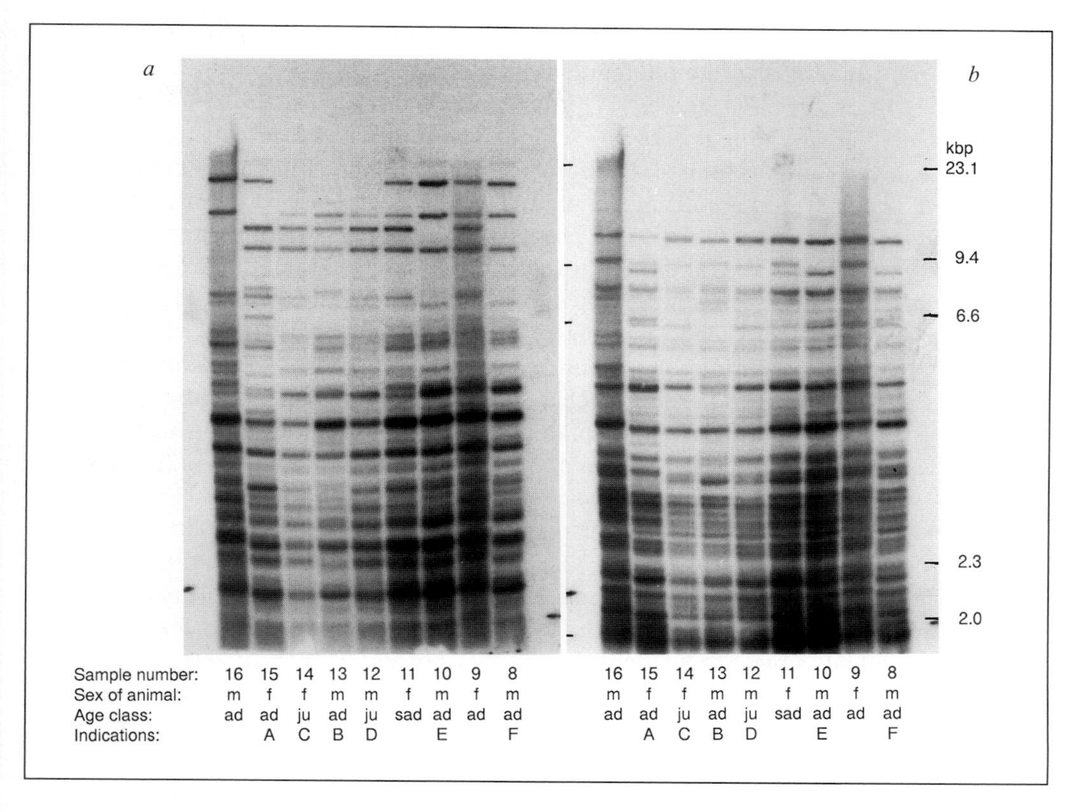

Fig. 2. Autoradiographs showing hybridization patterns in the marmoset group trapped at Dos Irmaos. A = An older breeding female (pregnant) with worn dentition; B = older male with well-worn dentition; C + D = twins; E + F = twins. m = Male; f = female; ad = adult; sad = subadult; ju = juvenile. *a* Using human minisatellite probe 33.6. *b* Using probe 33.15. kbp = Kilo base pairs.

DNA Fingerprints

Autoradiographs, showing DNA fingerprints of the two groups are shown in figures 1 (Tapacura) and 2 (Dos Irmaos). In preparing these fingerprints, the two hybridizations were carried out sequentially; the filters were first probed with minisatellite probe 33.6 and then with probe 33.15. The results obtained with probe 33.15 are complementary to those from probe 33.6 and there is no significant overlap in the variable bands detected by the two probes in each group. (Please note that the original autoradiographs available to us for data analysis and interpretation are of better quality than the photographs reproduced.)

From previous studies [17] it is known that probe 33.6 is more informative in revealing relatedness within captive marmoset groups, but that band variability is limited. Examination of the banding patterns of both the wild-caught groups indicates a similar limited degree of variability. The greatest variation in banding occurs within the high-molecular-weight-region (above 5 kbp) and there are clear differences between the two groups when this region of the autoradiographs is compared. However, within each group these bands tend to be shared by several individuals (fig. 1, 2).

With these limitations in mind, some analysis of familial relationships is possible. It is important to bear in mind the fact that dizygotic twins produced by marmosets share placental vascular anastomoses during fetal development and that the twins are haemopoietic chimaeras [22]. The DNA fingerprints of captive-born twins are virtually identical [17]. We therefore analyzed the data for the wild-caught groups for occurrence of twinning on this basis. In the TP group, in no case do adult individuals show virtually identical banding patterns. However, three juvenile animals are of particular interest (No. 6, 18, 19). The DNA fingerprints for these individuals appear to be identical and they are all juveniles of very similar body weight and physical development (table 1). As survival of triplets is known to be problematic with marmosets, one might have assumed that these would be the offspring of two different females, but their banding patterns suggest that these individuals may in fact be triplets. As mentioned above, bands tend to be shared within groups by several individuals, but there is one exception. Specimen 20 shows a unique band in the highest molecular-weight region not represented in any of the other individuals (fig. 1a). This result, together with the fact that this subadult male was caught singly, may be interpreted as an indication that he may not belong to that social unit.

In the DI group, two probable sets of twins are identifiable: the adults, male 8 and male 10, and the juveniles, male 12 and female 14. Because the breeding female of the group (female 15) was trapped, it is possible to conduct some analysis of parentage within the group. On the assumption that the 2 juveniles 12 and 14 might be previous offspring of this female, we compared the banding patterns of these animals in relation to those of the 4 adult males in the group (No. 8, 10, 13, and 16). On this basis, we are unable to exclude any of these males as possible fathers of this set of twins.

Discussion

DNA fingerprinting has been applied successfully to studies of individual identification, inheritance of banding pattern, and relatedness in free-ranging birds (e.g. sparrows [23, 24]) and to captive groups of various Old World monkey species [15, 16]. To date, the technique has rarely been applied to free-ranging primates and this report represents the first application of DNA fingerprinting to a free-ranging callitrichid species.

Sequential hybridization using two minisatellite probes, 33.6 and 33.15, produced minimal overlap in banding patterns for each of the two marmoset groups studied. The results obtained with the two probes are therefore complementary. However, as in previous studies of captive marmoset groups [17], probe 33.6 proved to be more informative in determining relatedness of individuals within groups. A major problem concerns the high degree of band sharing between all group members, with the major source of variability being in the higher molecular-weight region. The two groups studied inhabited isolated forests approximately 40 km apart. Comparison of the high-molecular-weight banding patterns shows some obvious differences between the two groups which may relate to the degree of genetic isolation between them. However, the limited variability of banding patterns within each group makes it exceedingly difficult to determine the biological relatedness of all its members. Low variability in DNA banding patterns within these free-ranging marmoset groups may be interpreted in two ways. Firstly, the populations studied might be subject to some degree of inbreeding. The forests at Tapacura and at Dos Irmaos, represent remnants of the once extensive Atlantic Coastal Forest of Brazil. At Tapacura the 392-ha forest contains a very high density of common marmosets [19]. Even allowing for intergroup migration [21], it is possible that this isolated population exhibits effects of inbreeding. The second possibility is that all group members are indeed very closely related as members of a family unit. This circumstance, coupled with the phenomenon of bone marrow chimaerism between twins, would result in great similarities between the banding patterns of many group members. Indeed, this picture also emerged in previous studies of captive family groups of marmosets [17].

DNA fingerprints of free-ranging marmoset groups are particularly useful for identification of sets of twins or triplets. Haemopoietic chimaerism of the offspring has long been known for marmosets [22, 25] and this results in almost identical DNA fingerprinting patterns when using blood samples [17]. In a recent study on common marmosets (n = 15; one triplet and 6 twin sets from 3 different captive families), all uterine siblings had identical blood DNA fingerprints [Anzenberger, Signer et al.; unpublished data]. This fact allowed us to identify provisionally two sets of twins and one set of triplets in the two groups studied. The finding of surviving triplets in the wild is most unusual, since a third offspring rarely survives, even under the favourable conditions provided in captive breeding colonies [26][1]. As the triplets weighed 140–160 g, they were probably at least 3–4 months old and were certainly fully weaned. The DNA fingerprinting analysis suggests that these individuals are triplets rather than being the offspring of two separate females.

The present study was of limited duration and it was not possible to trap all marmosets within a defined area of forest either at Tapacura or Dos Irmaos. From our field observations, it is certain that some adult group members were not trapped. These facts make it very difficult to analyse parentage within groups or to take account of migrations between groups which occur at Tapacura [21]. Fingerprinting techniques should be useful in future to determine whether sets of twins remain together or emigrate separately between groups. Given the fact that twin births are common in the species studied, we appear to have detected less twin pairs than might have been expected.

The potential of DNA fingerprinting as a means of understanding primate mating systems and of integrating laboratory and field approaches is now well recognized [27]. We believe that this approach will be useful in studies of social organization and mating systems in callitrichids and can contribute to resolving the debate about polyandry and monogamy in this primate family [11, 12, 26, 28]. However, to achieve this goal, longer-term field studies combining behavioural observations with trapping programs will be essential. Our studies show that the use of the existing human minisatellite probes 33.6 and 33.15 has limitations when applied to *Callithrix jacchus*. It would be valuable to attempt DNA fingerprinting on tissues other than white blood cells (e.g. skin biopsies) to avoid complications due to haemopoietic chimaerism. A variety of new probes and banding techniques is becoming available, so that these may also be tested in the hope of obtaining more informative DNA fingerprints, such as exist for various Old World primates [15, 16].

[1] There is only one proven record of *Callithrix jacchus jacchus* triplets surviving into adulthood under captive conditions [Anzenberger; unpubl. data].

Acknowledgments

A.F. Dixson is grateful to the Royal Society, British Council and Medical Research Council (UK) for travel funds and to Mark, Meg and Robbie Rouse for provision of much-needed dry ice at Heathrow Airport. G. Anzenberger gratefully acknowledges generous financial support from the Georges and Antoine Claraz-Schenkung, and also field assistance by Renata König-Anzenberger. A.J. Jeffreys is a Lister Institute Research Fellow and part of this work was supported by grants to him from the Medical Research Council (UK) and the Wolfson Foundation. Commercial enquiries regarding the use of the Minisatellite probes should be addressed to Cellmark Diagnostics, 8 Blacklands Way, Abingdon Business Park, Abingdon, Oxon OX14 1DY, UK. All authors wish to thank Dr. E. Signer (for assistance in interpreting the DNA fingerprint data), the Rural University of Pernambuco and Dr. S. Mendes (Director, Ecological Station, Tapacura), Dr. Elizabeth Chaves (LIKA, Federal University of Pernambuco), Senor Arlindo, Senor Antonio Rossano M. Pontes and the students of M.A.O. Monteiro da Cruz for assistance with fieldwork in Brazil.

References

1 Darwin C: The Descent of Man and Selection in Relation to Sex. London, Murray, 1871.
2 Clutton-Brock TH, Harvey PH: Evolutionary rules and primate societies; in Bateson PP, Hinde RA (eds): Growing Points in Ethology. Cambridge, Cambridge University Press, 1976, pp 195–237.
3 Short RV: Sexual selection and its component parts, somatic and genital selection, as illustrated by man and the great apes. Adv Stud Behav 1979;9:131–158.
4 Harcourt AH, Harvey PH, Larson SG, Short, RV: Testis weight, body weight and breeding system in primates. Nature 1981;293:55–57.
5 Dixson AF: Observations on the evolution of the genitalia and copulatory behaviour in male primates. J Zool 1987;213:423–443.
6 Møller AP: Ejaculate quality, testes size and sperm competition in primates. J Hum Evol 1988;17:479–488.
7 Cords M: Forest guenons and patas monkeys: Male-male competition in one group; in Smuts BB et al (eds.): Primate Societies. Chicago, Chicago University Press, 1986, pp 98–111.
8 Rowell TE: Beyond the one male group. Behaviour 1988;104:188–201.
9 Chism J, Rowell TE: Mating and residence patterns of male patas monkeys. Ethology 1986;72:31–39.
10 Kleiman DG: Monogamy in mammals. Q Rev Biol 1977;52:39–69.
11 Sussman RW, Garber PA: A new interpretation of the social organization and mating system of the Callitrichidae. Int J Primatol 1987;8:73–92.
12 Goldizen AW: Facultative polyandry and the role of infant carrying in wild saddleback tamarins *(Saguinus fuscicollis).* Behav Ecol Sociobiol 1987;20:99–109.
13 Jeffreys AJ, Wilson V, Thein SL: Hypervariable 'minisatellite' regions in human DNA. Nature 1985;314:67–73.

14 Jeffreys AJ, Wilson V, Thein SL: Individual specific 'fingerprints' of human DN Nature 1985;316:76–79.
15 Weiss ML, Wilson V, Chan C, Turner T, Jeffreys AJ: Application of DNA-finge printing probes to Old World Monkeys. Am J Primatol 1988;16:73–79.
16 Washio K, Misawa S, Ueda S: Individual identification of nonhuman primates usi DNA-fingerprinting. Primates 1988;30:217–221.
17 Dixson AF, Hastie N, Patel I, Jeffreys AJ: DNA 'fingerprinting' of captive famil groups of common marmosets. Folia Primatol 1988;51:52–55.
18 Hubrecht RC: Field observation on group size and composition of the commor marmoset *(Callithrix jacchus jacchus)* at Tapacura, Brazil. Primates 1984;25:13–21.
19 Stevenson MF, Rylands AB: The marmosets, genus *Callithrix;* in Mittermeier A, Rylands AB, Coimbra-Filho A, Fonseca GAB (eds): Ecology and Behaviour of Neotropical Primates. Washington, World Wildlife Fund, 1988, vol 2, pp 131–222.
20 Hearn JP: A device for restraining small monkeys. Lab Anim 1977;11:261–262.
21 Scanlon CE, Chalmers NR, Monteiro da Cruz MAO: Changes in the size, composition and reproductive condition of wild marmoset groups *(Callithrix jacchus jacchus)* in North East Brazil. Primates 1988;29:295–305.
22 Benirschke K, Anderson JM, Brownhill LE: Marrow chimerism in marmosets. Science 1962;138:513–515.
23 Wetton JH, Carter RE, Parkin DT, Walters D: Demographic study of a wild house sparrow population by DNA-fingerprinting. Nature, 1987;327:147–149.
24 Burke T, Davies NB, Bruford MW, Hatchwell BJ: Parental care and mating behaviour of polyandrous dunnocks *(Prunella modularis)* related to paternity by DNA fingerprinting. Nature 1989;338:249–251.
25 Wislocki GB: Observations on twinning in marmosets. Am J Anat 1939;664:445–483.
26 Hearn JP: Marmosets and tamarins; in Poole T (ed): The UFAW Handbook on the Care and Management of Laboratory Animals. Harlow, Longman Scientific and Technical, 1988, pp 568–581.
27 Rosenblum LA, Kummer H, Nadler RD, Robinson J, Suomi SJ: Interface of field and laboratory-based research in primatology. Am J Primatol 1989;18:61–64.
28 Ferrari SF, Lopes Ferrari MA: A re-evaluation of the social organisation of the Callitrichidae, with reference to the ecological differences between genera. Folia Primatol 1989;52:132–147.

Dr. A.F. Dixson, Centre International de Recherches Médicales de Franceville, B.P. 769, Franceville (Gabon)

Martin RD, Dixson AF, Wickings EJ (eds): Paternity in Primates:
Genetic Tests and Theories. Basel, Karger, 1992, pp 203–224

Monogamous Social Systems and Paternity in Primates

Gustl Anzenberger

Zoologisches Institut, Ethologie und Wildforschung, Universität Zürich, Schweiz

In the first study conducted on a monogamous neotropical primate, the titi monkey *Callicebus moloch*, Mason [1] already touched on the topics and problems to be dealt with in the following paper. Mason found that (a) titi monkey groups usually consist of an adult pair and one or more young, (b) there are many indications that the bond between mates is strong and enduring, and (c) the adult male carried the infant at virtually all times. This citation illustrates three levels at which we can deal with monogamy, namely the sociographic, the motivational, and the functional level. Traditionally, behavioral studies rely on all three levels, and commonly provide a basis from which the entire social system or mating system of a given species is inferred [for a detailed discussion of such problems when dealing with the topic of monogamy, see ref. 2]. There was, however, an additional observation reported in Mason's paper, namely that 'animals will occasionally copulate with members of adjacent groups', which brings us to the genetic level of monogamy and vividly illustrates the fact that grouping pattern and behavioral mechanisms do not necessarily reflect the genetic relationships within a social unit. In other words, it is almost impossible to establish genetic relationships in field studies by using behavioral observations alone. Wickler [3] wrote: 'If the term monogamy is to remain biologically meaningful it must imply that one or both partner(s) will produce offspring exclusively with the other – regardless by what behaviour mechanism this exclusiveness is ensured'. From this, it follows that not only the genetic level of monogamy but especially the motivational level is of crucial importance. Accordingly, this paper will first focus on behavioral mechanisms as encountered in monogamous primate species and then turn to their possible genetic consequences.

Taxonomy and General Remarks

There are four taxonomic groups within the order Primates in which monogamy is the common rather the exceptional social grouping pattern. These four groups, following the classification of Martin [4], are the Callitrichidae (marmosets and tamarins), the Callimiconinae (Goeldi's monkey), the Aotinae (owl monkeys and titi monkeys) and the Hylobatinae (gibbons and siamangs). Comparisons of the different taxa conducted here will vary according to the types of evidence available. For gibbons and siamangs, natural studies – including some experimental studies of wild-living groups – have provided the basis of our knowledge. There have been only a few studies on captive hylobatid groups. For callitrichids, the situation is quite the reverse because there is an overwhelming body of data from studies on captive groups, but only few from studies that have been conducted in the wild. The most balanced situation is encountered with the subfamily Aotinae, especially the genus *Callicebus,* where our knowledge rests on a balanced combination of studies conducted in the wild and studies conducted in the laboratory. Without question, *Callicebus* is the best-studied monogamous genus in primates. As representatives of their respective taxa, the following discussion will focus within Callitrichidae on common marmosets *(Callithrix jacchus),* saddle-back tamarins *(Saguinus fuscicollis)* and golden lion tamarins *(Leontopithecus rosalia),* within Aotinae on dusky titi monkeys *(Callicebus moloch),* and within Hylobatinae on several gibbon species (*Hylobates* spp). Although there are several studies dealing with the genera *Callimico* (reviewed in [5]) and *Aotus* (reviewed in ref. 6, 7), there are too few data available for a direct comparison to be made with the other experimental studies reviewed here.

Whereas the monogamous grouping pattern has not been seriously questioned for Aotinae and Hylobatinae, there is an ongoing discussion with respect to the Callitrichidae [8, 9], because recent field studies on saddle-back tamarins indicate that facultative polyandry may occur in some species or populations [10]. Nevertheless, Callitrichidae will be included in this review for a number of reasons. First, tamarins and marmosets should not be treated as a uniform taxonomic group with only one given social system [11]; different grouping patterns may be expected according to species. Second, there is an enormous bulk of data available from studies in captivity which indicate a monogamous lifestyle for at least some callitrichid species [12, 13]. Regardless of how carefully they are conducted, field studies per se have no more validity than studies under captive conditions, and only a combined appreciation can lead to a genuinely synthetic understanding [14–16]. Lastly, determination of genetic relationships through the use of new techniques such as DNA fingerprinting seems more likely to be achieved in the near future with wild-living populations of species in the Callitrichidae [17] than with species in the Aotinae or Hylobatinae.

Behavioral Monogamy

There are three major behavioral factors affecting paternity and its certainty. (1) First, there is the pair bond, a hypothetical construct which is normally expressed by a mutual attraction between pairmates. This attraction is operationalized by a close spatial relationship, by partner-specific behaviors and by signs of distress during separation from the pairmate [see also ref. 3]. (2) Second, there is mate 'fidelity' by females which is normally operationalized by a clear preference for the familiar male partner and by specific behaviors towards unfamiliar males, ranging from indifference to intersexual aggression. (3) Third, there is mate monopolization by males which is normally operationalized by behaviors such as mate guarding and male intrasexual aggression. Within family groups, social and sexual control of mature offspring might be achieved by these same behaviors. Note that the behavioral dispositions attributed to males and females, respectively, might be developed symmetrically in the two sexes. However, as male intersexual aggression, male 'fidelity', female intrasexual aggression and female mate guarding would have no immediate effects on the assurance of paternity, they will not be discussed. Because my main emphasis will be on proximate causes of behaviors, I will make no distinction between a specific function of a given behavior and any functional side effect of that behavior. For example, female 'fidelity' clearly enhances the assurance of paternity for the male partner, but it does so regardless of whatever selection pressures shaped her behavior. In addition, I will not deal with specific morphological or physiological features of the female that could affect the male's certainty of paternity, such as the lack of perineal swelling or other visual signs around the time of ovulation [the issue of 'concealed ovulation'; see ref. 106].

The Pair Bond as Indicated by Field and Captive Studies

In its simplest form, bonding behavior between two individuals under natural conditions is operationalized by spatial proximity over a period of time, i.e. an individually identifiable male and female will be found together for remarkably long periods. Various reports have confirmed this, e.g. for lar gibbon and siamang pairs that have remained stable for at least 8 years [18], and for *Hylobates muelleri*, with stability over 7 years [19]. For titi monkeys *(Callicebus moloch* and *Callicebus torquatus)* numerous pairs have been reported to stay together for periods from at least 11 months (the duration of the study) up to several years [1, 20]. The authors concerned assumed on this basis that titi monkey pairs mate for life. There is no reliable information on the duration or the quality of the pair bond of specific callitrichid pairs under natural conditions.

Apart from this spatial proximity between pairmates over time, there are numerous indications of a mutual attraction from the affiliative, amicable, and often strikingly coordinated pattern that is evident within pairs of monogamous primate species [1, 18, 20, 21]. Mason described for *Callicebus* various signs of distress after pairmates had lost contact with one another [1]. In addition, pairmates of monogamous primates seem to be codominant, a phenomenon first described by Carpenter [22] for *Hylobates lar*.

Finding the same two individuals together over time could be taken as a manifestation of an individualized, emotional bond between pairmates. But the same grouping pattern, following a more parsimonious interpretation, could also result from site attachment combined with intrasexual aggression in both sexes. We therefore have to turn, wherever possible, to experimental procedures that allow one to test male and female motivational inclinations and partner-specific contributions to the observed social grouping. Several experimental designs are available to test this (see below).

The Pair Bond and Duetting

Titi monkeys and lesser apes share a conspicuous form of vocal behavior, so-called duet singing [for functional aspects of duetting in birds and primates, see ref. 23, 24]. Owl monkeys have no loud calls, but pairmates utter 'resonant whoops' in a coordinated fashion [25], while antiphonal long calls of mated individuals have been reported for golden lion tamarins [26]. In *Callicebus* and *Hylobates,* male and female pairmates contribute sex-specific (the only exception being *Hylobates hoolock*) loud calls to the complex and tightly coordinated duet and it is presumed that these vocal performances originated from joint territorial displays. In addition to this apparent territorial function, there are also some indications that social information is conveyed by the duetting pair. In titi monkeys under natural conditions duetting brings pairmates together, keeps them close to each other during territorial encounters, and may hence also function to maintain the social integrity of the mated pair [1, 27, 28]. Similar interpretations have been made for hylobatids [19, 29]. Whether duetting in fact serves social functions such as strengthening and maintaining the pair bond is not easy to demonstrate, because possible social functions are confounded by the territorial character of this behavior. However, for both taxa – titi monkeys and gibbons – experimental playback experiments have elucidated possible social functions of duetting, as will be shown below.

Wickler [30] proposed a mechanism that linked pair-specific duetting to pair-bonding, based on the assumption that the development of a complex duet is a kind of mutual investment between pairmates that prevents mate desertion. There is some supporting evidence for this hypothesis from the literature on duetting in birds [31]. For primates, Robinson suggested that in *C. moloch* a pair-specific duet develops slowly and only after considerable practice [32]. Kinzey found for *C. torquatus* that one pair did not duet at the beginning of this study, and one possible explanation was that the pairmates were newly mated [20]. By contrast, three newly constituted pairs of *C. moloch* at the California Primate Research Center in Davis, Calif., USA, performed full duets within days (personal observations); but there is no information on whether and when those duets became pair-specific. Geissmann [33] reported for siamangs that newly paired or re-paired individuals need some time together before highly coordinated duets are sung. The issue of whether duetting has a direct effect on the pair bond must remain open until more evidence becomes available.

The Pair Bond as Tested by Separation-Reunion Experiments

In separation-reunion experiments, pairmates are separated for a period of time and then reunited. Behaviors shown by pairmates during separation are normally seen as appetitive behaviors directed towards reunion with the familiar partner or, in other words, are interpreted as an expression of a bonding drive. As pairmates normally cease their restless or agitated behavior upon reunion and show affiliative behaviors, it may be concluded that an individualized pair bond contributes to a homeostatic state within individuals and that such proximate causes keep the pair together.

It is clear that such kinds of experiments cannot be conducted under natural conditions, and we must rely on information from studies conducted in captivity. The most elaborate series of experiments has been performed by Mason and co-workers [34–37] on titi monkeys *(C. moloch).* All experiments were done in a comparative framework, in that the same types of experiments were conducted both with the monogamous titi monkey, *C. moloch*, and with the polygamous squirrel monkey, *Saimiri sciureus*. The rationale behind this procedure was that it would distinguish specific lifestyle characteristics typical of a monogamous versus a polygamous cebid species, and separate them from general characteristics deriving from a common cebid heritage. The studies revealed that titi monkeys, in contrast to squirrel monkeys, showed pronounced reactions when separated from the familiar companion [37]. Compared with the undisturbed situation, locomotion, vocalization and heart rate went up in both sexes. The increase in heart rate was not attributable to the higher locomotor activity and may hence be interpreted as an indication of physiological processes when coping with stress due to separation. Although this is the only study to date that demonstrates an emotional correlate of the pair bond, this correlation provides a possible causal mechanism for keeping pairmates together: separation leads to a deviation from physiological homeostasis, and only reunion with the familiar partner guarantees physiological well-being. This interpretation is supported by the titi monkeys' differential reaction to the following three experimental conditions: (a) subject alone, (b) subject together with an unfamiliar, opposite-sexed conspecific, and (c) subject reunited with the familiar pairmate. It might be expected that, for such a highly socially disposed species as the titi monkey, being in the company of a social substitute would be better than being alone. However, the opposite was found to be true. Subjects reacted to the presence of an unfamiliar opposite-sexed conspecific with a further increase in locomotion, vocalization and heart rate, compared to when they were alone. This underscores the power of the hypothesized causal mechanism for bonding.

Female 'Fidelity' as Tested by Intruder and Choice Experiments

The social integrity of a pair will be threatened by the appearance of a third adult conspecific, because it represents for one pairmate a social rival and for the other a social alternative [38]. This situation can be simulated experimentally by introducing an 'intruder' to a mated pair. Let us first consider that part of the intruder-experiment relevant to female 'fidelity', i.e. how females react towards unfamiliar males in the presence or absence of their own mates. These reactions allow the experimenter to assess the females' contributions to the strength and specificity of the pair bond.

When given a choice, female titi monkeys showed in paired comparisons a preference for their familiar pairmate over an unfamiliar male, by staying significantly closer to the screen separating them from their mates than to the screen separating them from strange males [35, 39]. In a situation in which unrestricted interactions were possible [for diagrams of the experimental set-ups, see ref. 39], females performed both huddling and tail-twining with their mates. By contrast, when unrestricted interactions with unfamiliar males were possible, proximity in such unfamiliar pairs was most frequently the result of the male following the female and never occurred in a relaxed manner. However, no intersexual aggression was shown by female titi monkeys towards unfamiliar males. Moreover, when their pairmates were absent during encounters, some sexual interactions between female titi monkeys and unfamiliar males were observed [39].

Results from studies on members of the family Callitrichidae revealed a heterogeneous array of reactions from females when given a choice between the pairmate and an unfamiliar male (in all three studies to be reviewed here [41–43], diagrams of the respective experimental set-ups are provided). In common marmosets *(C. jacchus),* females not only preferred their mates over unfamiliar males in terms of social distance [38, 40], but also physically attacked unfamiliar males. There were no amicable, soliciting or sexual behaviors, not even in the absence of their mates [41]. In saddle-back tamarins *(Saguinus fuscicollis),* females approached strange males almost as often as they approached their mates. Moreover, females tended to direct more tongue displays and to gaze significantly more at strange males than at their pairmates; both behaviors are clearly sexually motivated [42]. In golden lion tamarins *(L. rosalia),* females showed a social preference for their own mates, and they seemed to be more exclusive in this regard than males of this species. However, females did not actively attack or avoid unfamiliar males, and they tended to show increased interest in unfamiliar males as soon as their mates were absent.

Nevertheless, long calls given by females increased substantially at the same time, which can be interpreted as attempts to establish contact with their absent mates [43].

For several gibbon species, intrusions of unfamiliar conspecifics have been simulated with playback experiments. Playback of sex-specific calls or duets of unfamiliar animals to resident, territorial pairs provides a first indication of inclinations of each sex. In *H. muelleri*, females initiated duets only in response to playbacks of female solos or duets, but they followed their mates when they silently approached playbacks of male solo songs [44]. In *H. lar*, reactions of females to playbacks were similar [45]. They initiated vocalizations when female solos were played back, but they only occasionally followed their mates when the latter moved towards playbacks of male solos or duets. In *Hylobates agilis*, females never reacted to male solos, either by approaching or vocalizing; but they showed some reaction to female solos and duets [46]. If we accept these playback experiments as simulations of a real territorial and social intrusion, then the females of all tested gibbon species showed little if any attraction to unfamiliar males. However, they joined their partners vocally and physically in territorial disputes, and there are some reports of intersexual aggression in newly paired or widowed females of different gibbon species [19].

Female Extra-Pair Copulations and Deviations from Monogamy

There remains the puzzling result that female titi monkeys showed less exclusivity with regard to sexual behavior than females of the other monogamous species reviewed. Even though the captive female titis were determined – by birth records (4 animals) and by the lack of observed sexual behavior with the established pairmate (3 animals) – not to be in the ovulatory phase during the experiments [39], such that sexual interactions would have had no effect on the security of paternity of their mates, there is the report by Mason [1] that wild *Callicebus* females crossed borders 'especially during the seasonal peak of sexual activity' and occasionally copulated with male neighbors. In comparison with other studies on *Callicebus,* Mason found a very high population density in his study area [20, 25]. Although he spent nearly a year in the field, it is possible that the (unknown) previous history of his study objects may account for this peculiar behavior of extra-pair copulations. Ågren [47] described a striking parallel for another monogamous mammal. Under seminatural conditions, female gerbils *(Meriones unguiculatus)* trespassed into adjacent territories while in estrus and copulated there with unfamiliar males. This behavior is interpreted as an alternative female mating strategy to monogamy which occurs under high population densities, mainly in connection with inbreeding avoidance.

There is one report of a polygynous *Hylobates pileatus* group, and the authors presume that it arose from a family [48]. It is noteworthy that this deviation from the usual monogamous lifestyle also occurred in a population with a very high density. It is quite likely that the daughter did not disperse but stayed within her family unit. It could not be determined whether her offspring was sired by her father or by a neighboring male.

Mate Guarding by Males and Male Intrasexual Aggression

From an evolutionary point of view, it is clear that in species where males invest considerable amounts of time and energy indirectly or directly in their offspring [49], they should do their utmost to prevent themselves from being cuckolded [50, 51]. Behavioral mechanisms fulfilling this demand are mate guarding and male intrasexual aggression. These two terms are not completely exclusive of one another, but we will consider mate guarding as mainly mate-oriented behaviors and intrasexual aggression as clearly rival-oriented behaviors. [NB, with respect to *Homo sapiens*, seen as a facultatively monogamous primate, literature on mate guarding, mate retention, paternal confidence and paternal investment is provided in ref. 52–54.]

Mate guarding normally refers to behaviors such as consorting and/or herding imposed by the male during the estrus or the ovulatory period of the female in mammals or around the egg-laying period in birds [for a review of the abundant literature on birds, see ref. 55, 56]. If we take into account the fact that the fertile period of any female primate probably lasts only 3–4 days, mate guarding by males for this period twice every year in callitrichids, or only once per year in *Callicebus,* or once every second or third year in gibbons and siamangs, would ensure paternity. Although paternity must be protected during the fertile days of the female, there is another aspect to consider which, theoretically speaking, is simply a prolongation of mate guarding beyond this period. Assurance of paternity in its broadest sense includes protection of the sired offspring for a certain period of their early embryonic life. Records from our colony of common marmosets show that females (n = 5) that were pregnant from a previous mate obviously terminated pregnancy when a new male was introduced prior to the 8th week of pregnancy. Unfortunately, we have no endocrine records for these cases, but the early termination of pregnancy could be conclusively inferred from the next delivery date, which was in all cases almost exactly 150 days after pairing with the new male; according to Hearn [57], the gestation period for common marmosets is 144 ± 2 days, with a range from 141 to 146 days. These observations are reminiscent of the Bruce effect described for various rodents [58; for the effects of new males on females in several Old World monkey species, see ref. 59, 60].

It is noteworthy that, in a recent paper, van Schaick and Dunbar [61] proposed a new hypothesis regarding the evolution of hylobatid monogamy. They suggest that the possible biological advantage of monogamy to males may be that males are thus able to prevent infanticide of their offspring by other males. Under this assumption, the assurance of paternity is extended even further, relative to the time of conception.

Mate Monopolization by Males in Natural and Captive Studies

Descriptions of mate-guarding behaviors of males of free-living monogamous primate species are lacking, and direct observations of copulatory behavior are scarce. Mason [1] reported that wild male *C. moloch* clearly tried to prevent extra-pair copulations of their mates. During territorial encounters and while duetting, male titi monkeys often press against their mates and even hold them. Subsequent experimental studies confirmed this mate-monopolizing behavior of males. In tests where mated *Callicebus* males could observe their mates experimentally brought gradually closer to unfamiliar males, test males showed pronounced signs of behavioral agitation ('jealousy') [62]. In another experiment in which unfamiliar pairs were confronted with each other, males placed themselves between the mate and the intruders and showed physical restraint and mounting of their mates. In addition, males showed greater cardiac activity during confrontations than females [63]. Despite the fact that male titi monkeys show signs of agitation as soon as the social and territorial integrity of the pair is jeopardized, there is rarely overt physical aggression, and most spacing seems to be accomplished by the ritualized duet singing of the pair [1, 27, 28]. However, in an experimental study where males could watch through a Plexiglas screen as their mates interacted directly with an unfamiliar intruder, the male opponents showed clear fight intentions through the screen [39].

There are no reports on mate guarding or male intrasexual aggression in free-living Callitrichidae. Although there is detailed information available on wild saddle-back tamarins [64], there have been no reports of these two behaviors, probably because this species is facultatively polyandrous.

By definition, in polyandry 'two or more males must mate with one female, with neither male monopolizing the female around the time that she ovulates' [65]. By contrast, there are some reports from captivity of mate guarding by common marmosets, saddle-back tamarins and golden lion tamarins [12, 66]. For some callitrichid species it has been reported that males initiate more affiliative behaviors than females do [67–69]. This is commonly interpreted as a contribution to the pair bond, but can be seen at the same time as mate surveillance. In common marmosets, males seem to assure their paternity in two stages [pers. obs.]. First, males show pronounced mate guarding, as expressed by extremely close following of the female and by huddling. The extent of their guarding behavior seems unnecessary under captive conditions, but there are clear indications of elder sons showing an interest in their mothers around the time of ovulation. Secondly, males achieve a kind of sperm preponderance, because during the ovulatory phase of the female the first copulations on a given day almost always occur within 15 min after the pair (and other family members) have left the sleeping site.

With regard to male intrasexual aggression in callitrichid species, we must once again turn to intruder experiments. Male common marmosets, golden lion tamarins, cotton-top tamarins and saddle-back tamarins all showed overt intrasexual aggression or agonistic behaviors during these tests, but they differed in the following ways: when males of common marmosets *(C. jacchus)* were able to observe unrestricted interactions between their mates and strange males through wire mesh, they clearly tried to attack the rival. They spent more than one third of total observations clinging to the dividing screen, directing a high frequency of threat vocalizations at the intruder [38, 40]. In studies in which pairs of common marmosets [70, 71] as well as cotton-top tamarins *(Saguinus oedipus)* [71] were confronted with single unfamiliar intruders – separated by perspex screens [70] or wire mesh [71], respectively – males of both species showed high frequencies of intrasexual aggression. In addition, marmoset males also addressed affiliative behaviors towards their mates, which may be seen as a kind of simultaneous mate guarding, during encounters with unfamiliar males [71]. In saddle-back tamarins *(S. fuscicollis)* there was a less pronounced manifestation of intrasexual aggression in males when, together with their mates, they encountered an unfamiliar male; the intruder was again separated by wire mesh [12]. The most limited reaction to same-sexed intruders was found in male golden lion tamarins *(L. rosalia)*. When resident mated pairs were presented with a male intruder (separated by wire mesh), the resident males showed virtually no attack responses [72]. The findings for the latter two species are in agreement with the fact that, in both species, occasionally relatively stable trios of two males and one female have been formed [12, 66], indicating a higher compatibility among males than among females. The facultative polyandrous lifestyle found in some *S. fuscicollis* populations [64] is in accord with these experimental findings, but for *Leontopithecus* a statisfying explanation is lacking.

For practically all gibbon and siamang species there are numerous reports of male intrasexual aggression, which mainly has a territorial function but at the same time serves to protect the social integrity of the pair or group [19, 29, 61]. Again, playback experiments allow one to simulate territorial and social intrusions and reveal a great deal about the behavioral mechanisms ensuring monogamy in the lesser apes. There were clear reactions of males of *H. agilis* and *H. lar* to playbacks of male solos and duets, as expressed by approaching the source of vocalization as well as by vocal responses [45, 46]. In *H. muelleri*, males also approached sites of male song playbacks, but did so silently [44].

Mate Monopolization by Males within Family Groups

As time passes, a family group develops around the initial monogamous pair [13]. Because behavioral mechanisms effective within family groups seem less spectacular at first sight than overt male intrasexual aggression, they are sometimes less acknowledged. Nevertheless, within family groups there are two mechanisms at work which lead to mate monopolization [13, 73]. First, there is the possibility that maturing offspring are expelled from the family, or at least peripheralized, which results in the maintenance of one reproductively active pair per group. The second mechanism is the sexual suppression or inhibition of progeny as long as they stay within the family unit.

By and large, groups of Hylobatinae and Aotinae seem to be typified by the first mechanism, whereas Callitrichidae seem to be typified by the second. This could be inferred from group size and birth intervals alone, but in addition there exist detailed descriptions of gibbon and siamang males becoming aggressive towards their sons, especially during periods when parents were sexually active [18]. We encounter a parallel situation in the female sex, which is not at all surprising for monogamous species [13]. Same-sexed offspring in fact become social and sexual competitors on reaching maturity, as is shown by reports on incestuous mating and pair formation in hylobatids, especially after one parent had died or disappeared [18, 19]. Geissmann [74] presented data for different *Hylobates* species indicating that, under captive conditions, the onset of sexual maturity may be earlier than previously assumed. Although there were two other plausible explanations for these findings, it seems possible that the second mechanism of sexual suppression or inhibition by parents is also effective within family groups of gibbons. It is worth mentioning in this connection that infants within hylobatid families often join their duetting parents in vocalization. However, around the time of maturity, male offspring cease to sing with their parents whereas subadult females continue calling [19].

Mason [1] gave some indication that young adult *Callicebus* were occasionally targets of aggression from their parents, but other authors report that in both *Aotus* and *Callicebus* males left their natal group without agonistic interactions prior to their departure [25].

For Callitrichidae, there is an abundant literature dealing with sexual suppression or inhibition of subdominant individuals by dominants. These investigations have been conducted in captivity with 'naturally' developed family groups as well as with artificially constituted groups [75, 76]. It was initially assumed for subordinate females that there is total sexual inhibition both at the behavioral level and at the endocrinological level, given that these females do not ovulate. More recent papers have led to a more differentiated picture of the situation with respect to subdominant female callitrichids [77–83]. As far as the males are concerned, it was assumed that males are inhibited only at the behavioral level, because the onset and maintenance of spermatogenesis is unaffected [84, 85]. However, Abbott [75] found for subdominant males in artificial groups a tendency to have lower plasma testosterone levels than dominant males. Recently, we found in 'naturally' developed family groups in our colony that mature sons had significantly lower urinary testosterone levels than their fathers [Anzenberger et al., in preparation]. Although testosterone concentration is only part of the complicated endocrine feedback mechanism involved in controlling sexual behavior, there is a clear correlation between social status and sexual activity of males within families or artificial groups of callitrichids in all species studied [75, 86; see also 87].

Summary of Female and Male Motivational Inclinations

In almost all of the New World monkey species reviewed here, females show a clear predilection towards monogamy, and there are ample indications of a long and enduring pair bond. When the results of choice experiments are compared for the two sexes, females seem to contribute more to the pair bond than males do, because they clearly prefer their mates over male social alternatives, stay closer to them, are more agitated when separated from them, and even occasionally show aggression towards unfamiliar males. Nevertheless, there are indications that all of these reactions are less pronounced when the mate is absent. That is, the intrinsic female contribution to the pair bond is somehow reinforced by the presence of her mate. Of all females tested, *S. fuscicollis* showed the least exclusivity with regard to the established pair bond, i.e. females were obviously interested in unfamiliar males and even showed soliciting behaviors. This experimental result, in combination with another report indicating a higher degree of compatibility between males than between females [12, 88], provides confirmation from captivity for postulated facultative polyandry in free-living *S. fuscicollis*. For gibbon and siamang females, field studies have provided abundant evidence for a long and enduring exclusive pair bond and for joint territorial (and social) defense, including weak components of female intersexual aggression.

In males of almost all the monogamous species reviewed, high levels of intrasexual aggression have been observed under free-living conditions as well as in captivity. Monogamously mated males, with the exception of male *Leontopithecus*, are clearly more agitated by the appearance of a same-sexed conspecific than their female partners are in the reversed situation. The male's behavior can be seen as part of a male strategy aimed at assuring paternity by preventing any intrusions or immigrations into the social group.

In general, we find a behavioral syndrome among monogamous primate species which is typified by a strong female contribution to the pair bond and by pronounced mate monopolization by males, resulting from high male intrasexual aggression. One might expect that this convergence of female and male behaviors would almost automatically assure paternity within monogamously living species.

Genetic Monogamy

One of the first studies indicating that we have to differentiate between the demographic labelling of a given social system and its genetic consequences was conducted on red-winged blackbirds *(Agelaius phoeni-*

ceus). In this study, male territory-holders were vasectomized, but the females nesting within the territories of those males nevertheless produced fertile clutches [89]. Recently, for the same species, it has been shown by means of DNA fingerprinting that males realize an average of 21% of their reproductive success in nests outside their own territories [90]. An increasing body of data from studies on different bird species applying genetic techniques has revealed that extra-pair fertilizations of females can occur [91–94, 104]. That is to say, a monogamous social system is not necessarily equivalent to mating exclusivity between the two partners of a pair. The tremendous growth in data from ornithological studies in recent years is due to several advantages of birds for this kind of study. First, there are very effective procedures for capturing them. Second, immature offspring are assembled in a nest. Third, only a few drops of blood are needed for obtaining DNA, because in birds (in contrast to mammals) red blood cells contain nuclei. It is clear that none of these advantages holds for primates, and an application of the newly available genetic techniques, which require trapping of specimens, will not be easy to achieve [for logistical problems encountered when dealing with primates, see also ref. 95].

Determination of Paternity with New Techniques

There are no published reports of any attempts to determine kinship in social groups of Hylobatinae or Aotinae using modern genetic techniques based on analysis of DNA. Therefore, we will restrict ourselves to the situation within the family Callitrichidae. The discovery of hypervariable minisatellite sequences in the human genome [96, 97] was the starting point for the development of the DNA fingerprinting technique. In the meantime, for primates, this technique has been used to perform individual identification and to study familial relationships in humans [96, 97], in a few Old World monkey species [98, 99] and in the common marmoset [100]. So far, there has been only one study determining genetic relationships within two groups of free-living common marmosets [17].

Normally, marmosets and tamarins give birth to dizygotic twins and the placental discs of the twins are connected by vascular anastomoses leading to hemopoietic and bone marrow cell chimerism. This enigmatic fact has long been known for marmosets [101, 102] and it creates in uterine siblings practically identical DNA fingerprinting patterns when produced from blood samples [100; Signer, Anzenberger et al., in preparation].

Accordingly, DNA fingerprints of free-ranging common marmoset groups have been particularly useful for identification of sets of twins. On the other hand, the phenomenon of hemopoietic and bone marrow chimerism poses a problem which Dixson et al. [17] have addressed: 'It would be valuable to attempt DNA-fingrprinting on tissues other than white blood cells (e.g. skin biopsies) to avoid complications due to hemopoietic chimaerism.' In other words, from a given individual, one has to expect that DNA fingerprints obtained via blood samples versus those obtained via tissue samples will be different.

Over the last few years, we have taken blood samples as well as samples from different body tissues (including gonadic tissues) from any animal that died in our colony, e.g. animals that had to be euthanized due to sickness or old age and still-born fetuses resulting from cesarian sections. Preliminary results have revealed three important facts [Anzenberger, Signer et al., in preparation]: first, there is no single specimen (n = 8; 4 males, 4 females) in which the DNA fingerprint patterns obtained by blood samples are congruent with those obtained by tissues. Second, tissue samples from different organs of the same specimen, gonadic tissues included, lead to identical DNA fingerprint patterns (n = 8; 4 males, 4 females). Third, same-sexed twins (n = 4; 2 males, 2 females) show identical DNA fingerprints when using their blood samples, but show different ones when using other tissues, regardless of whether organ or gonadic tissues are used. In other words, only tissue samples reveal the true genotype of individuals. This fact is underlined by data from another study [Signer, Anzenberger et al., in preparation], in which uterine siblings of sets of twins or triplets had the same blood DNA fingerprint, regardless of whether the siblings were of the same or opposite sex (n = 15; 6 twin sets and one triplet from 3 different captive families).

This means that, within the family Callitrichidae, it is not sufficient to do kinship/paternity determination by the use of blood samples alone, because twins (and triplets) showing identical banding patterns in their blood DNA might contribute different banding patterns to the next generation via germ cells. Thus, considering the fact that the facultative polyandry described for *S. fuscicollis* might sometimes be a fraternal system [103], i.e. brothers remain together and pair up with the same female, determination of paternity cannot be performed reliably with blood samples [see ref. 11 for formation of a new group in a callitrichid species; for an example in birds, see ref. 105].

Even though the new technique of DNA fingerprinting, when using only blood samples, does not solve all problems in determining paternity in marmosets and tamarins, it is a powerful instrument in addition to conventional analyses of genetic variability and observational studies. In the only study so far conducted on free-living common marmosets [17], it was possible to identify one set of surviving triplets. Without the aid of this new technique, and relying only on behavioral observations, this fact certainly would have been taken as the first indication of two females simultaneously reproducing within the same social group.

Some people might be tempted to say that as long as we cannot get genetic information we cannot say anything about social systems and especially about mating systems. On the other hand, the evidence collected from field studies, behavioral investigations in captivity and experimental observations is merely different, not weaker. If we learn by genetic determination that mate guarding of males in a given species is not as effective as assumed or proposed, then we have learned more about animal behavior than banding patterns of DNA fingerprints alone would have told us. However, it is to be hoped that behavioral observations combined with genetic determination of kinship will be achieved in some monogamous primate species in the near future, thus shedding new light on our understanding of one of the most challenging social systems within mammals.

Acknowledgments

Dr. C.R. Menzel and Prof. R.D. Martin read and improved the manuscript substantially. The author's research on common marmosets was carried out at the Biomathematical Section of the Department of Psychology, University of Zürich.

References

1 Mason WA: Social organization of South American monkey, *Callicebus moloch:* A preliminary report. Tulane Stud Zool 1966;13:23–28.
2 Wickler W, Seibt U: Monogamy: An ambiguous concept; in Bateson P (ed): Mate Choice. Cambridge, Cambridge University Press, 1983, pp 33–50.
3 Wickler W: The ethological analysis of attachment. Z Tierpsychol 1976;42:12–28.
4 Martin RD: Primate Origins and Evolution. London, Chapman & Hall, 1990.
5 Heltne PG, Wojcik JF, Pook AG: Goeldi's monkey, Genus *Callimico;* in Coimbra-Filho AF, Mittermeier RA (eds): Ecology and Behavior of Neotropical Primates. Rio de Janeiro, Academia Brasileira de Ciencias, 1981, vol 1, pp 169–210.
6 Wright PC: The night monkeys, genus *Aotus;* in Coimbra-Filho AF, Mittermeier RA (eds): Ecology and Behavior of Neotropical Primates. Rio de Janeiro, Academia Brasileira de Ciencias, 1981, vol 1, pp 211–240.
7 Dixson AF: The owl monkey *(Aotus trivirgatus);* in Hearn JP (ed): Reproduction in New World Primates. Lancaster, International Medical Publishers, 1983, pp 181–215.

8 Sussman RW, Garber PA: A new interpretation of the social organization and mating system of the Callitrichidae. Int J Primatol 1987;8:73–92.

9 Sussman RW, Kinzey WG: The ecological role of the Callitrichidae: A review. Am J Phys Anthropol 1984;64:419–449.

10 Goldizen AW: Tamarin and marmoset mating systems: Unusual flexibility. Trends Evol Ecol 1988;3:36–40.

11 Ferrari SF, Lopes Ferrari MA: A re-evaluation of the social organisation of the Callitrichidae, with reference to the ecological differences between genera. Folia Primatol 1989;52:132–147.

12 Epple G: The behavior of marmoset monkeys (Callithricidae); in Rosenblum LA (ed): Primate Behavior. New York, Academic Press, 1975, pp 195–239.

13 Kleiman DG: Monogamy in mammals. Q Rev Biol 1977;52:39–69.

14 Kummer H, Dasser V, Hoyningen-Huene P: Exploring primate social cognition: some critical remarks. Behaviour 1990;112:84–98.

15 Rosenblum LA, Kummer H, Nadler RD, et al: Interface of field and laboratory-based research in primatology. Am J Primatol 1989;18:61–64.

16 Epple G, Küderling I, French JA, Tardif S: Social and sexual strategies in callitrichid monkeys. Primate Rep 1987;14:73.

17 Dixson AF, Anzenberger G, Monteiro Da Cruz MAO, et al: DNA fingerprinting of free-ranging groups of common marmosets *(Callithrix jacchus jacchus)* in NE Brazil; in Martin RD, Dixson AF, Wickings EJ (eds): Paternity in Primates: Genetic Tests and Theories. Basel, Karger, 1992, pp 192–202.

18 Chivers DJ, Raemaekers JJ: Long-term changes in behaviour; in Chivers DJ (ed): Malayan Forest Primates. New York, Plenum Press 1980, pp 209–260.

19 Leighton RL: Gibbons: Territorality and monogamy; in Smuts BB, Cheney DL, Seyfarth RM, et al (eds): Primate Societies, Chicago University of Chicago Press, 1986, pp 135–145.

20 Kinzey WG: The titi monkeys, genus *Callicebus;* in Coimbra-Filho AF, Mittermeier RA (eds): Ecology and Behavior of Neotropical Primates. Rio de Janeiro, Academia Brasileira de Ciencias, 1981, vol 1, pp 241–276.

21 Menzel CR: An experimental study of territory maintenance in captive titi monkeys *(Callicebus moloch);* in Else JG, Lee PC (eds): Primate Ecology and Conservation. Cambridge, Cambridge University Press, 1986, pp 133–143.

22 Carpenter CR: A field study in Siam of the behavior and social relations of the gibbon. Comp Psychol Monogr 1940:16(5).

23 Farabaugh SM: The ecological and social significance of duetting; in Kroodsma DE, Miller EH (eds): Acoustic Communication in Birds, 1982, vol 2.

24 Serpell JA: Duetting in birds and primates. Anim Behav 1981;29:963–965.

25 Robinson JG, Wright PC, Kinzey WG: Monogamous cebids and their relatives: Intergroup calls and spacing; in Smuts BB, Cheney DL, Seyfarth RM, et al (eds): Primate Societies, Chicago University of Chicago Press, 1986, pp 44–53.

26 McLanahan, Green KM: The vocal repertoire and an analysis of the contexts of vocalization in *Leontopithecus rosalia*; in Kleiman DR (ed): The Biology and Conservation of the Callitrichidae, Washington DC, Smithsonian Institution Press, 1977, pp 251–269.

27 Robinson JG: Vocal regulation of use of space by groups of titi monkeys *Callicebus moloch.* Behav Ecol Sociobiol 1979;5:1–15.

28 Robinson JG: Vocal regulation in inter- and intragroup spacing during boundary encounters in the titi monkey, *Callicebus moloch*. Primates 1981;22:161–172.
29 Brockelman WY, Srikosamatara S.: Maintenance and evolution of social structure in gibbons; in Preuschoft H, Chivers DJ, Brockelman WY, et al (eds:) The Lesser Apes. Edinburgh, University Press, 1984, pp 298–323.
30 Wickler W: Vocal dueting and the pair bond. Z Tierpsychol 1980;52:201–209.
31 Wickler W, Seibt U: Dueting and the pair bond. Anim Behav 1982;30:943–944.
32 Robinson JG: An analysis of the organization of vocal communication in the titi monkey *Callicebus moloch*. Z Tierpsychol 1979;49:381–405.
33 Geissmann T: Funktion der gesanglichen Lautäusserungen des Siamangs, *Hylobates syndactylus*, mit besonderer Berücksichtigung der Paarbindungs-Hypothese. Zool Gart NF, in press.
34 Mason WA: Comparative studies of social behavior in *Callicebus* and *Saimiri:* Behavior of male-female pairs. Folia Primatol 1974;22:1–8.
35 Mason WA: Comparative studies of social behavior in *Callicebus* and *Saimiri;* Strength and specificity of attraction between male-female cagemates. Folia Primatol 1975;23:113–123.
36 Phillips M, Mason WA: Comparative studies of social behavior in *Callicebus* and *Saimiri:* Social looking in male-female pairs. Bull Psychonomic Soc 1975;7:55–56.
37 Cubicciotti DD, Mason WA: Comparative studies of social behavior in *Callicebus* and *Saimiri:* Male-female emotional attachments. Behav Biol 1975;16:185–197.
38 Anzenberger G: Social conflict in two monogamous New World primates: pairs and rivals; in Mason WA, Mendoza SP (eds): Primate Social Conflict. New York, SUNY Press, in press.
39 Anzenberger G: The pair bond in the titi monkey *(Callicebus moloch):* Intrinsic versus extrinsic contributions of the pairmates. Folia Primatol 1989;50:188–203.
40 Anzenberger G: Bindungsmechanismen in Familiengruppen von Weissbüscheläffchen *(Callithrix jacchus)*. Zürich, Juris Druck und Verlag, 1983.
41 Anzenberger G: How stranger encounters of Common Marmoset *(Callithrix jacchus jacchus)* are influenced by family members: The quality of behavior. Folia Primatol 1985;45:204–224.
42 Epple G: Sex differences in partner preference in mated pairs of saddle-back tamarins *(Saguinus fuscicollis)*. Behav Ecol Sociobiol 1990;27:455–459.
43 Inglett BJ, French JA, Dethlefs TM: Patterns of social preference across different social contexts in golden lion tamarins *(Leontopithecus rosalia)*. J Comp Psychol 1990;104:131–139.
44 Mitani JC: The behavioral regulation of monogamy in gibbons *(Hylobates muelleri)*. Behav Ecol Sociobiol 1984;15:225–229.
45 Raemaekers JJ; Raemaekers PM: Field playback of loud calls to gibbons *(Hylobates lar):* territorial, sex-specific and species-specific responses. Anim Behav 1985;33: 481–493.
46 Mitani JC: Territoriality and monogamy among agile gibbons *(Hylobates agilis)*. Behav Ecol Sociobiol 1987;20:265–269.
47 Ågren G (1985): Alternative mating strategies in the mongolian gerbil. Behaviour 1984;91:229–244.
48 Srikosamatara S, Brockelman WY: Polygyny in a group of pileated gibbons via a familial route. Int J Primatol 1987;8:389–393.

49 Mendoza SP, Mason WA: Parental division of labour and differentiation of attachments in a monogamous primate *(Callicebus moloch)*. Anim Behav 1986;34:1336–1347.

50 Trivers R: Parental investment and sexual selection; in Campbell D (ed): Sexual Selection and the Descent of Man. Chicago, Aldine, 1972, pp 136–179.

51 Trivers RL: Social Evolution. Menlo Park, Benjamin/Cummings, 1985.

52 Flinn MV: Mate guarding in a Carribean village. Ethol Sociobiol 1988;9:1–28.

53 Buss DM: From vigilance to violence. Ethol Sociobiol 1988;9:291–317.

54 Gaulin SJ, Schlegel A: Paternal confidence and paternal investment: A cross cultural test of a sociobiological hypothesis. Ethol Sociobiol 1980;1:301–309.

55 Birkhead TR: Sperm competition in birds. Trends Ecol Evol 1987;2:268–272.

56 Lamprecht J: Mate guarding in geese: Awaiting female receptivity, protection of paternity or support female feeding?; in Rasa AE, Vogel C, Voland E (eds): The Sociobiology of Sexual and Reproductive Strategies. London, Chapman & Hall, 1989, pp. 48–60.

57 Hearn JP: The common marmoset *(Callithrix jacchus);* in Hearn JP (ed): Reproduction in New World Primates. Lancaster, MTP Press, 1983, pp 181–215.

58 Bruce HM: A block to pregnancy in the house mouse caused by the proximity of strange males. J Reprod Fertil 1960;1:96–103.

59 Agoramoorthy G, Mohnot SM, Sommer V, Srivastava: Abortions in free ranging Hanuman langurs *(Presbytis entellus)* – A male induced strategy? Hum Evol 1988;3: 297–308.

60 Colmenares F, Gomendio M: Changes in female reproductive condition following male take-overs in a colony of Hamadryas and hybrid baboons. Folia Primatol 1988; 50:157–174.

61 Van Schaick CP, Dunbar RIM: The evolution of monogamy in large primates: A new hypothesis and some crucial tests. Behaviour 1990;115:30–62.

62 Cubicciotti DD; Mason WA: Comparative studies of social behavior in *Callicebus* and *Saimiri:* Heterosexual jealousy behavior. Behav Ecol Sociobiol 1978;3:311–331.

63 Anzenberger G, Mendoza SP, Mason WA: Comparative studies of social behavior in *Callicebus* and *Saimiri:* Behavioral and physiological responses of established pairs to unfamiliar pairs. Am J Primatol 1986;11:37–51.

64 Wilson Goldizen A: Facultative polyandry and the role of infant-carrying in wild saddle-back tamarins *(Saguinus fuscicollis)*. Behav Ecol Sociobiol 1987;20:99–109.

65 Wilson Goldizen A: Tamarins and marmosets: Communal care of offspring; in Smuts BB, Cheney DL, Seyfarth RM, et al (eds): Primate Societies. Chicago, University of Chicago Press, 1986, pp 34–43.

66 Kleiman DG: Characteristics of reproduction and sociosexual interactions in pairs of lion tamarins *(Leontopithecus rosalia rosalia)* during the reproductive cycle; in Kleiman DR (ed): The Biology and Conservation of the Callitrichidae, Washington, Smithsonian Institution Press, 1977, pp 181–190.

67 Box H, Morris JM: Behavioural observations on captive pairs of wild caught tamarins *(Saguinus mystax)*. Primates 1980;21:53–65.

68 Evans S, Poole TB: Pair bond formation and breeding success in the common marmoset *Callithrix jacchus*. Int J Primatol 1983;4:83–97.

69 Ruiz JC: Comparison of affiliative behaviors between old and recently established pairs of golden lion tamarin, *Leontopithecus rosalia.* Primates 1990;31:197–204.
70 Evans S: The pair-bond of the common marmoset, *Callithrix jacchus jacchus:* An experimental investigation. Anim Behav 1983;31:651–658.
71 Harrison ML, Tardif SD: Species differences in response to conspecific intruders in *Callithrix jacchus* and *Saguinus oedipus.* Int J Primatol 1989;10:343–362.
72 French JA, Inglett BJ: Female-female aggression and male indifference in response to unfamiliar intruders in lion tamarins. Anim Behav 1989;37:487–497.
73 Bischof N: Comparative ethology of incest avoidance; in Fox R (ed): Biosocial Anthropology, ASA Studies. London, Malaby Press, 1975, pp 37–67.
74 Geissmann T: Reassessment of age of sexual maturity in gibbons (*Hylobates* spp). Am J Primatol 1991;23:11–22.
75 Abbott DH: Behavioral and physiological suppression of fertility in subordinate marmoset monkeys. Am J Primatol 1984;6:169–186.
76 Abbott DH, McNeilly AS, Lunn SF, et al: Inhibition of ovarian function in subordinate female marmoset monkeys *Callithrix jacchus jacchus.* J Reprod Fertil 1981; 63:335–345.
77 Evans S, Hodges JK: Reproductive status of adult daughters in family groups of common marmosets *(Callithrix jacchus jacchus).* Folia Primatol 1984;42:127–133.
78 Tardif SD: Social influences and sexual maturation of female *Saguinus oedipus oedipus.* Am J Primatol 1984;6:199–210.
79 French JA, Inglett BJ, Dethlefs TM: The reproductive status of nonbreeding group members in captive golden lion tamarin social groups. Am J Primatol 1989;18: 73–86.
80 French JA, Abbott DH, Snowdon CT: The effect of social environment on oestrogen excretion, scent marking, and sociosexual behavior in tamarins *Saguinus oedipus.* Am J Primatol 1984;6:155–167.
81 Ziegler TE, Savage A, Scheffler G, Snowdon CT: The endocrinology of puberty and reproductive functioning in female cotton-top tamarins *(Saguinus oedipus).* Biol Reprod 1987;37:618–627.
82 Abbott DH, George LM, Barrett J, et al: Social control of ovulation in marmoset monkeys: A neuroendocrine basis for the study of infertility; in Ziegler TE, Bercovitch FB (eds): Socioendocrinology of Primate Reproduction. New York, Wiley-Liss, 1990, pp 135–158.
83 Ziegler TE, Snowdon CT, Uno H: Social interactions and determinants of ovulation in tamarins *(Saguinus);* in Ziegler TE, Bercovitch FB (eds): Socioendocrinology of Primate Reproduction. New York, Wiley-Liss, 1990, pp 113–134.
84 Dixson AF: Plasma testosterone concentrations during postnatal development in the male common marmoset. Folia Primatol 1986;47:166–170.
85 Abbott DH, Hearn JP: Physical, hormonal and behavioural aspects of sexual development in the marmoset monkey, *Callithrix jacchus.* J Reprod Fertil 1978;53:155–166.
86 Epple G, Küderling I, Belcher AM, et al: Estimation of immunoreactive testicular androgen metabolites in the urine of saddle-back tamarins. Am J Primatol 1991;23: 87–98.
87 Fuchs E, Rosenbusch J, Anzenberger G: Urinary protein pattern reflects social rank in male common marmosets *(Callithrix jacchus).* Folia Primatol, in press.

88 Epple G: Effect of pair-bonding with adults on the ontogenetic manifestation of aggressive behavior in a primate, *Saguinus fuscicollis*. Behav Ecol Sociobiol 1981;8: 117–123.
89 Bray OE, Kennelly JJ, Guarino JL: Fertility of eggs produced on territories of vasectomized red-winged blackbirds. Wilson Bull 1975;87:187–195.
90 Gibbs HL, Weatherhead PJ, Boag PT, et al: Realized reproductive success of polygynous red-winged blackbirds revealed by DNA markers. Science 1990;250:1394–1397.
91 Gowaty PA: Multiple parentage and apparent monogamy in birds; in Gowaty PA, Mock DW (eds): Avian Monogamy. Ornithol Monogr. Washington, The American Ornithologists Union, 1985, vol 37, pp 11–21.
92 Burke T, Davies NB, Bruford MW, Hatchwell BJ: Parental care and mating behaviour of polyandrous dunnocks *(Prunella modularis)* related to paternity by DNA fingerprinting. Nature 1989;338:249–251.
93 Rabenold PP, Rabenold KN, Piper WH, et al: Shared paternity revealed by genetic analysis in cooperatively breeding tropical wrens. Nature 1990;348:538–540.
94 Sherman PW, Morton ML: Extra-pair fertilizations in mountain white-crowned sparrows. Behav Ecol Sociobiol 1988;22:413–420.
95 Wilson Goldizen A: A comparative perspective on the evolution of tamarin and marmoset social systems. Int J Primatol 1990;11:63–83.
96 Jeffreys AJ, Wilson V, Thein SL: Hypervariable 'minisatellite' regions in human DNA. Nature 1985;314:67–73.
97 Jeffreys AJ, Wilson V, Thein SL: Individual specific 'fingerprints' of human DNA. Nature 1985;316:76–79.
98 Weiss ML, Wilson V, Chan C, Turner T, Jeffreys AJ: Application of DNA fingerprinting probes to Old World Monkeys. Am J Primatol 1988;16:73–79.
99 Washio K, Misawa S, Ueda S: Individual identification of nonhuman primates using DNA fingerprinting. Primates 1988;30:217–221.
100 Dixson AF, Hastie N, Patel I, Jeffreys AJ: DNA 'fingerprinting' of captive family groups of common marmosets. Folia Primatol 1988;51:52–55.
101 Benirschke K, Anderson JM, Brownhill LE: Marrow chimerism in marmosets. Science 1962;138:513–515.
102 Wislocki GB: Observations on twinning in marmosets. Am J Anat 1939;664:445–483.
103 McGrew WC: Kinship terms and callitrichid mating patterns: A discussion note. Primate Eye 1986;30:25–26.
104 Lifjeld JT, Slagsvold T, Lampe HM: Low frequency of extra-pair paternity in pied flycatchers revealed by DNA fingerprinting. Behav Ecol Sociobiol 1991;29:95–101.
105 Maynard Smith J, Ridpath MG: Wife sharing in the Tasmanian native hen *(Tribonyx mortierii)*: A case of kinship selection? Am Nat 1972;106:447–452.
106 Martin RD: Female cycles in relation to paternity in primate societies; in Martin RD, Dixson AF, Wickings EJ (eds): Paternity in Primates: Genetic Tests and Theories. Basel, Karger, 1992, pp 238–274.

Dr. Gustl Anzenberger, Zoologisches Institut, Ethologie und Wildforschung, Universität Zürich-Irchel, Winterthurerstrasse 190, CH–8057 Zürich (Switzerland)

Martin RD, Dixson AF, Wickings EJ (eds): Paternity in Primates:
Genetic Tests and Theories. Basel, Karger, 1992, pp 225–237

Sperm Competition, Reproductive Tactics, and Paternity in Savanna Baboons and Rhesus Macaques

Fred B. Bercovitch

Caribbean Primate Research Center, University of Puerto Rico,
Medical Sciences Campus, Sabana Seca, Puerto Rico

'Yakov ... I came to say I've given birth to a child ...'

'So what do you want from me? ...'

'Yakov ... it might make things easier if you wouldn't mind saying you are my son's father ...'

'Who's the father ...'

'... He came, he went, I forgot him. He fathered the child but he's not the father. Whoever acts the father is the father ...'

B. Malamud, *The Fixer*

For every savanna baboon offspring produced, somewhere between 20 billion and 700 billion sperm have glided into the reproductive tract of a female. Competition occurs when resources are inequitably divided among those entities seeking the resource, so the existence of sperm competition for a single baboon ovum is axiomatic. However, the mechanics of sperm competition and the potential interconnections among sperm competition, testis size, mating systems, mating strategies and paternity represent areas in which research is at a very early stage. The dominant problem addressed in this contribution is how to assess potential relationships between sperm competition and paternity in savanna baboons *(Papio cynocephalus)* and rhesus macaques *(Macaca mulatta)*.

Sperm Competition: Evolutionary Framework

Sperm competition is an unusual form of male-male rivalry because it entails conflicts not only among males but also among the gametes of a single individual. The latter mode of competition can be approached from the perspective of the Elm-Oyster Model for the evolution of sexual reproduction [1]. This model starts from the observation that, in both elms and oysters, an enormous number of gametes are shed into the environment, but only one of them succeeds in occupying the space available to mature into adulthood. Producing a genetically diverse batch of gametes increases the likelihood of success for any given individual [1]. Paradoxically, under conditions of restricted resource availability, if each genotype possesses slightly different abilities to exploit the limited resource(s), then competition is more intense among similar genotypes than among different genotypes [2]. It is reasonable to assume that release of a single ovum, coupled with the microphysiological heterogeneity of a female's reproductive tract, creates conditions that favor the development of genetic heterogeneity to maximize a male's likelihood of fertilization. Those same conditions pose a conundrum for male sperm competition because a male's own sperm may find themselves more in conflict with one another than with sperm from a different male.

Meiosis ensures that the spermatozoa of an individual will contain genetically variable attributes, but disentangling the impact of chance factors from nonrandom components is a major predicament confronting evolutionary biologists interested in sperm competition. The evolutionary issue was recognized in *The Origin of Species* [3, p. 68] '... with all beings there must be fortuitous destruction, which can have little, or no influence on the course of natural selection. For instance, a vast number of eggs or seeds are annually devoured, and these could be modified through natural selection only if they varied in some manner which protected them from their enemies. Yet many of these eggs or seeds would perhaps, if not destroyed, have yielded individuals better adapted to their conditions of life than any of those which happened to survive ...'.

Traits that are closely related to fitness have a low heritability [4], but the heritability of sperm morphology is extremely high [5, 6], which confronts us with a second principal dilemma. If fitness is associated with traits having a low heritability, and if sperm morphology has a high heritability, what is the extent to which sperm morphology influences variation in male reproductive success? Evolutionary inferences regarding sperm competition usually resort to analysis of relative testis size rather than spermatozoal traits, but testis size also has a high heritability [7].

Across a variety of mammals, relatively large testes can be found in multimale species, seasonally breeding species or both [8–12], supporting the idea that sperm competition is more intense in these species. Sperm production is correlated with testis size [13–15], and it is likely that evolution has favored the development of relatively large testes among males that have high levels of sexual activity, such as those residing in multimale groups. This reasoning has been expanded to suggest that deposition of large quantities of sperm increases the chances of siring offspring, with testis size expected to be related to male reproductive success [16–18]. Although this latter suggestion is dubious [19], the initial premise remains a major assumption that needs to be verified or refuted.

Sperm Competition: The Fertilization Gauntlet

Following ejaculation, mammalian sperm are confronted by a veritable obstacle course on their journey towards the ovum. In human beings, the depletion in sperm number between deposition in the vagina and arrival at the ampulla, the site of fertilization, is prodigious. Of the 200 to 300 million sperm emitted in a single ejaculation, less than 10% reach the oviduct, and only about 200 come close to the egg [20]. Seminal plasma remains in the vagina, with sperm transport to the oviduct due to uterine contractions. Some sperm arrive at the oviduct within 30 min of sexual intercourse in human beings, and uterine contractions may be facilitated by the prostaglandins in seminal fluid. Movement within the oviduct is dependent primarily upon the ciliary action and muscular contractions of the oviduct simultaneously transporting the ovum in one direction and the sperm in the opposite direction. Within the ampulla, multiple spermatozoa contact the ovum. Prior to fertilization, the sperm must furrow through the cumulus cells, corona radiata, zona pellucida, and plasma membrane surrounding the target cell. The acrosome reaction required for egg penetration is initiated after binding to the zona pellucida, with zona penetration dependent upon sperm motility. Sperm motility is acquired in the epididymis and influenced by male accessory gland secretions, with progressive motility possibly facilitating passage through the cervix. Capacitation within the female reproductive tract induces hypermotility in sperm, and the increased velocity favors rapid zona penetration.

The fertilization gauntlet in mammals is almost entirely directed by the female reproductive tract, with three possible exceptions relevant to the mechanics of sperm competition: (1) after sexual intercourse, seminal plasma remaining in the vagina may form a cervical barrier, acting as a type of copulatory plug [21, 22], that can disrupt the entrance of sperm deposited later by a competitor, as well as facilitating movement of the first male's sperm; (2) after attachment to the egg, sperm motility, a key factor influencing male fertility [23, 24], is required for burrowing through the zona pellucida, so sperm competition can be mediated by differential sperm motility, and (3) deposition of large quatities of abnormal sperm may function in a 'Kamikaze' fashion to thwart the ability of competing sperm to fertilize the egg [25], although the validity of this proposal has been strongly disputed [26]. The reduced chances of deformed sperm traveling to the oviduct may be due to factors intrinsic to the sperm, enhanced by the female reproductive tract functioning as a sieve.

Overall, the probability of fertilization seems to depend more upon the microphysiology of a female's reproductive tract than upon spermatozoal traits [27]. One might expect that the chances of fertilization can be modified by alterations in the female reproductive tract, with such modifications functioning as a form of 'passive female choice' in the context of sexual selection [28, 29]. This suggestion is an oxymoron contradicting Darwin's [30, p. 894] reasoning that '... [what] is required [for sexual selection to operate] ... is that choice should be exerted *before* [my emphasis] the parents unite ...' Sexual selection is a concept explaining the origins of secondary sexual traits, and spermatozoa are not embellished with secondary sexual characteristics.

Haploid gene expression, i.e., mRNA synthesis in spermatids after meiosis, would seem to be a necessary condition for female selectivity of specific sperm, but current evidence indicates no haploid gene activation in mammals [5, 20]. If haploid gene expression can be documented in mammalian sperm, then sperm competition is a form of gametic selection, where the probability of fertilization is a consequence of differences in haploid genotype [31]. If spermatozoa are genetically quiescent, even though subjected to intense competition, the role of chance in fertilization probabilities seems more pronounced, and it becomes difficult to envisage how the outcome of sperm competition can be linked to variation among male rivals in sperm traits. Differential fertility in female rabbits inseminated with equal numbers of spermatozoa from two males has been attributed to selective transport in the female tract and speed of capacitation

[32], rather than to haploid gene expression. Identification of the role of chance factors in sperm transport is essential for an understanding of the nature of sperm competition. Even if rapidity of capacitation and motility influence the probability of fertilization, both characteristics are not fully activated until *after* sperm have reached the oviduct.

One possible exception to this condition is the potential for facultative sex ratio adjustment by females. If the minuscule size differential between sperm bearing the X or Y chromosome can be distinguished by the female reproductive tract, and if modifications can be made to alter the probability of fertilization according to spermatozoal size, then sex ratio adjustment functions as a form of female sperm selectivity.

If fluctuations within the female reproductive tract modulate the chances of fertilization by particular sperm, then males may increase their likelihood of siring offspring by saturating a female's reproductive tract with their own gametes. Ejaculatory rates are not correlated with testis size in savanna baboons [19], and sperm output is not accurately predicted on the basis of testis size [13]. A prevailing assumption requiring critical evaluation is that males that ejaculate more often than conspecifics are more likely to sire offspring.

Reproductive Tactics in Savanna Baboons and Rhesus Macaques

The mating system of savanna baboons and rhesus macaques is described as female defense polygyny [33, 34]. In both species, sexually receptive females form temporary mating bonds, or consortships, with males [savanna baboons: ref. 35–44; rhesus macaques: ref. 45–51], but male mate-guarding tactics differ between the two species. Conceptions in the absence of consortships are extremely rare in baboons [52, 53], but not in rhesus macaques [54]. Table 1 summarizes some factors associated with sperm competition and reproductive tactics in savanna baboons and rhesus macaques.

Determination of probable paternity in both species is clouded by imprecise external signs of ovulation coupled with a pattern of multiple male partners. Additional complexities are introduced if spermatozoal longevity has been underestimated among savanna baboons and rhesus macaques [79]. Mating with multiple males has been interpreted as a female reproductive tactic designed to obscure paternity, decrease the risk of infanticide, and increase the chances of paternal care [55–57]. An alterna-

Table 1. Sperm competition and reproductive tactics in savanna baboons and rhesus macaques

	Baboons	Macaques
Body weight, kg	27	9
Testes weight, g	72	46
Gonadosomatic index (testes/body weight), %	0.26	0.51
Ejaculate volume, ml	3.5	1.3
Sperm concentration/ml $\times 10^6$	420	600
Sperm count/ejaculate $\times 10^6$	1469	780
Ejaculatory rate/h	0.91 (avg)	0.45 (max)
Masturbation rate/h	0.10	?
Reproductive periodicity	nonseasonal	seasonal
Number of preconception cycles (average)	3–5	0–1
Mating days per cycle (average)	5–7	6–9
Mating partners per cycle (average)	3	3
Aggressive competition for females	common	rare
Male coalitions for females	yes	no
'Sneaky' male matings	very rare	common
Male copulatory vocalizations	very rare	sometimes
Female copulatory vocalizations	common	none
Female sexual swellings	puberty–death	puberty
Pregnancy sign	obvious	uncertain
Postconception sexual consortships	no	yes
Male-infant associations	common	rare

Data are presented as a general overview of parameters associated with reproduction. Values should be treated cautiously due to differences in sample sizes between studies, combination of captive and wild data, differences in methodological techniques for deriving both physiological and behavioral data, and a large variance in many parameters. Key references consulted: 8, 12, 19, 40, 41, 43–49, 52, 71–78.

tive suggestion is that mating with multiple males improves the probability of conception and reduces the amount of sperm available to other females [58]. Both explanations require that females control the number of mating partners. Female savanna baboons influence the likelihood of consortship formation with particular males, but ejaculatory rates and male tenure in consort are not a function of female cooperation [52, 59]. In rhesus macaques, male mate selection probably plays a greater role in consortship formation than does female mate selection [49]. The suggestion that increased reproductive activity enhances the likelihood of conception is discredited by data demonstrating that no significant differences exist between the conception cycle and the penultimate cycle in savanna baboons in the number of consort partners, the number of days in consort, or ejaculatory rates on the most likely days of ovulation [43].

Paternity confusion, whether a cause or a consequence of multiple male partners, is a double-edged sword. Females are as ignorant of offspring paternity as are their mates unless they are capable of modifying their reproductive tract to favor the sperm of particular males, which is unlikely. Although paternity is camouflaged in both savanna baboons and rhesus macaques, formation of friendships between male and infant baboons can be linked to the likelihood of paternity [41, 59, 60]. Baboon sexual swellings are not precise indicators of ovulation, but males ejaculate more often and compete more intensely for access to females on the 4 most likely days of ovulation [19, 44]. The timing of mating and the fertilizing capacity of spermatozoa combine with social strategies related to mate acquisition to influence the likelihood of putative male reproductive success [19]. Patterns of reproductive tactics among male baboons allow the investigator to restrict potential paternity to a subset of males, whereas rhesus macaque reproductive tactics pose a more formidable challenge to assessing potential paternity. Rhesus macaque females not only lack pronounced signs of ovulation, they also participate in postconception sexual activity (table 1).

Savanna baboons and rhesus macaques differ in reproductive tactics pertinent to gaining and maintaining access to sexually receptive females. Relative testis size can be linked to variation in mating systems, but not to variation in mating success or mating tactics. Methods for the determination of paternity in multimale primate societies should come to terms with dissimilarities between species in modes of mate acquisition and reproductive behavior before they are applied to identify potential male progenitors.

Paternity in Savanna Baboons and Rhesus Macaques

A male that acts as the father is not necessarily the father, but such a male probably has a greater likelihood of being the father. In both savanna baboons and rhesus macaques, mate selectivity limits consortship formation [42, 49, 52]. Notwithstanding the multimale nature of mating activity, a single male performs over 60% of ejaculations on the 4 most likely days of ovulation in close to one-third of conception cycles among savanna baboons [43]. One or two males monopolize a female rhesus macaque during her receptive period in close to 50% of cycles [47]. In both species, male reproductive success has been gauged on the basis of at least six different criteria, all of which are correlated [43], but the potential association with paternity is unknown. The fundamental goal in paternity determination is to uncover at least one accurate behavioral indicator of male reproduction, and this quest has led to the use of DNA fingerprinting to assess paternity.

At least four field studies are investigating the potential for DNA fingerprinting in savanna baboons, but results are still unavailable. A pilot study of DNA fingerprinting in free-ranging rhesus macaques revealed that furtive matings can result in successful fertilizations [54]. A few studies of captive macaques have sought to evaluate the relationship between behavioral estimates of paternity and actual paternity based upon genetic polymorphisms [61–64], with a consensus developing that paternity is not accurately deduced from behavioral records. However, even though this assessment may be correct, the sampling protocols used in the captive studies do not permit a firm conclusion. For example, one study found a lack of correlation between sexual activity and paternity, but only 28% of females that conceived were observed to copulate during the probable days of conception [63]. Because of differences in male mate-guarding patterns between savanna baboons and rhesus macaques, and because of the limited opportunity for furtive matings in captivity, conclusions regarding the accuracy of behavioral indices as mirrors of paternity in captive macaques may not be applicable either to savanna baboons or to free-ranging macaques.

Among captive rhesus macaques, adult males associate more often with their own offspring than with unrelated individuals [65]. Immature savanna baboons may spend more time with adult males than with their mothers as they advance beyond the weaning period [66]. Occasionally, male baboons that are not likely to have sired a particular infant form a friendship with that infant [41, 67]; but in most cases male savanna baboons form affiliations only with infants that they could have fathered [41, 59, 60, 68, 69].

Speculation about paternity in savanna baboons and rhesus macaques has been based on sexual activity around the presumed time of conception, association patterns between males and infants, or both. However, the actual bond between paternity and male behavior remains to be assessed.

Reconnoitering the Future

There are parallels between the testis and the brain. Measurements of both organs have been used in interspecific comparisons to generate inferences about the structure of mammalian societies. It has been proposed that relatively large brains tend to be associated with species residing in complex social systems and relatively large testes tend to be associated with species residing in multimale mating systems. However, within a species, brain size is an inaccurate reflection of intelligence and testis size is probably an inaccurate index of fitness. Sperm competition is reflected in testis size, but the impact of sperm competition on paternity depends upon males gaining access to females. Male reproductive tactics modulating sperm competition include the timing of insemination, the frequency of ejaculation, the tenure of mate guarding, as well as sperm count and quality. The foundation for advancement in our understanding male and female reproductive tactics needs to be built upon a framework that welds paternity to behavior.

Determination of paternity in natural populations of savanna baboons and rhesus macaques includes the adoption of field logistics requiring the capture of target infants, their mothers, and all potential fathers. In addition, only anecdotal information is currently available on night-time copulatory activity, but detailed information is necessary for inferring male paternity. Monitoring receptive females throughout the day would appear to be no more than a minimum requirement for accurately partitioning the likelihood of fertilization based upon male sexual activity. Sufficient sample sizes will be required to map paternity with male traits, since male reproductive tactics alter with age, rank, troop tenure, and social relationships [19, 41, 44, 70]. Specific information regarding the outcome of sperm competition for paternity requires experimental laboratory procedures that compare the quality and quantity of different male sperm with each male's reproductive output in a free-ranging situation.

No panacea exists for uncovering the labyrinthine relationship between sperm competition and paternity in savanna baboons and rhesus macaques. Splicing these two entities together requires additional knowledge spanning the gamut from sperm motility and longevity to alternative male reproductive tactics. The web connecting sperm competition with reproductive tactics and paternity possesses the same characteristics that Winston Churchill attributed to the Soviet Union at the outbreak of the Second World War: it is a riddle wrapped in a mystery inside an enigma.

Acknowledgments

Research funds have been provided by NSF Grant BNS7806914 to R.H. Byles, S.C. Strum, and T.J. Olivier, NIMH Grant 5T32MH15133 to R.H. Byles, D.G. Lindburg, and B.J. Williams, NIH Grant RR03640 to the CPRC, and the Harry Frank Guggenheim Foundation. Suggestions for improvement in the text have been made by S.M. Schwartz and R.D. Martin.

References

1 Williams GC: Sex and Evolution. Princeton, Princeton University Press, 1975.
2 Lewontin RC: The Genetic Basis of Evolutionary Change. New York, Columbia University Press, 1974.
3 Darwin CR: On the Origin of Species. London, Murray, 1859.
4 Falconer DS: Introduction to Quantitative Genetics. Edinburgh, Oliver & Boyd, 1960.
5 Beatty RA: Genetics of animal spermatozoa; in Mulcahy DL (ed): Gamete Competition in Plants and Animals. Amsterdam, North Holland, 1975, pp 61–68.
6 Afzelius BA: Abnormal human spermatozoa including comparative data from apes. Am J Primatol 1981;1:175–182.
7 Schinkel AP, Johnson RK, Kittok RJ: Testicular development and endocrine characteristics of boars selected for either high or low testis size. J Anim Sci 1984;58: 675–685.
8 Harcourt AH, Harvey PH, Larson SG, et al: Testis weight, body weight and breeding system in primates. Nature 1981;293:55–57.
9 Harvey PH, Harcourt AH: Sperm competition, testes size, and breeding systems in primates; in Smith RL (ed): Sperm Competition and the Evolution of Animal Mating Systems. New York, Academic Press, 1984, pp 589–600.
10 Kenagy GJ, Trombulak SC: Size and function of mammalian testes in relation to body size. J Mammal 1986;67:1–22.
11 Dixson AF: Observations on the evolution of the genitalia and copulatory behaviour in male primates. J Zool 1987;213:423–443.
12 Moller AP: Ejaculate quality, testes size and sperm competition in primates. J Hum Evol 1988;17:479–488.
13 Amann RP: Sperm production rates; in Johnson AP, Gomes WR, Vandemark NL (eds): The Testis. New York, Academic Press, 1970, vol 1, pp 433–482.
14 Amann RP: A critical review of methods for evaluation of spermatogenesis from seminal characteristics. J Androl 1981;2:37–58.
15 Johnson L, Petty CS, Neaves WB: Influence of age on sperm production and testicular weights in men. J Reprod Fert 1984;70:211–218.
16 Popp JL, DeVore I: Aggressive competition and social dominance theory: Synopsis; in Hamburg DA, McCown ER (eds): The Great Apes. Menlo Park, Benjamin/Cummings, 1979, pp 317–338.
17 Short RV: Sexual selection and its component parts, somatic and genital selection, as illustrated by man and the great apes. Adv Stud Behav 1979;9:131–158.
18 Short RV: Sexual selection in man and the great apes; in Graham CE (ed): Repro-

ductive Biology of the Great Apes. New York, Academic Press, 1981, pp 319–341.
19 Bercovitch FB: Body size, sperm competition, and determinants of reproductive success in male savanna baboons. Evolution 1989;43:1507–1521.
20 Edwards RG: Conception in the Human Female. New York, Academic Press, 1980.
21 Devine MC: Copulatory plugs in snakes: Enforced chastity. Science 1975;187:844–845.
22 Mathews M, Adler NT: Facilitative and inhibitory influences of reproductive behavior on sperm transport in rats. J Comp Physiol Psychol 1977;91:727–741.
23 Mortimer D, Templeton AA: Sperm transport in the human female reproductive tract in relation to semen analysis characteristics and the time of ovulation. J Reprod Fert 1982;64:401–408.
24 Mahadevan MM, Trounson AO: The influence of seminal characteristics on the success rate of human in vitro fertilization. Fert Steril 1984;42:400–405.
25 Baker RR, Bellis MA: 'Kamikaze' sperm in mammals? Anim Behav 1988;36:936–939.
26 Harcourt AH: Sperm competition and the evolution of nonfertilizing sperm in mammals. Evolution 1991;45:314–328.
27 Quiatt D, Everett J: How can sperm competition work? Am J Primatol 1982;suppl 1:161–169.
28 Thornhill R: Cryptic female choice and its implications in the scorpionfly *Garpobittacus nigriceps.* Am Nat 1983;122:765–788.
29 Sullivan BK: Passive and active female choice: A comment. Anim Behav 1989;37:692–694.
30 Darwin CR: The Descent of Man and Selection in Relation to Sex. London, Murray, 1871.
31 Lewontin RC: The units of selection. Annu Rev Ecol Syst 1970;1:1–18.
32 Parrish JJ, Foote RH: Fertility differences among male rabbits determined by heterospermic insemination of fluorochrome-labeled spermatozoa. Biol Reprod 1985;33:940–949.
33 Emlen ST, Oring LW: Ecology, sexual selection, and the evolution of mating systems. Science 1977;197:215–223.
34 Berenstain L, Wade TD: Intrasexual selection and male mating strategies in baboons and macaques. Int J Primatol 1983;4:201–235.
35 Saayman GS: The menstrual cycle and sexual behaviour in a troop of free ranging chacma baboons. Folia Primatol 1970;12:81–110.
36 Hausfater G: Dominance and Reproduction in Baboons *(Papio cynocephalus).* Contrib Primatol. Basel, Karger, 1975, vol 7.
37 Seyfarth RM: Social relationships among adult male and female baboons. I. Behaviour during sexual consortship. Behaviour 1978;64:204–226.
38 Rasmussen KLR: Consort Behaviour and Mate Selection in Yellow Baboons *(Papio cynocephalus)*; PhD thesis, Cambridge, 1980.
39 Rasmussen KLR: Spatial patterns and peripheralisation of yellow baboons *(Papio cynocephalus)* during sexual consortships. Behaviour 1986;97:161–180.
40 Collins DA: Social Behaviour and Patterns of Mating among Adult Yellow Baboons (*Papio c. cynocephalus* L. 1776); PhD thesis, University of Edinburgh, 1981.

41 Smuts BB: Sex and Friendship in Baboons. New York, Aldine, 1985.
42 Bercovitch FB: Male rank and reproductive activity in savanna baboons. Int J Primatol 1986;7:533–550.
43 Bercovitch FB: Reproductive success in male savanna baboons. Behav Ecol Sociobiol 1987;21:163–172.
44 Bercovitch FB: Coalitions, cooperation, and reproductive tactics among adult savanna baboons. Anim Behav 1988;34:1198–1209.
45 Carpenter CR: Sexual behavior of free ranging rhesus monkeys *(Macaca mulatta)*. I. Specimens, procedures and behavioral characteristics of estrus. J Comp Psychol 1942;33:113–142.
46 Carpenter CR: Sexual behavior of free-ranging rhesus monkeys *(Macaca mulatta)*. II. Periodicity of estrus, homosexual, autoerotic and non-conformist behavior. J Comp Psychol 1942;33:143–162.
47 Kaufmann JH: A three-year study of mating behavior in a freeranging band of rhesus monkeys. Ecology 1965;46:500–512.
48 Lindburg DG: The rhesus monkey in North India: An ecological and behavioral study; in Rosenblum LA (ed): Primate Behavior. New York, Academic Press, 1971, pp 1–106.
49 Lindburg DG: Mating behavior and estrus in the Indian rhesus monkey; in Seth PK (ed): Perspectives in Primate Biology. New Delhi, Today & Tomorrow's Printers & Publishers, 1983, pp 45–61.
50 Chapais B: Reproductive activity in relation to male dominance and the likelihood of ovulation in rhesus monkeys. Behav Ecol Sociobiol 1983;12:215–228.
51 Hill DA: Social relationships between adult male and female rhesus macaques. I. Sexual consortships. Primates 1987;28:439–456.
52 Bercovitch FB: Reproductive Tactics in Adult Female and Adult Male Olive Baboons; PhD thesis, University of California at Los Angeles, 1985.
53 Altmann J, Hausfater G, Altmann SA: Determinants of reproductive success in savannah baboons; in Clutton-Brock TH (ed): Reproductive Success. Chicago, University of Chicago Press, 1988, pp 403–418.
54 Berard J, Schmidtke J, McGeehan L: Male reproductive success in a free-ranging colony of rhesus macaques. Am J Primatol 1990;20:173.
55 Hrdy SB: The Woman That Never Evolved. Cambridge, Harvard University Press, 1981.
56 Stacey PB: Female promiscuity and male reproductive success in social birds and mammals. Am Nat 1982;120:51–64.
57 Davies EM, Boersma PD: Why lionesses copulate with more than one male. Am Nat 1984;123:594–611.
58 Small MF: Female primate sexual behavior and conception. Are there really sperm to spare? Curr Anthropol 1988;29:81–100.
59 Bercovitch FB: Mate selection, consortship formation, and reproductive tactics in adult female savanna baboons. Primates 1991;32, in press.
60 Stein DM: The Sociobiology of Infant and Adult Male Baboons. Norwood, Ablex, 1984.
61 Duvall SW, Bernstein IS, Gordon TP: Paternity and status in a rhesus monkey group. J Reprod Fert 1976;47:25–31.

62 Curie-Cohen M, Yoshihara D, Luttrell L, et al: The effects of dominance on mating behavior and paternity in a captive troop of rhesus monkeys *(Macaca mulatta)*. Am J Primatol 1983;5:127–138.
63 Stern BR, Smith DG: Sexual behaviour and paternity in three captive groups of rhesus monkeys *(Macaca mulatta)*. Anim Behav 1984;32:23–32.
64 Shively C, Smith DG: Social status and reproductive success of male *Macaca fascicularis*. Am J Primatol 1985;9:129–135.
65 Berenstain L, Rodman PS, Smith DG: Social relations between fathers and offspring in a captive group of rhesus monkeys *(Macaca mulatta)*. Anim Behav 1981;29:1057–1063.
66 Nicolson NA: Weaning and the Development of Independence in Olive Baboons, PhD thesis, Harvard University, 1982.
67 Strum SC: Why males use infants; in Taub DM (ed): Primate Paternalism. New York, Van Nostrand Reinhold, 1984, pp 146–185.
68 Packer C: Male care and exploitation of infants in *Papio anubis*. Anim Behav 1980; 28:512–520.
69 Busse C, Hamilton WJ III: Infant carrying by male chacma baboons. Science 1981; 212:1281–1283.
70 Strum SC: Almost Human. New York, Random House, 1987.
71 Schultz AH: The relative weight of the testes in primates. Anat Rec 1938;72:387–394.
72 Kinsky M: Quantitative Untersuchungen an äthiopischen Säugetieren. Anat Anz 1960;108:65–82.
73 Kraemer DC, Vera Cruz NC: Collection, gross characteristics and freezing of baboon semen. J Reprod Fert 1969;20:345–348.
74 Hill WCO: Primates. Comparative Anatomy and Taxonomy. VIII. Cynopithecinae. New York, Wiley-Interscience, 1970.
75 Dukelow WR: Semen and artificial insemination; in Hafez ESE (ed): Comparative Reproduction of Nonhuman Primates. Springfield, Thomas, 1971, pp 115–127.
76 Flechon J-E, Kramer DL, Hafez ESE: Scanning electron microscopy of baboon spermatozoa. Folia Primatol 1976;26:24–35.
77 Wildt DE: Spermatozoa: Collection, evaluation, metabolism, freezing, and artificial insemination; in Dukelow WR, Erwin J (eds): Comparative Primate Biology. New York, Liss, 1986, vol 3, pp 171–193.
78 Bornman MS, van Vuuren M, Meltzer DGA, et al: Quality of semen obtained by electroejaculation from chacma baboons *(Papio ursinus)*. J Med Primatol 1987;17: 57–61.
79 Martin RD: Female cycles in relation to paternity in primate societies; in Martin RD, Dixson AF, Wickings EJ (eds): Paternity in Primates: Genetic Tests and Theories. Basel, Karger, 1992, pp 238–274.

Fred B. Bercovitch, Caribbean Primate Research Center, P.O. Box 1053,
Sabana Seca, PR 00952 (Puerto Rico)

Martin RD, Dixson AF, Wickings EJ (eds): Paternity in Primates: Genetic Tests and Theories. Basel, Karger, 1992, pp 238–274

Female Cycles in Relation to Paternity in Primate Societies

R.D. Martin

Anthropologisches Institut und Museum, Universität Zürich-Irchel, Schweiz

It is widely assumed that overt behaviour of sexually mature males in primate social groups is directly linked to paternity of any offspring that are born. This assumption underlies certain interpretations of primate social organization. In particular, it is often held that dominance relationships among males in primate social groups directly determine priority of mating access to females and hence paternity of offspring. But until recently, observers of primate behaviour could do no more than identify mothers of infants directly; the only guides to paternity of offspring were (1) the apparently exclusive presence of just one sexually mature male in unimale groups or (2) observed copulatory behaviour in multimale groups. The context of the latter was succinctly stated by Sade [1]:

> 'It may be more common for a female to mate with several males during her oestrous cycle with perhaps one particular male mating with her most frequently or exclusively during the few days of the cycle on which she is most likely to ovulate. Some females, however, may mate with several males at any time during the cycle. In order to estimate paternity from observation of the mating behavior, one must draw inferences from the frequency of copulation and the most likely time of conception.'

For primates living in groups containing several sexually mature males, the inference of paternity from observed copulation can be a particularly risky undertaking. With the advent of DNA fingerprinting, supplementing the commonly inadequate technique of blood protein electrophoresis, it has now become possible to infer paternity in primate groups relatively reliably. Surprisingly, no strong correlation has been found between observed patterns of male sexual behaviour and paternity in several studies of macaques in captivity [e.g. 2–7], although there are exceptions [e.g. 8]. It is possible, however, that poor correlations emerge because most studies have been conducted with primate groups living under artificial captive conditions. For this reason, it is especially important to examine the relationship between observed sexual behaviour and paternity in wild-living primate groups. A field study of *Macaca fascicularis* [9] has now yielded evidence that the highest-ranking males in a social group are most likely to father offspring within the group. Nevertheless, with some species at least, male rank may be poorly correlated with paternity even in groups living under natural conditions. This is the case, for instance, with three groups of seasonally breeding vervet monkeys *(Cercopithecus aethiops),* studied by Andelman [10]. For the mating season as a whole, males in the top third of the dominance hierarchy performed significantly more successful copulations than lower-ranking males; but during the week when conception was most probable (as indicated by backdating from birth dates) there was no correlation between rank and successful copulations. As most males mated with most females in their group, it is impossible to infer paternity of offspring from observations of copulation. Such findings suggest that the relationship between copulatory behaviour and paternity in primate social groups may be more complex than is generally supposed.

There has, in fact, been a major limitation in many discussions of the potential relationship between copulatory behaviour and reproductive success in that hypotheses have been predominantly centred on males. It is entirely possible that females living in social groups containing several mature males pursue their own strategies and that for this reason male copulatory activity is poorly correlated with paternity. Indeed, there are indications from an entirely different direction that this may be so. General comparisons of female cycles, copulation and pregnancy in mammals, combined with consideration of the timing of ovulation, copulation and fertilization, suggest that an unusual development took place in the course of primate evolution with far-reaching implications for female mating strategies. The starting-point for discussion of this alternative evidence is the oft-repeated but unfounded claim that the 'loss of oestrus' is unique to the human female.

Loss of Oestrus in Humans

Speculative accounts of human behavioural evolution often list apparent distinctions between humans and other primates and then suggest an explanation for their origin. One widely cited distinction is apparent 'loss of oestrus' in the human female, associated with the occurrence of copulation at times when fertilization cannot occur (both during the menstrual cycle and during pregnancy). Despite the fact that many non-human primate species lack obvious external signs of ovulation, the absence of any such sign in the human female has often been cited as a unique feature, even in serious academic texts [e.g. 11], and the invention of novel explanations has been rife in the literature on human ethology [e.g. 12–14]. It has also been suggested [e.g. 13] that human beings are unique in that copulation may take place during pregnancy. Accordingly, scenarios of human evolution are commonly designed to explain the following two related points:

(1) Copulation commonly takes place at times when the 'obvious' biological function (i.e. fertilization) is precluded.

(2) There are no obvious indicators of ovulation, which is hence 'concealed'.

Various hypothetical explanations have been proposed to account for 'prolonged sexual receptivity' and 'concealed ovulation':

(1) Loss of externally visible indicators of oestrus reduces competition between males [e.g. 15, for non-human primates] and may reduce the risk of infanticide [10] or enhance cooperation between males [e.g. 16].

(2) Copulation at times other than close to ovulation reinforces the pair-bond between male and female [e.g. 13, 17].

(3) 'Concealment' of ovulation from the male represents a female strategy to promote increased paternal investment in offspring [e.g. 14].

(4) 'Concealment' of ovulation from the female's own awareness prevents her from taking measures to avoid pregnancy [18].

Such hypotheses are largely irrelevant, however, because the basic assumption of human uniqueness is flawed. In fact, this myth has hindered investigation of underlying processes. Comparison of non-human primates and other mammals reveals a very different picture.

Copulatory Behaviour in Mammals

It is necessary, first of all, to distinguish sharply between copulation during pregnancy (which cannot be fertile) and copulation that is thought to be remote from ovulation during the non-pregnant cycle (which need not necessarily be infertile). As will be shown below, the implications are quite different in the latter case.

A general comparison of mammals reveals that copulation during pregnancy is widespread, occurring not only in many primates but also in numerous other mammals. Heape [19] – author of the term 'oestrus' (see below) – specifically acknowledged that mating could occur at unusual times, particularly during pregnancy, referring to this as 'abnormal oestrus' [see also 20]. Heape noted that copulation during pregnancy has 'been noticed in most species of domestic mammals' and proposed the following explanation for 'abnormal oestrus':

> 'When oestrus occurs during pregnancy it is probably due to a temporary diversion of a superabundant supply of placental blood; when it occurs at other times, the highly nutritious food, with which the animals which experience it appear to be generally supplied, or the condition resulting therefrom, is possibly largely responsible for it.'

Zuckerman [20] later proposed a more general explanation. According to these views, copulation during pregnancy amounts to an 'error' in a system otherwise adapted to ensure coincidence of copulation and ovulation.

Hammond and Marshall [21] observed mating during pregnancy in the rabbit and erroneously thought that this was the only non-human species to show this. In fact, mating during pregnancy is widespread among mammals and may be the rule rather than the exception. Kleiman and Mack [22], for example, report mating during pregnancy in carnivores and ungulates as well as in various simian primates (table 1). As mating during pregnancy may be of little biological significance, insofar as it does not adversely affect the developing embryo, it is possible that interpretation of such mating as 'error' or as a special behavioural mechanism (e.g. to enhance male-female bonding) is justified. However, this is certainly not the case for mating during the non-pregnant cycle.

There is hence no basis whatsoever for the claim that copulation during pregnancy is unique to humans. The reason for the occurrence of such infertile copulation in many mammalian species remains obscure, but it is clear that it has no special relevance to human evolution.

Turning to mating during the non-pregnant female cycle, it is necessary as a first step to define the term 'oestrus', for which Young [23] provided a useful definition:

> 'The period of oestrus or heat is the time of sexual activity in the female, occurring either rhythmically throughout the year or at one or more seasons. At the time of oestrus the eggs are shed from the ovary and the females desire and will receive the males. The word oestrus is derived from the Greek name for the gadfly or warble fly, whose sting drives cattle crazy.'

Table 1. Occurrence of mating in simian primates[1]

Species	Context[2]	Cycle[3]	Pregnancy[4]	Source
Platyrrhines				
Aotus trivirgatus	CAPT	DIFF	EARL	30
Callithrix jacchus	CAPT	DIFF	–	31
Callithrix jacchus	CAPT	–	EARL	32
Leontopithecus rosalia	CAPT	EXTE	–	33
Leontopithecus rosalia	CAPT	–	EARL	22
Saguinus oedipus	CAPT	DIFF	THRO	34, 35
Saimiri sciureus	CAPT	CONF	EARL	36,37
Catarrhines				
Cercocebus albigena	WILD	EXTE	–	38
Cercocebus atys	CAPT	–	EARL	39
Cercopithecus aethiops	WILD	DIFF	EARL	10, 40
Cercopithecus ascanius	WILD	DIFF	EARL	41
Cercopithecus mitis	CAPT	EXTE	EARL	42
Erythrocebus patas	CAPT	DIFF	EARL	26
Gorilla gorilla	CAPT	CONF	–	43, 44, 45
Gorilla gorilla	WILD	CONF	THRO	46, 47
Hylobates lar	CAPT	DIFF	EARL	48
Hylobates lar	WILD	–	EARL	49
Mandrillus spp.	CAPT	–	EARL	39
Macaca fuscata	CAPT	DIFF	–	50
Macaca fuscata	CAPT	–	EARL	6
Macaca mulatta	CAPT	DIFF	–	51, 52, 53
Macaca mulatta	FREE	EXTE	EARL	54, 55
Macaca mulatta	WILD	EXTE	EARL	56
Macaca nemestrina	CAPT	–	EARL	39
Macaca sylvanus	WILD	EXTE	–	57
Macaca sylvanus	FREE	EXTE	EARL	58
Miopithecus talapoin	CAPT	EXTE	–	59
Miopithecus talapoin	WILD	EXTE	–	60
Pan paniscus	CAPT	DIFF	–	61
Pan troglodytes	CAPT	EXTE	EARL	61, 62, 63
Pan troglodytes	WILD	DIFF	–	64, 65
Papio ursinus	WILD	DIFF	EARL	66, 67
Pongo pygmaeus	CAPT	DIFF	–	68
Pongo pygmaeus	CAPT	EXTE	THRO	69
Pongo pygmaeus	CAPT	–	EARL	70
Pongo pygmaeus	WILD	EXTE	–	71
Presbytis entellus	CAPT	EXTE	EARL	72, 73
Theropithecus gelada	WILD	EXTE	EARL	74, 75

[1] Table compiled partly with reference to Andelman [10] and Loy [26].

[2] Context: CAPT = captive; FREE = captive, but free-ranging; WILD = wild-living.

[3] Cycle: CONF = copulation confined to ≤ 3 days; EXTE = copulation lasts for several days (> 3); DIFF = diffuse copulation more or less throughout the cycle.

[4] Pregnancy; EARL = copulation at least during early pregnancy; THRO = copulation throughout pregnancy.

The term was originally coined by Heape [19] and defined as follows:

> '... it is the special period of sexual desire of the female; it is during oestrus, and only at that time, the female is willing to receive the male and fruitful coition rendered possible in most, if not all, mammals.'

Among mammals, copulation is indeed typically confined to a short period, often less than 24 h and at most 72 h, around the time of ovulation. Females commonly exhibit recognizable behavioural and other indications of readiness to mate. Nevertheless, it should be noted that female copulatory behaviour potentially includes at least three separable components [24, 25]:

(1) proceptivity: female behaviour that serves to initiate copulation;

(2) receptivity: the readiness of the female to permit the male to copulate;

(3) attractiveness: a combination of behavioural and non-behavioural cues that increase the probability of male mounting attempts.

As Dixson [25] noted, a mating peak may originate in different ways, for example through increased attractiveness (with no change in proceptivity or receptivity) or through increased proceptivity and receptivity (with no change in attractiveness). Presumably, all three components usually coincide in mammals with a well-defined oestrus [see also 26]; but it is particularly important to bear in mind the distinctions between them if copulation occurs at times of the cycle other than close to ovulation. In what follows, attention will be largely focussed on the presence or absence of *copulation,* as it is the probability of fertilization rather than the behavioural mechanism leading to copulation that is paramount. Nevertheless, possible underlying differences is mechanism should not be forgotten as they are particularly relevant to female strategies.

Information on primates indicates that there is indeed a distinction among them with respect to the relationship between ovulation and copulatory activity. However the major distinction is between prosimians (lemurs, lorises and tarsiers) and simians (monkeys, apes and humans), rather than between all non-human primates and the human species [see also 26].

Studies on prosimians, both in the laboratory and in the field, clearly indicate that copulatory activity is typically confined to a period of a few days, associated with definite signs of oestrus in the female. Madagascar lemurs all exhibit strictly seasonal patterns of reproduction [27]. In most cases, there is only a single, brief mating period each year. Females exhibit a well-defined oestrus of very limited duration. The reproduction of lemurs and lorises has been effectively reviewed by Van Horn and Eaton [28], who showed that members of the loris group resemble lemurs in exhibiting very confined periods of mating activity during the non-pregnant cycle of the female. Finally, Izard et al. [29] have reported that Bornean tarsiers *(Tarsius bancanus)* mate only during a very restricted period in each non-pregnant cycle. The term 'oestrous cycle' would thus seem to be appropriate for the prosimians, as for non-primate mammals generally.

On the other hand, there appears to be a definite contrast between prosimians and simians, as indicated by Van Horn and Eaton [28]:

> 'In some ways, the reproductive physiology and behaviour of prosimians resemble those of other mammals more than they do those of simians. Prosimians, for example, are sexually receptive only for a brief period during their ovarian cycle, whereas at least some simians are potentially receptive during all phases of the cycle.'

A review of simian primates (table 1) [see also 10] clearly confirms this distinction. Although copulatory frequency may vary during the non-pregnant cycle, most simian primates exhibit copulation at times when fertilization seems unlikely. There are some exceptions (e.g. the squirrel monkey, *Saimiri,* and the gorilla), and there is also some indication that the spread of copulations over the cycle may typically be greater in captivity than in the wild, but in many primates copulation is clearly not confined to the peri-ovulatory period even in the wild. This basic distinction between simian primates and other mammals was recognized by both Heape [19] and, particularly, by Zuckerman [20]: 'The matings of the lower mammal are confined to short oestrous periods, towards the end of the follicular phase. The matings of the primate ... are diffused over the entire cycle ...'. When this was written, very little was known about reproduction of prosimians, so 'primate' here actually refers only to monkeys and apes. Recent evidence has abundantly confirmed Zuckerman's statement as far as diffuse copulation in many simian primates is concerned.

The distinction between prosimians and simians, with restriction of copulation to a limited period around the time of ovulation in the former, is interesting in the context of primate evolution (fig. 1). It would seem that the primitive mammalian pattern characterized by a distinctive oestrus was initially retained in ancestral primates but subsequently lost during the evolution of simians. Although it is possible that the loss of a distinctive

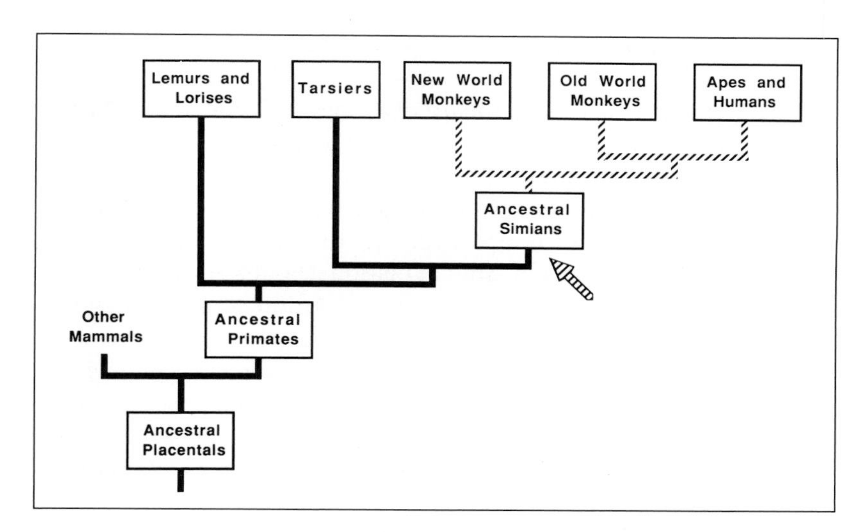

Fig. 1. Outline phylogenetic tree of the primates [based on 141], showing lineages characterized by the occurrence of oestrus (black branches) and those characterized by loss of oestrus in all or most species (hatched branches). An economical interpretation is that the loss of typical mammalian oestrus occurred in the ancestral stock of the simian primates (arrowed).

oestrus occurred independently in various lineages of monkeys and apes, the simplest hypothesis is that this change took place in the common ancestor of simians. According to this interpretation, 'loss of oestrus' is a basic feature of simian primates, not just of humans. It is therefore best to avoid the term 'oestrus' when referring to simian primates [26, 76]. Although the condition in humans may represent an extreme development, we need to explain why such a fundamental change with major implications (see below) occurred in ancestral simians.

It is possible that copulation during the female cycle at times other than close to ovulation may serve some special function in simian primates. Zuckerman [20] suggested that 'diffusion' of mating throughout the menstrual cycle of monkeys and apes was related to the emergence of a higher form of social organization. This interpretation was explicitly adopted, for example, by Sahlins [77]. In fact, however, all such interpretations fail to acknowledge a basic problem that arises when copulation is not tightly synchronized with ovulation, in stark contrast to the typical mammalian condition.

Desynchronization of Copulation and Ovulation

The absence of clearly identifiable 'oestrus' and the occurrence of copulation outside the apparent time of ovulation during the female cycle have been reliably reported only for simian primates. These unusual characteristics are hence far less widespread among mammals than copulation during pregnancy. Indeed, it is surprising that they occur at all, as desynchronization of copulation and ovulation should theoretically lead to major problems. Synchronization of copulation with ovulation is obviously advantageous, in that fertilization of the egg(s) soon after ovulation by relatively fresh spermatozoa is thereby guaranteed. Although there is some variation between mammalian species, it is generally accepted that eggs typically have a limited lifespan of just a few hours, with a maximum of less than 24 h, while sperm usually survive for up to 48 h (although some bats are exceptional). Fertilization would hence normally be expected to take place within a period of about 72 h surrounding the time of ovulation, which agrees closely with the maximum duration of typical mammalian oestrus. As soon as mating becomes more 'diffused' through the non-pregnant cycle, however, mating prior to or following the time of ovulation (fig. 2) can potentially lead to : (1) fertilization of a freshly ovulated egg by degenerating sperm; (2) fertilization of a degenerating egg by fresh sperm. In both cases, abnormal development is likely to follow such untimely fertilization. Experimental work on guinea pigs, rats and rabbits [78–82] has shown that gamete aging is associated with reduced survival of embryos following fertilization. In the case of the egg, an increased frequency of multiple fertilization (polyspermy) may occur following aging. Further, completion of meiosis does not take place until after fertilization, so penetration of a degenerating egg by a single fresh sperm can also be associated with various chromosomal abnormalities [83, 84].

At the very least, fertilization involving aging gametes will lead to early reproductive failure and a consequent delay in reproduction. It may be that there is usually an 'all-or-none' effect in that the development of zygotes with abnormalities arising from degeneration of gametes is arrested at an early stage. It is also possible that some zygotes on the boundary between 'normality' and 'abnormality' slip through the net to continue their development. Orgebin-Christ [85] suggests that there is some evidence both for humans and other mammals that the latter occurs, although the human evidence (necessarily circumstantial) is not conclusive. Iffy [86] and Iffy et al. [87] have presented data indicating that certain abnormali-

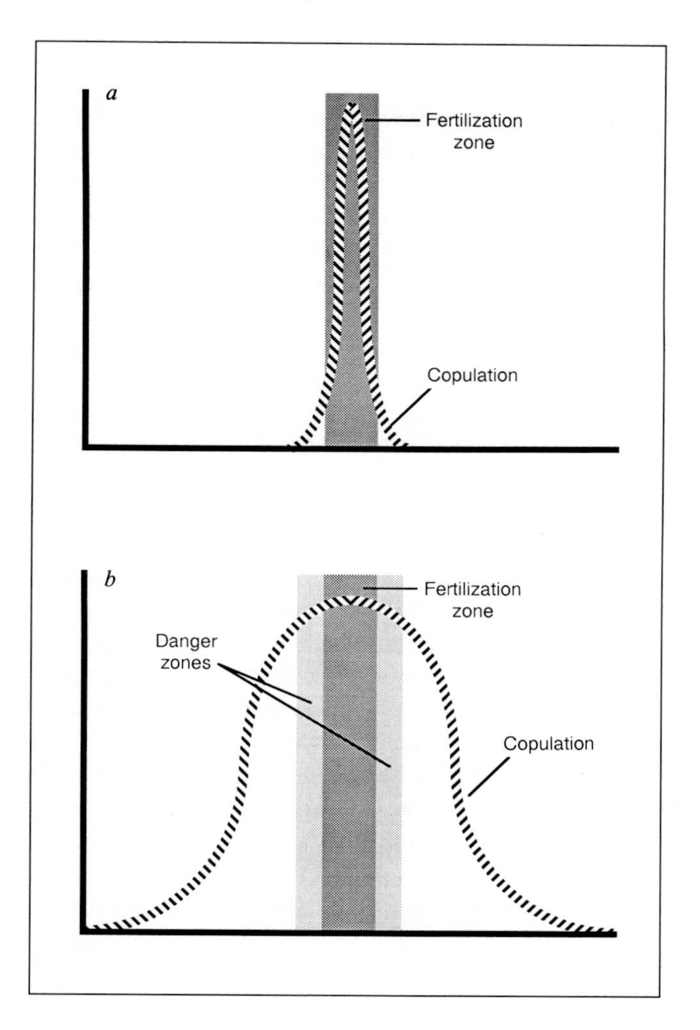

Fig. 2. Contrast between the typical mammalian non-pregnant cycle, with tight synchronization of copulation with ovulation (*a*), and the cycle generally recognized for simian primates (*b*), in which ovulation occurs at a specific point in the cycle, while copulation can occur at other times. If copulation occurs at times other than close to ovulation, two danger zones arise in which degenerating sperm may fertilize a recently ovulated egg or a degenerating ovum can be fertilized by fresh spermatozoa. Zygotes arising from fertilization in these danger zones are likely to exhibit abnormalities.

ties in human foetal development are associated with conception at time other than close to mid-cycle, particularly in the concluding phase. It has also been proposed [84] that the association between Down's syndrome and maternal age may be due to reduced copulatory frequency per cycle and a resulting higher incidence of delayed fertilization. Be that as it may, the 'danger zones' illustrated for the simian model in figure 2 should theoretically lead to gamete loss and postponement of pregnancy, if not to birth of deformed offspring, unless special mechanisms exist to preclude fertilization involving aging gametes.

Viewed in this light, the common 'diffusion' of mating through the non-pregnant cycle in simian primates assumes a special significance. In the typical mammalian condition, the female exhibits a well-defined oestrus, with mating confined to a relatively short peri-ovulatory period, such that the 'danger zones' in figure 2 are essentially avoided. Given such a basic pattern, it is to be expected that natural selection would eliminate deviations, as females that permit mating outside oestrus or exhibit ovulation out of synchrony with other components of the cycle would suffer a delay in reproduction. How, then did the condition in most simian primates evolve? In the absence of special mechanisms to prevent untimely fertilization, the selective advantage of developing diffuse copulation would have to outweigh the risk of heavy reproductive costs. Any explanation of the diffusion of mating through the non-pregnant cycle in simian primates must either identify special mechanisms that eliminate untimely fertilization or specify major advantages that offset the potential disadvantages of such fertilization, which would be particularly acute for simian species given their very slow rate of reproduction.

The theoretical existence of 'danger zones' before and after ovulation in the standard model of the simian female cycle, with the accompanying possibility of untimely fertilization with heavy reproductive costs, is particularly relevant to dominance among males. In simian primate species that live in muti-male groups, it is common to find a dominance hierarchy among sexually mature males and it is generally assumed that male dominance rank is positively correlated with reproductive success. Dominant males are presumed to have preferential or exclusive access to females at ovulation and hence to father a greater proportion of offspring in the group. It is well known that females in multi-male groups can mate with several sexually mature males in the course of a single cycle (e.g. baboons, macaques, chimpanzees), and various authors have therefore suggested that dominant males have priority of access to females at ovulation. A particular form of behaviour that is thought to ensure such priority of access is the formation of an exclusive 'consortship' between a particular male and a particular female, lasting for a period of several days that is presumed to include the time of ovulation. In many Old World simian species that live in multi-male groups (e.g. *Cercocebus* spp., *Macaca nemestrina, Macaca nigra, Macaca sylvanus, Miopithecus talapoin, Papio anubis, Papio ursinus, Pan troglodytes*), the female has a readily visible sexual swelling that changes in size and coloration during the non-pregnant cycle [88]. Maximal development of the sexual swelling typically occurs at mid-cycle and it is commonly accepted that ovulation occurs during maximal swelling, at some time close to the onset of deflation. It is therefore often assumed that in such species a dominant male can recognize the approach of ovulation and behave appropriately to monopolize copulatory access to the female at this time.

If it is genuinely possible for a given male to ensure exclusive access to a female around the time of ovulation, it might be thought that the potential problem of 'danger zones' illustrated in figure 2 will not arise. It is, however, necessary to consider what happens when a single male copulates with the female in the 'danger zone' prior to ovulation and then copulates with her again close to the time of ovulation. The sperm from the earlier (untimely) copulation will already be far up in the female tract when sperm from the subsequent (peri-ovulatory) copulation are introduced. Is there some mechanism to ensure that in any resulting competition between fresh and aging sperm the latter are consistently excluded from fertilization? (Incidentally, a similar question also arises with respect to human reproduction.) At present, there would seem to be no answer to this question. A similar consideration applies when a subordinate male copulates with a female in the 'danger zone' a few days prior to ovulation and a dominant male subsequently mates with her close to the time of ovulation (fig. 3). Aging sperm from the subordinate male may be close to the egg at the time of ovulation and sperm from the dominant male will enter the female tract at a later stage. The hypothesis that a dominant male can ensure paternity by monopolizing copulation with a female close to the time of ovulation therefore requires either (1) total exclusion of all other males during a period of several days (with ovulation occurring toward the end of that period) and a mechanism to ensure that only fresh sperm from the dominant male will fertilise the egg, or (2) a reliable mechanism to ensure that the dominant male's fresh sperm will out-compete any aging sperm from a subordinate male.

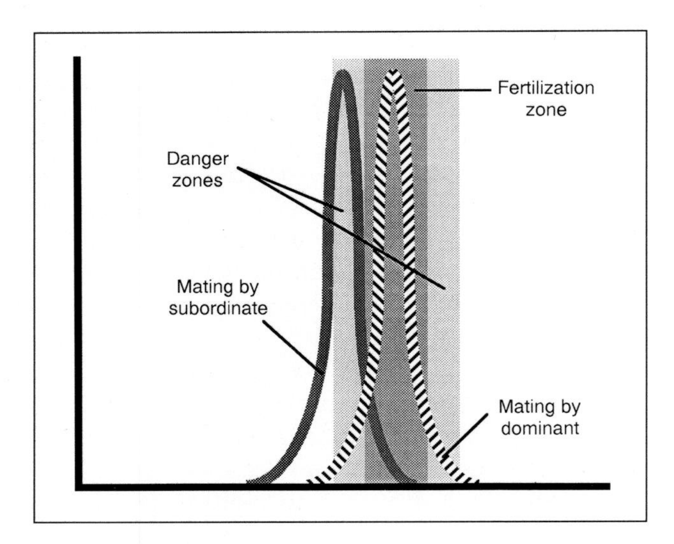

Fig. 3. Illustration of the cycle generally recognized for simian primates, showing occurrence of mating by subordinate and dominant males relative to ovulation at mid-cycle. The dominant male is presumed to concentrate his copulation around the time of ovulation, while subordinates males may mate at other times. If a subordinate male mates with the female prior to ovulation, copulation in the pre-ovulatory 'danger zone' may occur and in the event of fertilization this could incur reproductive costs.

The commonly postulated relationship between male dominance and preferential access to females at the time of conception also suffers from another drawback in that the hypothesis is almost entirely male-oriented. It is clearly advantageous for a male to seek to monopolize a female when ovulation is likely to occur so that he can maximize his own reproductive output. It is therefore to be expected that males will compete for access to females at ovulation (assuming that they can recognize when this will occur) and the formation of a dominance hierarchy among males may be the outcome of this competition. On the other hand, it is by no means so obvious that it is in the female's interest to have her offspring fathered by the top-ranking male in the dominance hierarchy. It may be in her interest to copulate most with that male and to create the impression that he is likely to be the father of her offspring, for example in order to reduce aggressive interactions. Yet it may be that a lower-ranking (perhaps less aggressive and more cooperative) male would be genetically more suitable as a father of her offspring.

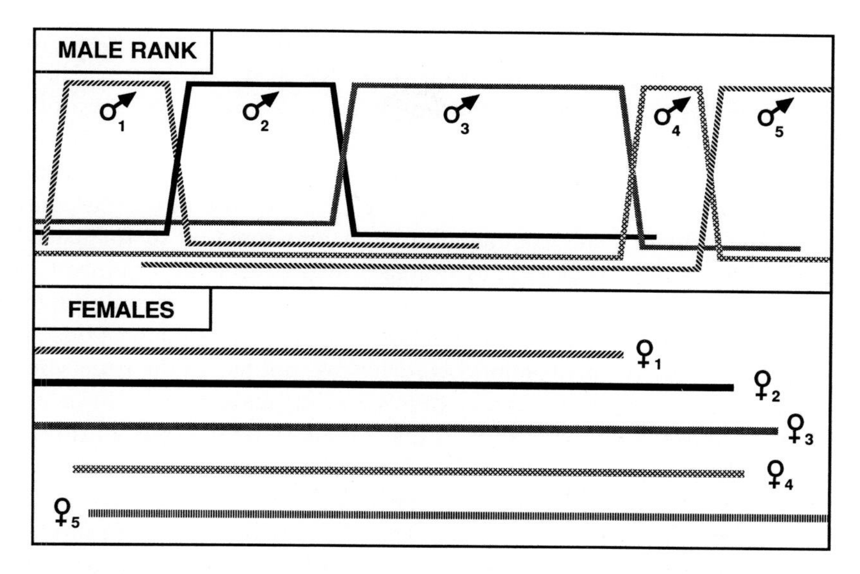

Fig. 4. Diagram of male rank relationships in a multi-male simian primate group containing 5 sexually mature males and 5 sexually mature females, showing the dynamic nature of dominance. In this hypothetical example, each male achieves the alpha position in the group at some time, but males differ in their length of tenure. Male 3 has the longest tenure as alpha male in the group. If length of tenure is linked to desirable genetic properties, it would be in the interests of the females to bear offspring fathered by this male both before and after his occupation of the top rank and not merely during his tenure as alpha male.

It is widely assumed that dominant males are in some way 'genetically superior' and that it is necessarily in a female's interest to have offspring of dominant males, but this assumption is by no means compelling. Further, although it might be argued that breeding with the dominant male could be generally advantageous for the female within the social group even without a genetic benefit (e.g. because of protection provided by that male), it is actually only necessary for the female to establish a positive relationship with the dominant male; he need not necessarily be the father of her offspring.

In fact, even if it is assumed that dominant males are in some way preferable on genetic grounds as fathers of offspring, there is a basic flaw in arguments linking paternity to male dominance rank in that dominance relationships undergo dynamic change in social groups of simian primates (fig. 4). As a general rule, a given male will achieve dominance relatively late in life and will then remain dominant for a limited period, subsequently declining in rank following a change in the hierarchy. Although certain males may never achieve dominance and may hence be regarded as in some way inferior, a number of males in a given group will be dominant at some time in their lives. Among males that do achieve a high dominance rank, there are important differences in the period of tenure. It is conceivable that a male that becomes dominant and remains so for a lengthy period might possess genetic advantages. From a female's point of view, however, he would be genetically 'superior' at all times. It could therefore be in the interest of females to have offspring fathered by such a male both prior to and following his tenure of high rank, not just at the time when he happens to be dominant. Choice of a 'superior' male prior to his achievement of dominance would, of course, require female recognition of a male's superiority from criteria other than temporary dominance rank, but this is not beyond the bounds of possibility.

At this point, it must be noted that the generally accepted model for the non-pregnant cycle of simians illustrated in figures 2 and 3 incorporates the interpretation that ovulation takes place spontaneously at a well-defined point (e.g. close to mid-cycle), in response to a marked surge in circulating levels of the hormone LH. The inference that diffuse copulation might lead to untimely fertilization (fig. 2b, 3) directly depends on this standard interpretation. However, data on the variability of gestation periods recorded for simian primates indicate a special condition in comparison to other mammals.

Analysis of Mammalian Gestation Periods

As a rule, the gestation period shows suprisingly little variability within any given mammalian species. It was noted by Kiltie [89], however, that in simian primates the coefficient of variation for recorded gestation periods seems to be greater than in other mammals. Kiltie included no prosimians in his sample, but a preliminary survey of primates [70] indicated that, in comparison to monkeys and apes, lemurs and lorises typically exhibit lower values for the coefficient of variation in gestation period. Analysis of a carefully checked set of data confirms all of these findings. For a sample of 27 non-primates (table 2), coefficients of variation for gestation period are 0.8–3.5% (average: 2.1%). For a sample of 12 prosim-

Table 2. Variation in inferred gestation periods for non-primate species

Species	Gestation period				Source
	mean	n	SD	CV, %	
Sylvilagus aquaticus	36.8	18	± 1.0	2.7	90
Neotoma mexicana	32.7	8	± 0.8	2.4	91
Zygodontomys microtinus	25.0	29	± 0.2	0.8	92
Spermophilus richardsonii	22.0	9	± 0.3	1.4	93
Peromyscus leucopus	23.0	16	± 0.5	2.2	94
Peromyscus maniculatus	23.0	18	± 0.6	2.6	94
Mesocricetus auratus	15.0	71	± 0.4	2.7	95
Mus musculus	20.0	93	± 1.1	3.0	96
Oryctolagus cuniculus	31.5	569	± 0.6	3.4	97
Capra hircus	150.8	6,761	± 3.3	2.2	98
Ovis aries	151.0	1,584	± 2.2	1.5	99
Oreamnos americanus	185.8	9	± 1.4	0.8	100
Tragelaphus spekei	250.9	10	± 5.8	2.3	101
Lama pacos	341.6	169	± 5.3	1.5	102
Cervus elaphus	231.0	13	± 4.5	2.0	103
Odocoileus virginianus	205.3	10	± 5.0	2.4	104
Bos taurus	278.2	384	± 4.8	1.7	105
Camelus dromedarius	390.0	33	± 12.1	3.1	106
Giraffa camelopardalis	458.4	12	± 13.0	2.8	107
Sus scrofa (wild)	115.2	41	± 2.4	2.1	108
Sus scrofa (domestic)	113.5	1,450	± 1.7	1.5	109
Tayassu tajacu	144.9	9	± 2.3	1.6	110
Hyaena hyaena	90.0	9	± 1.8	2.0	111
Bassariscus astutus	51.4	19	± 0.7	1.3	112
Felis geoffroyi	65.3	9	± 2.3	3.5	113
Panthera uncia	98.9	8	± 1.1	1.1	114, 115
Rhinoceros unicornis	478.5	27	± 7.1	1.5	116

ians (table 3), coefficients of variation for gestation period are 0.5–2.9%, with a similar average of 1.9%. By contrast, for a sample of 15 simians (table 3), coefficients of variation for gestation period are 2.5–5.4% (average: 4.0%). As figure 5, shows, there is a clear difference between prosimians and simians.

The greater variability in gestation periods recorded for simian primates, compared to other mammals, can be explained in one of two ways. One possibility is that gestation periods are inherently more variable in

Table 3. Variation in inferred gestation periods for primate species

Species	Gestation period				Source
	mean	n	SD	CV, %	
Prosimian primates					
Galago crassicaudatus	136	21	± 3.7	2.7	117
Galago garnettii	130	12	± 2.5	1.9	117
Galago moholi	125	28	± 2.4	1.9	117
Galago senegalensis	141	23	± 1.7	1.2	118
Lemur catta	136	14	± 1.6	1.2	28
Lemur coronatus	125	6	± 2.9	2.3	119
Lemur fulvus	121	7	± 3.5	2.9	70
Lemur macaco	128	9	± 0.7	0.5	70
Loris tardigradus	168	10	± 3.8	2.3	120–123
Microcebus murinus	60	33	± 1.7	2.8	124
Nycticebus coucang	192	9	± 3.6	1.9	125
Varecia variegata	102	45	± 1.7	1.7	126–128
Simian primates					
Callithrix jacchus	148	17	± 4.3	2.9	129
Callimico goeldii	149	6	± 5.2	3.5	130
Cercopithecus aethiops	163	38	± 6.2	3.8	131
Cercopithecus ascanius	172	8	± 8.2	4.8	41
Erythrocebus patas	164	10	± 6.7	4.1	132
Gorilla gorilla	260	21	± 7.7	3.0	70
Homo sapiens	280[1]	15,861	± 15.0	5.4	133
Macaca arctoides	177	25	± 5.1	2.9	134
Macaca cyclopis	165	8	± 6.4	3.9	135
Macaca fascicularis	160	109	± 7.0	4.4	136
Macaca fuscata	173	17	± 6.9	4.0	137
Macaca mulatta	164	561	± 6.0	3.7	136
Macaca radiata	162	16	± 4.0	2.5	136
Papio anubis	173	78	± 9.4	5.4	138
Pan troglodytes	226	116	± 11.8	5.2	139

[1] For human beings, it is customary to define 'gestation periods' with respect to the last menstrual period, as there is no simple indicator of the actual time of conception. If it is assumed that ovulation normally occurs at mid-cycle, with a typical cycle length of 28 days, then 14 days must be subtracted from this average figure of 280 days to give the true period from conception to birth, i.e. approximately 266 days.

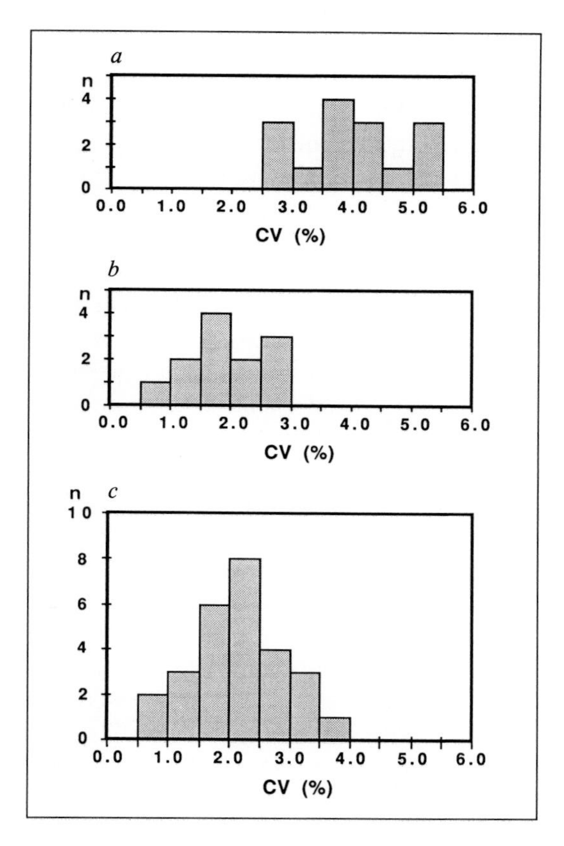

Fig. 5. Distributions of values for the coefficient of variation (CV) in gestation period for simian primates (*a*), prosimian primates (*b*) and non-primate mammals (*c*). The distributions for prosimians and non-primates are indistinguishable and the average values (1.9% vs. 2.1%) are very similar. By contrast, the values for simian primates generally exceed those for the other mammals shown and the average value (4.0%) is twice as high.

simians. Given that gestation periods show a very low level of variability in all other mammals, however, this explanation is unacceptable without a special argument to account for the increased variability of gestation in simians. The other possibility is that gestation periods recorded for simian primates are generally less reliable than for other mammals and that natural variability in gestation is exaggerated through errors in determining the time of conception.

It must be emphasized that the figures in tables 2 and 3 generally refer to *inferred* gestation periods. The standard practice is to take some standard, externally visible reference point in the non-pregnant cycle as an indicator of the likely time of ovulation and to calculate the gestation period from that point in what appears to be the cycle of conception. With both non-primates and prosimians, this is relatively unproblematic because there is usually a quite well-defined period of oestrus with copulation confined to that period. For this reason, coefficients of variation probably reflect intrinsic variability in the gestation period itself rather than any significant error in inference of the time of fertilization. (Nonetheless, it should be remembered that many mammals mate during pregnancy and that simply calculating gestation periods from the last observed copulation is an unreliable procedure for mammals generally. Figures for 'gestation periods' must always be carefully checked.) With simian primates, in contrast to other mammals, there is usually no well-defined oestrous period. The date of conception must be inferred from observed copulation, perhaps with the aid of additional guides such as changes in a sexual swelling (if present) and the common assumption for Old World simians that ovulation takes place at approximately mid-cycle.

As simian primates commonly lack a definite restriction of copulation to the peri-ovulatory period of the cycle, determination of gestation periods is problematic. For example, with certain simian species listed in table 3 *(Macaca mulatta, Papio anubis, Pan troglodytes)* it has been customary to take some point in the sexual swelling cycle, in some cases maximum turgescence, as an indicator of the approximate time of ovulation. But, with many other species, including *Homo sapiens,* even this procedure is ruled out. In the human case, the 'gestation period' is often calculated as the interval from the last menstrual period (LMP) to birth, with the result that any variability in the interval from LMP to the next ovulation is included in the overall variability of inferred lengths of pregnancy, which are hence not directly comparable with the actual gestation periods of any other mammal. This could partly account for the fact that human beings exhibit the greatest variation (in days) among the species listed in table 3, although this itself implies that the standard model invoking ovulation some 14 daxs after LMP must be relatively inaccurate. Two other simian species in table 3 are of special interest. The common marmoset *(Callithrix jacchus)* exhibits no readily visible external signs of the time of ovulation and copulation can occur at any time of the cycle (table 1). The gestation periods included in table 3 were, however, calculated on the basis of hormonal evidence (LH peak) of the time of ovulation, and it is noteworthy that the recorded coefficient of variation (2.9%) is one of the lowest for all of the simian species listed, falling at the upper end of the general mammalian range. Indeed, for *Callithrix jacchus,* a later study also timing conception from the LH peak [140] indicated a gestation period of 144 ± 1.5 days, with a coefficient variation of only 1.0%, although the sample was small (n = 5). The evidence for *Callithrix* therefore indicates that hormonally timed gestation periods may be less variable than those determined from external indicators of ovulation. The second case is that of the gorilla *(Gorilla gorilla),* which is relatively unusual among the simian primates surveyed in that during the non-pregnant cycle copulation – both in captivity and in the wild – is typically restricted to a period of a few days in mid-cycle, such that the approximate time of ovulation can be inferred fairly reliably from copulatory activity. This species also exhibits a relatively low coefficient of variation for recorded gestation periods (3.0%). It is species of the genera *Cercopithecus, Macaca, Papio, Pan* and *Homo* that generally exhibit the greatest variation among the simian primates surveyed (table 3). This itself suggests that ovulation may not be so rigidly confined to mid-cycle in *Macaca, Papio, Pan* and *Homo* as is commonly implied.

There is, incidently, a serious problem with the common procedure in field studies of backdating from a birth date to infer the likely time of conception, taking a standard figure for the gestation period. For a monkey with a gestation period of 6 months and a coefficient of variation of ± 2% (the low average figure for mammals other than simians, tables 2 and 3), the 99% confidence limits on the inferred time of ovulation would be ± 10 days, so conception could in fact at any time within a period of 20 days. For a great ape with a gestation period of 8.5 months, the 99% confidence limits on the inferred time of ovulation would be ± 15 days, so conception could occur at any time within a period of 30 days. As the cycle length for most simian primates is about 28 days [141], this means that the cycle of conception can usually be reliably identified, but that conception can be allocated to almost any point in the cycle. If the higher average coefficient of variation of ± 4% for variation in gestation periods of simian primates (table 3) is a true reflection of variability in the time from conception to birth, the 99% confidence limits on the time of ovulation would be ± 20 days for the monkey and ± 30 days for the great ape. This would mean that conception could occur at any time within a total period of 40 days and 60 days, respectively, and it would hence be impossible even to identify the cycle of conception with any confidence.

Time of Conception in the Simian Cycle

As noted above, human gestation is usually timed relative to LMP and it is generally assumed that conception occurs about 14 days after LMP. Yet there are numerous indications that mid-cycle ovulation is at most a statistical probability [142, 143]. For 34 cases in which mothers were reasonably confident that they were able to limit the time of probable conception to a single period of 1–3 days, Finnström [144] calculated a mean interval of 12.6 ± 3.8 days between LMP and conception, indicating 95% confidence limits of 7.6 days and 99% confidence limits of 11.4 days (assuming normality of distribution). In other words, although conception is most likely to occur at about mid-cycle, it may in fact occur on at least 23 days of a 28-day cycle. It has also been suggested that ovulation, in addition to occurring spontaneously in the manner typical of primates, may sometimes be induced in the course of the human menstrual cycle [e.g. see 145], but the evidence for this is tentative (see later).

According to the widely accepted model for the human female, ovulation typically occurs at mid-cycle and that there is hence an extensive 'safe period' during which ovulation and hence fertilization do not occur. This concept was championed particularly by Knaus [146, 147] and Ogino [148, 149] in the face of a considerable body of conflicting evidence. Numerous studies have compiled data for conceptions attributable to a limited time in the cycle: from conceptions arising during brief periods of leave of soldiers during wartime [e.g. 150], from carefully checked accounts of single cohabitation during the cycle of conception [e.g. 151–153] and from incidences of rape [e.g. 145]. As can be seen from figure 6, all studies agree in showing that conception can occur on virtually any day of the human cycle. There is no evidence of a distinct mid-cycle peak, although there is a consistent trend for conception to occur far more often in the first half of the cycle than in the second half. Indeed, some of the studies indicate a

---

Fig. 6. Distributions of human conceptions relative to day of the menstrual cycle indicated by 5 separate studies (Pryll [150]: n = 723; Manulkin, [151]: n = 104; Dyroff [152]: n = 1,022; Hollenweger-Mayr [153]: n = 928; Jöchle [145]: n = 543). In order to match the graph for Pryll's data, a maximal cycle length of 40 days has been indicated in all cases although other authors cited data only for cycle lengths up to 31 days. It can be seen that conceptions can occur at any time during the cycle, although most tend to occur during the first 20 days. Peak conception is indicated at various times between day 8 and day 14 of the cycle, according to the study concerned.

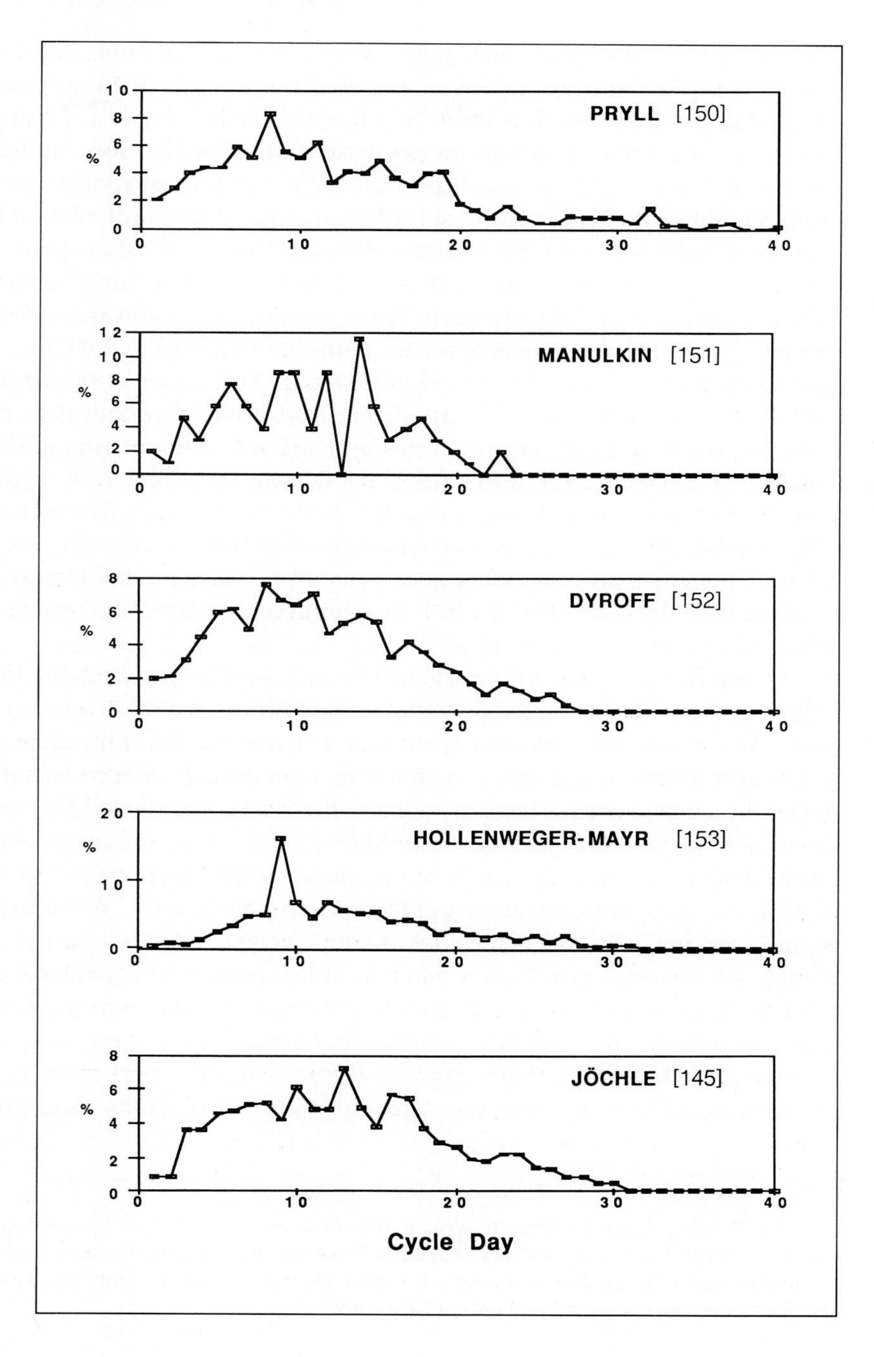
PRYLL [150]
MANULKIN [151]
DYROFF [152]
HOLLENWEGER-MAYR [153]
JÖCHLE [145]
%
Cycle Day

peak in conceptions before mid-cycle, on days 8–10. Numerous authors [e.g. ref. 147] have dismissed such evidence on the grounds that it is based on unreliable testimony, but there is little justification for this. Firstly, there is a remarkable agreement in the pattern of results between studies (fig. 6). Secondly, if the Knaus-Ogino model is correct, unreliable testimony should surely lead to increased scatter of conceptions around a mid-cycle peak and not to the biased pattern shown in figure 6. Thirdly, human gestation periods calculated from Dyroff's data for 1,150 human pregnancies with known dates of conception based on testimony from individual women [152] yield a mean value for the gestation period of 267.5 ± 10.7 days, with a coefficient of variation of ± 4.0%. This figure is markedly lower than that of 5.4% given in table 3, which is derived from calculations based on dates of LMP, and indicates a greater accuracy in timing the duration of pregnancy. It is also noteworthy that the same pattern is found regardless of whether data were collected in the normal marital context, with reference to brief periods of leave or exclusively from rape cases. There is therefore no compelling need to invoke induced ovulation, as a response to violence or great arousal, to account for conceptions occurring other than at mid-cycle [e.g. 145].

In parallel with the Knaus-Ogino model generally accepted for the human cycle, it is often held that ovulation also occurs at a relatively fixed point in the cycle of simians, taking place close to the middle of the cycle in Old World monkeys and apes. At first sight, this would seem to be supported by widely cited laboratory data for *Macaca mulatta* [154] and *Papio cynocephalus* [155], as shown in figure 7. The histograms as shown, however, give a misleading visual impression as the authors in both studies arranged matings primarily close to the mid-point of the cycle. When conceptions are expressed as a percentage of matings (fig. 8), it can be seen that there is no clear peak at mid-cycle and that little information is available at the beginning and the end of the cycle because very few matings were arranged at those times. Further, a direct laparoscopic study of the ovary in baboons [156] has shown that – although there is a marked peak at the end of maximal swelling of the sex skin – mature and rupturing follicles can be

---

Fig. 7. Histograms showing the distribution of conceptions during the cycle for *Macaca mulatta* [154] and *Papio cynocephalus* [155], giving a visual impression of a pronounced peak at mid-cycle. In fact, for both species the investigators primarily arranged matings close to the mid-point of the cycle.

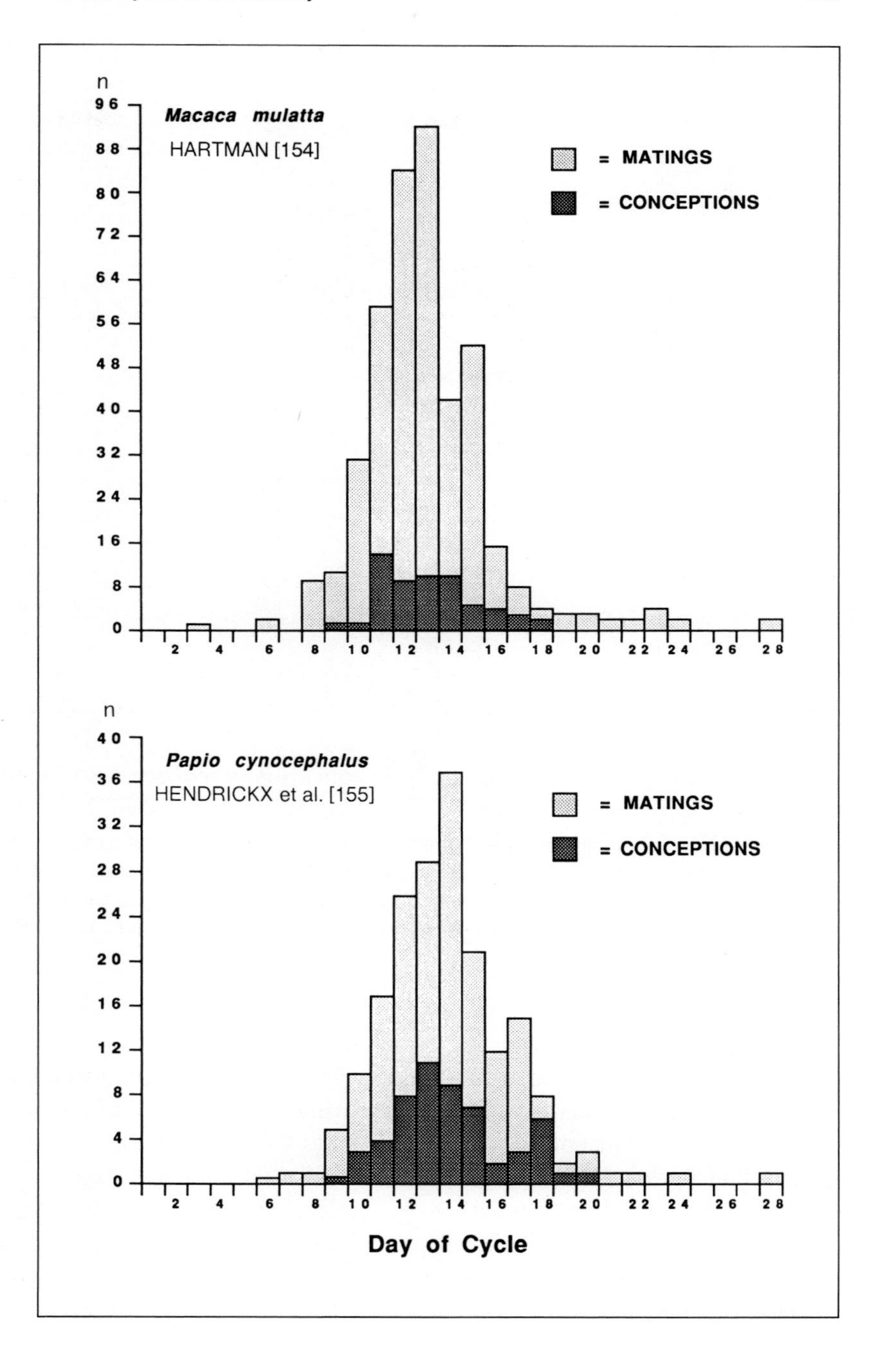
n
96
88
80
72
64
56
48
40
32
24
16
8
0
Macaca mulatta
HARTMAN [154]
= MATINGS
= CONCEPTIONS
2 4 6 8 10 12 14 16 18 20 22 24 26 28
n
40
36
32
28
24
20
16
12
8
4
0
Papio cynocephalus
HENDRICKX et al. [155]
= MATINGS
= CONCEPTIONS
2 4 6 8 10 12 14 16 18 20 22 24 26 28
Day of Cycle

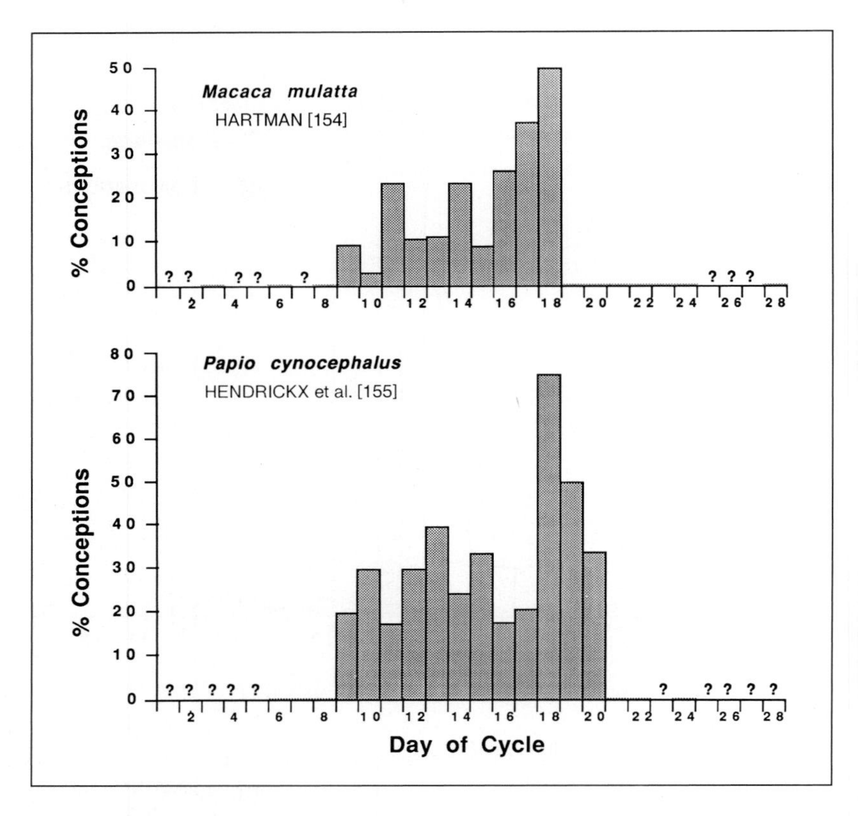

Fig. 8. Histograms showing conceptions as a proportion of mating during the cycle for *Macaca mulatta* [154] and *Papio cynocephalus* [155]. The question marks indicate days on which no matings were conducted. There is no clear mid-cycle peak in either species and the initial and terminal phases of the cycle have not been adequately tested for the occurrence of conceptions.

found over an 8-day period around mid-cycle. A study of the patas monkey *(Erythrocebus patas),* using single timed matings on days 9–19 of the cycle, also showed that conceptions occurred on days 9–16 with no obvious peak during that period [157]. As no matings were attempted for days 1–8 or days 20–30, it is unknown whether conception can occur at the beginning or end of the cycle. There is therefore no convincing evidence for a tightly restricted mid-cycle peak of conception in *Erythrocebus patas, Macaca mulatta* or *Papio cynocephalus.*

There is, however, one exceptional case for which reliable data now indicate a close relationship between sexual swelling, ovulation and successful copulation. The mandrill *(Mandrillus sphinx)* has a particularly well-developed consortship system involving 'mate-guarding' [158] in which a dominant male can effectively monopolize a female at ovulation and can demonstrably ensure paternity of offspring [159]. Gestation periods inferred both from a study of free-ranging mandrills in a large enclosure [160] and from breeding in captivity [161] show surprisingly limited variation. For the total sample (n = 11 + 5 = 16), the average figure for the gestation period is 176 ± 2.8 days, with a coefficient of variation of ± 1.6%, which lies well below the typical range of simian values (table 3) and falls within the general mammalian range (table 2). This indicates that, for this exceptional species, the time of conception can indeed be accurately inferred from behavioural observations despite diffuse mating and that the sexual swelling provides a reliable external indication of the time of ovulation, thus permitting successful mate-guarding.

Tests with Timed Fertilization

By using data for accurately timed fertilizations (from single timed matings, artificial insemination or embryo transfer following in vitro fertilization) it is possible to test whether the greater variability recorded for simian gestation periods is due to greater inherent variability in the true period of gestation or to inaccuracies in inferring times of conception.

Analysis of data from 43 normal human pregnancies resulting from single insemination and leading to single births [Prof. Y. Rumpler, Strasbourg, pers. commun.] yields a less variable estimate of the gestation period than that shown in table 3: 263.0 ± 11.6 days, with a coefficient of variation of ± 4.4%. This degree of variation is similar to that obtained by Dyroff [152] with gestation periods inferred from data on single cohabitations. An even more accurate figure was obtained by taking data from 26 normal pregnancies resulting from embryo transfer and leading to single births [Prof. Y. Rumpler, pers. commun.]: 262.3 ±7.2 days, with a coefficient of variation of ± 2.7%, which lies within the general range for mammals other than simian primates. (NB. There should in fact be a small difference between the average figure for embryo transfer and the true period between conception and birth, due to the fact that embryos are introduced to the womb some 4 days after fertilization of the egg. The average figure from embryo transfer therefore corresponds to a true gestation period of about 266 days for humans.) It is therefore clear that the true variability in the human gestation period is much less than that indicated by data based on LMP dates.

There are similar indications for rhesus macaques that recorded gestation periods are less variable when the time of fertilization is in some way controlled, in comparison to figures based on indirect indications of the time of conception. In Hartman's original study of *Macaca mulatta* [154], the average gestation period for 30 cases was 163.9 ± 8.0 days, with a coefficient of variation of 4.9%. In a subsequent study in which a female was placed with a male for a period of only 72 h in any given cycle [136], the coefficient of variation was 3.7%, as given in table 3. Finally, for 53 single timed matings in which each female was exposed to a male for only a very short period on a single day (12 min in 51 cases; 4 h in 2 cases), the gestation period was found to be 166.7 ± 5.2 days, with a coefficient of variation of only 3.1% [Prof. R.W. Goy, Wisconsin, pers. commun.]. This latter figure lies at the upper end of the range for mammals other than simian primates and should provide a relatively accurate indication of the true period from conception to birth.

Data are also available from work on the artificial insemination of chimpanzees [162]; Dr. K.G. Gould, Atlanta, Ga., pers. commun.]. The average of 9 figures for the gestation periods following artificial insemination is 237.4 ± 7.9 days, giving a coefficient of variation of ± 3.3%. This agrees well with a figure of 231.1 ± 5.9 days (coefficient of variation: ± 2.6%) for 7 single, timed matings cited by Martin [162], excluding one obviously aberrant figure of 203 days. In the latter instance, the coefficient of variation is within the usual range of variation in mammals, in marked contrast to the figures given in table 1, which are based on less precise estimates of gestation period obtained from records of menstrual cycles and copulation in a captive colony of chimpanzees.

From the limited evidence available, it can therefore be concluded that in several simian primates, at least, the natural timing of ovulation is probably rather variable with respect to any externally visible indicator of the non-pregnant cycle, whereas the level of intrinsic variation in the gestation period itself probably does not differ from that in other mammals. It therefore seems that simian primates may be characterized both by variation in copulation with respect to any fixed point in the non-pregnant cycle and by variation in ovulation in relation to that fixed point. Hence synchronization of an encounter between relatively fresh gametes (ova and spermatozoa) is rendered even more problematic.

Conclusions

It is clear that copulation in the human species is by no means as distinct from that of other mammals as has often been claimed. Mating during pregnancy is so widespread among mammals generally that it certainly cannot be regarded as a special human adaptation. Apart from copulation during pregnancy, it has often been claimed that the human female is unique among primates in exhibiting no clear external indication of the time of ovulation and in engaging in copulation at any stage of the non-pregnant cycle. Claims for human uniqueness in this respect are also unjustified, as diffusion of copulation through at least part of the cycle is common among simian primates and probably evolved in their common ancestor (fig. 1). This phenomenon is of fundamental significance because failure to synchronize copulation with ovulation can potentially lead to fertilization involving aged gametes (fig. 2).

It is a general rule among monkeys and apes that copulation can occur at times other than close to ovulation during the female cycle, although there are exceptions. The gorilla and the squirrel monkey, for instance, seem to exhibit a confined mating period coinciding with ovulation under most circumstances, although female squirrel monkeys may mate on any day of the cycle during relatively brief mating tests [163]. It is noteworthy that simian primates generally exhibit considerable flexibility in copulatory behaviour, which can be modified in various ways according to the management regime in captivity, and this supports Beach's [24] proposal that in simian primates there is greater control of sexual behaviour by the brain.

Dissolution of tight synchrony between ovulation and copulation raises a fundamental question about the origins of a pattern that may potentially lead to reproductive failure (fig. 2). In view of the selective disadvantage of potential gamete loss and occasional foetal wastage, it is unlikely that desynchronization of copulation and ovulation in simian primates has occurred for any of the relatively weak reasons given in the literature (error; diffuse response to stimuli; increased intelligence of simian primates; reinforcement of pair-bonds). Instead, it seems likely that such desynchronization has evolved as a special mechanism in a group of mammals characterized by heavy investment in single offspring and by greatly extended reproductive periods. Such mechanism would be expected to enhance rather than diminish reproductive output.

Evidence from certain simian primates, including humans, indicates that the time of copulation leading to conception can vary markedly within the non-pregnant cycle, perhaps with a peak *probability* of ovulation at mid-cycle, but with conception also possible following copulation at other times. The findings presented above could possibly be explained in a number of ways, including the following (singly or in combination):

(1) Spermatozoa of simian primates can in fact survive for significantly longer than the generally accepted period of 48 h in the female tract.

(2) Ovulation can occur over a much longer period in the female cycle than is generally accepted.

(3) More than one ovulation can occur during the cycle ('paracyclic ovulation').

(4) Ovulation is not (only) spontaneous, but (may also be) induced.

At present, it is impossible to make a clear choice between these alternatives. Similarly, it is unknown whether there is a special mechanism that usually prevents fertilization involving aged spermatozoa or ova, as would be theoretically expected as an outcome of diffuse mating during the female cycle. A solution must await the outcome of specifically designed research projects, but there is a distinct possibility that the phenomena presented above reflect a sophisticated system in which the female simian exerts special control over the timing of fertilization and hence over the paternity of her offspring.

Acknowledgments

Some basic concepts in this paper were presented in the 11th Curl Lecture (Royal Anthropological Institute, London, 1983) and further developed in a special guest lecture at the symposium 'Fertility in the Great Apes' (Atlanta, Ga., 1989). Thanks are due to the organizers of those lectures and also to following people for valuable comments, advice or assistance at various stages: Dr. H.M. Brand, Dr. A.F. Dixson, Dr. K.G. Gould, Dr. S.R. Kingsley, Dr. A.M. MacLarnon. Special thanks are due to Prof. Y. Rumpler for data on human artificial insemination/embryo transfer, to Dr. K.G. Gould for data on artificial insemination of chimpanzees and to Prof. R.W. Goy for data on timed matings in macaques.

References

1 Sade DS: Can fitness be measured in primate populations?; in Markl H (ed): Evolution of Social Behavior: Hypotheses and Empirical Tests. Weinheim, Verlag Chemie, 1980, pp 97–114.
2 Duvall SW, Bernstein IS, Gordon TP: Paternity and status in a rhesus monkey group. J Reprod Fertil 1976;47:25–31.

3 Curie-Cohen M, Yoshihara B, Blystad C, et al: The effects of dominance on mating behavior and paternity in a captive group of rhesus monkeys. Am J Primatol 1983;5: 127–138.

4 Stern BR, Smith DG: Sexual behavior and paternity in three captive groups of rhesus monkey *(Macaca mulatta).* Anim Behav 1984;32:23–32.

5 Inoue M, Takenaka A, Tanaka S, et al: Paternity discrimination in a Japanese macaque group by DNA fingerprinting. Primates 1990;31:563–570.

6 Inoue M, Mitsunaga F, Ohsawa H, et al: Male mating behaviour and paternity discrimination by DNA fingerprinting in a Japanese macaque group. Folia Primatol 1991;56:202–210.

7 Inoue M, Mitsunaga F, Ohsawa H, et al: Paternity testing in captive Japanese macaques *(Macaca fuscata)* using DNA fingerprinting; in Martin RD, Dixson AF, Wickings EJ (eds): Paternity in Primates: Genetic Tests and Theories. Basel, Karger, 1992, pp 131–140.

8 Witt R, Schmidt C, Schmitt J: Social rank and darwinian fitness in a multimale group of Barbary macaques (*Macaca sylvana* Linnaeus, 1758). Folia Primatol 1981; 36:201–211.

9 de Ruiter JR, Scheffrahn W, Trommelen GJJM, et al: Male social rank and reproductive success in wild long-tailed macaques. Paternity exclusions by blood protein analysis and DNA fingerprinting; in Martin RD, Dixson AF, Wickings EJ (eds): Paternity in Primates: Genetic Tests and Theories. Basel, Karger, 1992, pp 175–191.

10 Andelman SJ: Evolution of concealed ovulation in vervet monkeys *(Cercopithecus aethiops).* Am Nat 1987:129:785–799.

11 Wilson EO: Sociobiology: The New Synthesis. Cambridge, Belknap Press of Harward University Press, 1975.

12 Lancaster JB, Lee RB: The annual reproductive cycle in monkeys and apes; in DeVore I (ed): Primate Behavior: Field Studies of Monkeys and Apes. New York, Holt, Rinehart & Winston, 1965, pp 486–513.

13 Morris D: The Naked Ape: A Zoologist's Study of the Human Animal. London, Jonathan Cape, 1967.

14 Alexander R, Noonan KM: Concealment of ovulation, parental care, and human social evolution; in Chagnon NA, Irons WG (eds): Evolutionary Biology and Human Social Organization. North Scituate, Duxbury Press, 1979, pp 436–453.

15 Hrdy SB: Infanticide among animals: A review, classification, and examination of the implications for the reproductive strategies of females. Ethol Sociobiol 1979;1: 13–40.

16 Etkin W: Social behavioral factors in the emergence of man. Hum Biol 1963;35: 299–310.

17 Etkin W: Social behavior and the evolution of man's mental faculties. Am Nat 1954; 88:129–142.

18 Burley N: The evolution of concealed ovulation. Am Nat 1979;114:835–858.

19 Heape W: The 'sexual season' of mammals and the relation of the 'pro-oestrum' to menstruation. Quart J Microsc Sci 1900;44:1–70.

20 Zuckerman S: The Social Life of Monkeys and Apes. London, Kegan Paul, Trench & Trubner, 1932.

21 Hammond J, Marshall FHA: Reproduction in the Rabbit. London, Olivier and Boyd, 1925.

22 Kleiman DG, Mack DS: A peak in sexual activity during mid-pregnancy in the golden lion tamarin, *Leontopithecus rosalia* (Primates, Callitrichidae). J Mammal 1977;58:657–660.
23 Young JZ: The Life of Mammals. Oxford, Clarendon Press, 1957.
24 Beach FA: Sexual attractivity, proceptivity, and receptivity in female mammals. Horm Behav 1976;7:105–138.
25 Dixson AF: The hormonal control of sexual behaviour in primates. Oxf Rev Reprod Biol 1983;5:131–219.
26 Loy J: The sexual behavior of African monkeys and the question of estrus; in Zucker E (ed): Comparative Behavior of African Monkeys. New York, Liss, 1987, pp 175–195.
27 Martin RD: Adaptive radiation and behaviour of the Malagasy lemurs. Philos Trans R Soc Lond [B] 1972;264:295–352.
28 Van Horn RN, Eaton GG: Reproductive physiology and behavior in prosimians; in Doyle GA, Martin RD (eds): The Study of Prosimian Behavior. New York, Academic Press, 1979, pp 79–122.
29 Izard MK, Wright PC, Simons EL: Reproductive cycles in *Tarsius bancanus.* Am J Phys Anthropol 1985;66:184.
30 Dixson AF: The owl monkey *(Aotus trivirgatus);* in Hearn JP (ed): Reproduction in New World Primates. Lancaster, MTP Press, 1983, pp 69–113.
31 Kendrick KM, Dixson AF: The effect of the ovarian cycle on the sexual behaviour of the common marmoset *(Callithrix jacchus).* Physiol Behav 1983;30;735–742.
32 Rothe H: Some aspects of sexuality and reproduction in groups of captive marmosets *(Callithrix jacchus).* Z Tierpsychol 1975;37:255–273.
33 Kleiman DG: Characteristics of reproduction and social interactions in pairs of lion tamarins *(Leontopithecus rosalia)* during the reproductive cycle; in Kleiman DG (ed): The Biology and Conservation of the Callitrichidae. Washington, Smithsonian Institution Press, 1977, pp 181–190.
34 Brand HM, Martin RD: The relationship between female urinary estrogen excretion and mating behaviour in cotton topped tamarin *(Saguinus oedipus).* Int J Primatol 1983;4:275–290.
35 French JA, Abbott DH, Showdon CT: The effect of social environment on estrogen excretion, scent marking, and sociosexual behavior in tamarins. Am J Primatol 1984;6:155–167.
36 Travis JC, Holmes WN: Some physiological and behavioural changes associated with oestrus and pregnancy in the squirrel monkey *(Saimiri sciureus).* J Zool, Lond 1974;174:41–66.
37 Wilson MI: Characterisation of the estrous cycle and mating season of squirrel monkeys from copulatory behaviour. J Reprod Fertil 1977;51:57–63.
38 Wallis SJ: Sexual behavior and reproduction of *Cercocebus albigena johnstonii* in Kibale Forest, Western Uganda. Int J Primatol 1983;4:153–166.
39 Hadidian J, Bernstein IS: Female reproductive cycles and birth data from an Old World monkey colony. Primates 1979;20:429–442.
40 Gartlan JS: Sexual and maternal behaviour of the vervet monkey, *Cercopithecus aethiops.* J Reprod Fertil 1969;6:(suppl):137–150.
41 Cords M: Mating patterns and social structure in redtail monkeys *(Cercopithecus ascanius).* Z Tierpsychol 1984;64:313–329.

42 Rowell TE: Reproductive cycles of two *Cercopithecus* monkeys. J Reprod Fertil 1970;22:321–328.
43 Nadler RD: Cyclicity in tumescence of the perineal labia of female lowland gorillas. Anat Rec 1975;181:791–797.
44 Nadler RD: Laboratory research on sexual behavior of the great apes; in Graham CE (ed): Reproductive Biology of the Great Apes. New York, Academic Press , 1981, pp 192–238.
45 Hess JP: Some observations on the sexual behaviour of captive lowland gorillas; in Michael RP, Crook JH (eds): Comparative Ecology and Behaviour of Primates. London, Academic Press, 1973, pp 507–581.
46 Harcourt AH, Stewart KJ: Sexual behaviour of wild mountain gorillas; in Chivers DJ, Herbert J (eds): Recent Advances in Primatology. London, Academic Press, 1978, vol 1: Behaviour.
47 Harcourt AH, Fossey D, Stewart KJ, et al: Reproduction in wild gorillas and some comparisons with chimpanzees. J Reprod Fertil 1980;28(suppl):59–70.
48 Berkson G, Chaicumpa V: Breeding gibbons *(Hylobates lar entelloides)* in the laboratory. Lab Anim Care 1969;19:808–811.
49 Ellefson JO: A natural history of white-handed gibbons in the Malayan peninsula; in Rumbaugh DM (ed): Gibbon and Siamang. Basel, Karger, 1974, vol 3, pp 1–136.
50 Enomoto T, Seiki K, Haruki Y: On the correlation between sexual behavior and ovarian hormone level during the menstrual cycle in captive Japanese monkeys. Primates 1979;20:563:570.
51 Michael RP, Zumpe D: Rhythmic changes in the copulatory frequency of rhesus monkeys *(Macaca mulatta)* in relation to the menstrual cycle and a comparison with the human cycle. J Reprod Fertil 1970;21:199–201.
52 Keverne EB: Sexual receptivity and attractiveness in the female rhesus monkey. Adv Stud Behav 1976;7:155–200.
53 Johnson DF, Phoenix CH: Sexual behavior and hormone levels during the menstrual cycle of rhesus monkeys. Horm Behav 1978;11:160–174.
54 Conaway CH, Koford CB: Estrous cycles and mating behavior in a free-ranging band of rhesus monkeys. J Mammal 1964;45–577–588.
55 Kaufmann J: A three-year study of mating behaviour in a free-ranging band of rhesus macaques. Ecology 1965;46:500–512.
56 Lindburg DG: Mating behavior and estrus in the Indian rhesus monkey; in Seth PK (ed): Perspectives in Primate Biology. New Delhi, Today and Tomorrow's Printers & Publishers, 1983, pp 45–61.
57 Taub D: Female choice and mating strategies among wild Barbary macaques (*Macaca sylvanus* L.); in Lindburg DG (ed): The Macaques: Studies in Ecology, Behavior and Evolution. New York, Van Nostrand Reinhold, 1980.
58 Küster J, Paul A: Female reproductive characteristics in semifree ranging Barbary macaques (*Macaca sylvanus* L. 1758). Folia Primatol 1984;43:69–83.
59 Scruton DM, Herbert J: The menstrual cycle and its effect upon behaviour in the talapoin monkey *(Miopithecus talapoin)*. J Zool, Lond 1970;162:419–436.
60 Rowell TE, Dixson AF: Changes in social organization during the breeding season of wild talapoin monkeys. J Reprod Fertil 1975;43:419–434.
61 Savage-Rumbaugh ES, Wilkerson BJ: Socio-sexual behavior in *Pan paniscus* and *Pan troglodytes:* A comparative study. J Hum Evol 1978;7:327–344.

62 Yerkes RM, Elder JH: Oestrus receptivity and mating in chimpanzees. Comp Psychol Monogr 1936;13:1–39.
63 Yerkes RM: Sexual behavior in the chimpanzee. Hum Biol 1983;11:78–111.
64 Tutin CEG, McGinnis PR: Chimpanzee reproduction in the wild; in Graham CE (ed): Reproductive Biology of the Great Apes. New York, Academic Press, 1981, pp 239–264.
65 Goodall J: The Chimpanzees of Gombe: Patterns of Behavior. Cambridge, Belknap Press, Harvard University Press, 1986.
66 Saayman GS: The influence of hormonal and ecological factors upon sexual behavior and social organization in Old World primates.; in Tuttle RH (ed): Socioecology and Psychology of Primates. The Hague, Mouton, 1975, pp 181–204.
67 Saayman GS: The menstrual cycle and sexual behaviour in a troop of free-ranging chacma babbons *(Papio ursinus)*. Folia Primatol 1970;12:81–110.
68 Nadler RD: Sexual behavior of captive orangutans. Arch Sex Behav 1977;6:457–475.
69 Kingsley SR: The Reproductive Physiology and Behaviour of Captive Orang-Utans *(Pongo pygmaeus);* PhD thesis, London, 1981.
70 Cross JF, Martin RD: Calculation of gestation period and other reproductive parameters for primates. Dodo, J Jersey Wildl Pres Trust 1981;18:30–43.
71 Galdikas BMF: Orangutan sexuality in the wild; in Graham CE (ed): Reproductive Biology of the Great Apes. New York, Academic Press, 1981, pp 281–300.
72 Hrdy SB: The Langurs of Abu. Cambridge, Harvard University Press, 1977.
73 Sommer V: Kindestötungen bei indischen Langurenaffen *(Presbytis entellus)* – eine männliche Reproduktionsstrategie? Anthrop Anz 1984;42:177–183.
74 Dunbar RIM: Sexual behaviour and social relationships among gelada baboons. Anim Behav 1978;26:167–178.
75 Mori U: Reproductive behaviour. Contrib Primatol. Karger, Basel, 1979, vol 16, pp 183–197.
76 Keverne EB: Do Old World primates have oestrus? Malays Appl Biol 1981;10:119–126.
77 Sahlins MD: The social life of monkeys, apes and primitive man; in Spuhler JN (ed): The Evolution of Man's Capacity for Culture. Detroit, Wayne State University Press, 1959, pp 54–73.
78 Austin CR: Ageing and reproduction: Post-ovulatory deterioration of the egg. J Reprod Fertil 1970;(suppl):39–53.
79 Blandau RJ, Young WC: The effects of delayed fertilization on the development of the guinea pig ovum. Am J Anat 1939;64:303–329.
80 Blandau RJ, Jordan ES: The effect of delayed fertilization on the development of the rat ovum. Am J Anat 1941;68:275–291.
81 Blandau RJ: Aging ova and fertilization; in Uricchio WA (ed): Proceedings of a Conference on Natural Familiy Planning. Washington, Human Life Foundation, 1973, pp 81–84.
82 Blandau: Aging Gametes. Basel, Karger, 1975.
83 Austin CR: Chromosome deterioration in ageing eggs of the rabbit. Nature, Lond 1967;213:1018–1019.
84 German J: Mongolism, delayed fertilization and human sexual behaviour. Nature, Lond 1968;217:516–518.

85 Orgebin-Christ: Sperm age: effects on zygote development; in Uricchio WA (ed): Proceedings of a Conference on Natural Family Planning. Washington, Human Life Foundation, 1973, pp 85–95.
86 Iffy L: The time of conception in pathological gestations. Proc R Soc Med 1963;56: 1098–1100.
87 Iffy L, Wingate MB: Risks of the rhythm method of birth control. J Reprod Med 1970;5:11–17.
88 Dixson AF: Observations on the evolution and behavioral significance of 'sexual skin' in female primates. Adv Stud Behav 1983;13:63–106.
89 Kiltie RA: Intraspecific variation in the mammalian gestation period. J Mammal 1982;63:646–652.
90 Sorenson MF, Rogers JP, Baskett TS: Reproduction and development in confined swamp rabbits. J Wildl Manage 1968;32:520–531.
91 Olsen RW: Gestation period in *Neotoma mexicana.* J Mammal 1968;49:533–534.
92 Aguilera MM: Growth and reproduction in *Zygodontomys microtinus* from Venezuela in a laboratory colony. Mammalia 1985;49:75–83.
93 Michener GR: Estrous and gestation periods in Richardson's ground squirrels. J Mammal 1980;61:531–534.
94 Svilha A: A comparative life history study on mice of the genus *Peromyscus.* Misc Publ Mus Zool Michigan 1932;24:1–39.
95 Kupperman HS, Greenblatt RB, Hair LQ: The sexual cycle and reproduction in the golden hamster *(Cricetus auratus).* Anat Rec 1944;88:441–442.
96 Snell GD: Reproduction.; in Snell GD (ed): Biology of the Laboratory Mouse. New York, Dover, 1941, pp 55–88.
97 Rosahn PD, Greene HSH, Hu C: Observations on the gestation period of the rabbit. J Exp Zool 1935;72:195–212.
98 Lush JL: Animal Breeding Plans. Ames Iowa State College Press, 1945.
99 Terrill CE, Hazel LN: Length of gestation in range sheep. Am J Vet Res 1947;8: 66–72.
100 Hutchins M, Thompson G, Sleeper B, et al: Management and breeding of the Rocky Mountain goat *Oreamnos americanus* at Woodland Park Zoo. Int Zoo Yearb 1987; 26:297–308.
101 Densmore MA: Reproduction of sitatunga *Tragelaphus spekei* in captivity. Int Zoo Yearb 1980;20:227–229.
102 San-Martin M, Copaira M, Zuniga J, et al: Aspects of reproduction in the alpaca. J Reprod Fertil 1968;16:395–399.
103 Guinness F, Lincoln GA, Short RV: The reproductive cycle of the female red deer, *Cervus elaphus* L. J Reprod Fertil 1971;27:427–438.
104 Verme LJ: Reproduction studies on penned white-tailed deer. J Wildl Manage 1965; 29:74–79.
105 Jafar SM, Chapman AB, Casida LE: Cases of variation in length of gestation in dairy cattle. J Anim Sci 1950;9:592–601.
106 Mehta VS, Prakash AHA, Singh M: Gestation period in camels. Ind Vet J 1962;39: 387–389.
107 Skinner JD, Hall-Martin AJ: A note on foetal growth and development of the giraffe *Giraffa camelopardalis giraffa.* J Zool, Lond 1975;177:73–79.

108 Henry VG: Length of oestrous cycle and gestation in European wild hogs. J Wildl Manage 1968;32:406:408.

109 Cox DF: Relation of litter size and other factors to the duration of gestation in the pig. J Reprod Fertil 1964;7:405–407.

110 Sowls LK: Reproduction of the collared peccary *(Tayassu tajacu)*. Symp Zool Soc Lond 1966;15:155–172.

111 Rieger I: Breeding the striped hyaena *Hyaena hyaena* in captivity. Int Zoo Yearb 1979;19:193–198.

112 Poglayen-Neuwall I, Poglayen-Neuwall I: Gestation period and parturition of the ringtail, *Bassariscus astutus* (Lichtenstein, 1830). Z Säugetierk 1980;45:73–81.

113 Anderson D: Gestation period of Geoffroy's cat *Leopardus geoffroyi* bred at Memphis Zoo. Int Zoo Yearb 1977;17:164–166.

114 Marma BB, Yunchis VV: Observations on the breeding, management and physiology of snow leopards *Panthera u. uncia* at Kansas Zoo from 1962–1967. Int Zoo Yearb 1968;8:66–74.

115 Frueh RJ: A note on breeding snow leopards *Panther spekei* at St. Louis Zoo. Int Zoo Yearb 1968;8:74–76.

116 Lang EM, Leutenegger M, Tobler K: Indian rhinoceros, *Rhinoceros unicornis,* birth in captivity. Int Zoo Yearb 1977;17:237.

117 Izard MK, Simons EL: Infant survival and litter size in primigravid and multigravid Galagos. J Med Primatol 1986;15:27–35.

118 Zimmermann E: Aspects of reproduction, behavioral and vocal development in Senegal bushbabies (*Galago senegalensis)*. Int J Primatol 1989;10:1–16.

119 Kappeler PM: Reproduction in the crowned lemur *(Lemur coronatus)* in captivity. Am J Primatol 1987;12:497–503.

120 Nicholls L: Period of gestation of *Loris.* Nature, Lond 1939;143:246.

121 Manley GH: Gestation periods in the Lorisidae. Nature, Lond 1967;7:80–81.

122 Nieschalk U, Meier B: Haltung und Zucht von Schlankloris *(Loris tardigradus)*. Z Köln Zoo 1984;27:95–100.

123 Izard MK, Rasmussen DT: Reproduction in the slender loris *(Loris tardigradus malabaricus)*. Am J Primatol 1985;8:153–165.

124 Perret M: Influence de la captivité et du groupement social sur la physiologie du microcèbe (*Microcebus murinus* – Cheirogaleinae – Primates); thèse Université Paris-Sud, 1980.

125 Izard MK, Weisenseel K, Ange R: Reproduction in the slow loris *(Nycticebus coucang)*. Am J Primatol 1988;16:331–339.

126 Foerg R: Reproductive behavior in *Varecia variegata.* Folia Primatol 1982;38:108–121.

127 Shideler SE, Lindburg DG: Selected aspects of *Lemur variegatus* reproductive biology. Zoo Biol 1982;1:127–134.

128 Hick U: Haltung, Zucht und künstliche Aufzucht von Varis *(Varecia variegata)* im Kölner Zoo. Z Köln Zoo 1984;27:121–137.

129 Hearn JP, Lunn SF, Burden FJ, et al: Management of marmosets for biomedical research. Lab Anim 1975;9:125–134.

130 Ziegler TE, Snowdon CT, Warneke M: Postpartum ovulation and conception in Goeldi's monkey, *Callimico goeldii.* Folia Primatol 1989;52:206–210.

131 Johnson PT, Valerio DA, Thompson GE: Breeding the African green monkey, *Cercopithecus aethiops,* in a laboratory environment. Lab Anim Sci 1973;23:355–359.
132 Hartwell KM: Reproductive Cycles of Patas Monkeys *(Erythrocebus patas),* with Special Reference to Breeding Seasons; PhD thesis, Berkeley, 1972.
133 Gibson JR, McKeown T: Observations on all births (23,970) in Birmingham, 1947. I. Duration of gestation. Br J Soc Med 1950;4:221–233.
134 Nieuwenhuijsen K, Lammers AJJC, de Neef KJ, et al: Reproduction and social rank in female stumptail macaques (*Macaca arctoides*). Int J Primatol 1985;6:77–99.
135 Peng M-T, Lai Y-L, Yang C-S, et al: Reproductive parameters of the Taiwan monkey *(Macaca cyclopsis).* Primates 1973;14:201–213.
136 Valerio DA, Pallotta AJ, Courtney KD: Experiences in large-scale breeding of simians for medical experimentation. Ann NY Acad Sci 1969;162:282–296.
137 Nigi H: Some aspects related to conception of the Japanese monkey *(Macaca fuscata).* Primates 1976;17:81–87.
138 Kriewaldt FH, Hendrickx AG: Reproductive parameters of the baboon. Lab Anim Care 1968;18:361–370.
139 Peacock LJ, Rogers CM: Gestation period and twinning in chimpanzees. Science 1959;129:959.
140 Torii R, Nigi H, Koizumi H, et al: Serum chorionic gonadotropin, progesterone, and estradiol-17β during pregnancy in the common marmoset, *Callithrix jacchus.* Primates 1989;30:207–215.
141 Martin RD: Primate Origins and Evolution: A Phylogenetic Reconstruction. London/New Jersey, Chapman Hall/Princeton University Press, 1990.
142 Bell ET, Loraine JA: Time of ovulation in relation to cycle length. Lancet 1965;i: 1029–1030.
143 Vollman RF: Über Fertilität und Sterilität der Frau innerhalb des Menstruationszyklus. Arch Gynäk 1953;182:602.
144 Finnström O: Studies on maturity in newborn infants. VI. Comparison between different methods for maturity estimation. Acta Pediat Scand 1972;61:33–41.
145 Jöchle W: Coitus-induced ovulation. Contraception 1973;7:523–564.
146 Knaus H: Über den Zeitpunkt der Konzeptionsfähigkeit des Weibes im Intermenstruum. Münch med Wochenschr 1929;76:1152–1160.
147 Knaus H: Die periodische Fruchtbarkeit und Unfruchtbarkeit des Weibes. Wien, Maudrich, 1934.
148 Ogino N: Histologic studies on corpora lutea, period of ovulation, relation between corpora lutea and cyclic changes in uterine mucosa and the period of fertilization. Jpn Med World 1928;8:147–148.
149 Ogino K: Ovulationstermin und Konzeptionstermin. Zentralbl Gynäk 1929;8:464–479.
150 Pryll W: Kohabitationstermin und Kindgeschlecht. Münch Med Wochenschr 1916; 1916:1579–1582.
151 Manulkin AE: Über das zyklische Verhalten der Konzeptionen. Zentralbl Gynäk 1936;60:15–19.
152 Dyroff R: Beiträge zur Frage der physiologischen Sterilität. Zentralbl Gynäk 1939; 31:1717–1721.
153 Hollenweger-Mayr B: Die menschliche Schwangerschaftsdauer. Z Geburtsh Gynaek 1950;132:297–314.

154 Hartman CG: Studies in the reproduction of the monkey *Macaca (Pithecus) rhesus,* with special reference to menstruation and pregnancy. Contrib Embryol 1932;23: 1–161.
155 Hendrickx AG, Houston ML, Kraemer DC, et al: Embryology of the Baboon. Chicago, University of Chicago Press, 1971.
156 Wildt DE, Doyle LL, Stone SC, et al: Correlation of perineal swelling with serum ovarian hormone levels, vaginal cytology, and ovarian follicular development during the baboon reproductive cycle. Primates 1977;18:261–270.
157 Sly DL, Harbaugh SW, London WT, et al: Reproductive performance of a laboratory breeding colony of patas monkeys *(Erythrocebus patas).* Am J Primatol 1983;4: 23–32.
158 Feistner ATC: The Behaviour of a Social Group of Mandrills, *Mandrillus sphinx;* PhD thesis, Stirling, 1989.
159 Wickings EJ, Dixson AF: Application of DNA fingerprinting to familial studies of Gabonese primates; in Martin RD, Dixson AF, Wickings EJ (eds): Paternity in Primates: Genetic Tests and Theories. Basel, Karger, 1992, pp 113–130.
160 Feistner ATC: Aspects of reproduction in female mandrills (*Mandrillus sphinx*). Int Zoo Yearb, in press.
161 Carman M: The gestation and rearing periods of the mandrill, *Mandrillus sphinx,* at the London Zoo. Int Zoo Yearb 1979;19:159–160.
162 Martin DE: Breeding great apes in captivity; in Graham CE (ed): Reproductive Biology of the Great Apes. New York, Academic Press, 1981, pp 343–373.
163 Clewe TH: Mating of squirrel monkeys in captivity. Am Zool 1966;6:343–344.

Prof. Dr. Robert D. Martin, Anthropologisches Institut, Universität Zürich-Irchel, Winterthurerstrasse 190, CH–8057 Zürich (Switzerland)

Martin RD, Dixson AF, Wickings EJ (eds): Paternity in Primates: Genetic Tests and Theories. Basel, Karger, 1992, pp 275–280

Appendix: Methodology Associated with DNA Typing

Based upon the individual contributions and discussions generated at the symposium, an overview of the methods applicable to paternity exclusion (PE) methods in nonhuman primates seemed essential. (As the term 'DNA fingerprint' does not describe the allele patterns generated by RFLPs and single-locus VNTR probes, 'DNA typing' is a preferable descriptor.) The field of PE has moved forward very rapidly over the past seven years, from the time when mass screening of large series of protein markers was the only possibility. With the advent of DNA fingerprinting and the associated molecular technology, geneticists are now able to determine familial relationships with a high degree of accuracy. Since this time, almost as many different methods have been developed for PE as there are laboratories applying these techniques. It would have been very tempting to discard the classical PE protein analysis methods in the initial euphoria which greeted DNA fingerprinting. However, these traditional methods have proved indispensable as an initial screening step, especially in studies of large populations of primates for which moderate levels of allelic variation have been established. DNA analysis can then be employed to resolve any remaining ambiguities. Indeed, a combined approach using protein and DNA typing is probably as efficient as any using DNA typing alone and is certainly more cost-effective.

There is clearly no lack of choice of method for determining PE (table 1), and decisions on which approach to adopt will depend entirely on local facilities and expertise. There appears to be no reason why DNA typing should not be applicable to all nonhuman primate species, given that single-locus probes and multilocus probes (MLPs) have been applied successfully throughout the mammalian world. Difficulties experienced

Table 1. Compendium of the techniques discussed in preceding chapters for PE

Method	Sample	Procedure (once established)	Availability (system/probes/primers)	Source of error	Number of polymorphic loci for PE	Status of method
Protein markers	blood (I)	easy	<100 systems	protein degradation variable band intensity post-translational modification	8–10	classic
DNA polymorphisms						
RFLPs	blood (I)	medium complexity	hundreds of probes (human)	partial DNA digestion	14–16	applied
MLP	blood/tissue (I)	complex	6 probes	partial DNA digestion band shifts cross-gel differences variable band intensity	N/A	applied
SLP	blood (I)	medium complexity	hundreds of probes (human)	as for MLPs	4–5	applied
VNTR-PCR	blood/tissue (I) hair (NI)	easy	species-variable primers	degraded DNA contamination differential amplification	4–5	developmental
VDTR-PCR	blood/tissue (I) hair (NI)	easy	species-variable primers	as for VNTR-PCR	4–5	developmental

I = Invasive sampling technique; NI = non-invasive sampling technique; N/A = not applicable to MLPs since an indeterminate number of loci are recognised by these probes, but PE can usually be made using 2 MLPs; SLP = single-locus probe.

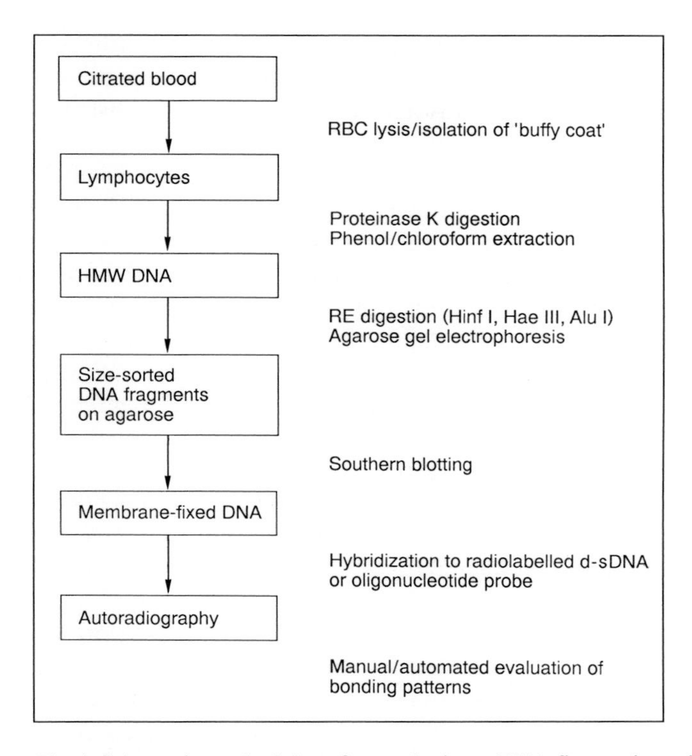

Fig. 1. Schematic methodology for producing a DNA fingerprint using MLPs.

with any given combination of restriction enzyme and DNA probe can often be overcome by trying other combinations until a suitable match is found. It may be necessary to try up to 16 different RFLPs to obtain the same degree of probability of PE as that attained with 2 MLPs (table 1). In some study populations, high band-sharing frequencies were found with the MLP approach, and it remains unclear whether this apparent loss of genetic variability results from inbreeding of isolated populations or is a function of the populations concerned. One considerable drawback to the use of MLP in PE is the inability to extract additional genetic information on the study population, because the Mendelian basis for the inheritance of individual bands of the fingerprint cannot be readily determined.

Shared bands are not limited to father-offspring relations, and unshared bands do not preclude paternity, so determining a minimum number of shared bands is a prerequisite to establishing paternity within a study. But the incidence of shared bands may be species- or population-specific, so a large number of samples are needed to establish baseline levels of variability.

The source of DNA has until recently been almost exclusively blood, but non-invasive sampling techniques (such as collection of shed hairs from resting sites) offer field workers the possibility of performing genetic analyses on their study populations without disturbing the habituation process. Clearly, DNA amplification and analysis from such material will become very popular, but its development and use at present is hampered by the need to develop and optimize polymerase chain reaction (PCR) primers for each species. In this and other aspects of DNA typing, technological advances (e.g. use of non-radioactive methods for visualization of bands on autoradiographs) will probably remove such reservations. The combination of minisatellite (and the microsatellite repeat sequences) with PCR will provide a rapid, accurate and inexpensive means of DNA typing, but the tremendous developmental investment needed for such techniques must also be taken into account.

Figure 1 shows a simplified flowchart of DNA fingerprinting using MLPs or synthetic oligonucleotides. It is not comprehensive, but is provided as guidance to those unfamiliar with the technique. An introduction to the methodology underlying DNA typing can be pursued by consulting the non-exhaustive recommended reading list given below.

Recommended Reading

Boerwinkle E, Xiong W, Fourest E, et al: Rapid typing of tandemly repeated hypervariable loci by the polymerase chain reaction: Application to the apolipoprotein B 3' hypervariable region. Proc Natl Acad Sci USA 1989;86:212–216.

Budowle B, Giusti AM, Waye JS, et al: Fixed-bin analysis for statistical evaluation of continuous distributions of allelic data from VNTR loci, for use in forensic comparisons. Am J Hum Genet 1991;48:841–855.

Devlin B, Risch N, Roederer K: Estimation of allele frequencies for VNTR loci. Am J Hum Genet 1991;48:662–676.

Dowling TE, Moritz C, Palmer JD: Nucleic acids. II. Restriction site analysis; in Hillis DM, Moritz C (eds): Molecular Systematics. Sunderland. Sinauer Associates, 1990. pp 250–317.

Dyke B, Williams-Blangero S, Dyer TD, et al: Use of isozymes in genetic management of nonhuman primate colonies; in Ogita ZI, Markert CL (eds): Isozymes: Structure. Function and Use in Biology and Medicine. New York, Wiley-Liss, 1990, pp 563–574.

Ely J. Ferrell RE: DNA fingerprints and paternity ascertainment in chimpanzees *(Pan troglodytes).* Zoo Biol 1990:9:91–98.

Harris H, Hopkinson DA: Handbook of enzyme electrophoresis in human genetics. Amsterdam. North-Holland, 1978.

Higuchi R, von Beroldingen, CH, Sensabaugh CF, et al: DNA typing from single hairs. Nature 1988:332:543–546.

Hillis DM, Larson A. Davis SK et al: Nucleic acids. III. Sequencing; in Hillis DM, Moritz C (eds): Molecular Systematics. Sunderland, Sinauer Associates, 1990, pp 318–370.

Innis MA, Gelfand H, Sninsky JJ, et al (eds): PCR protocols: A guide to methods and applications. San Diego, Academic Press, 1990.

Jeffreys AJ, Turner M, Debenham P: The efficiency of multilocus DNA fingerprint probes for individualization and establishment of family relationships, determined from extensive case work. Am J Hum Genet 1991;48:824–840.

Jeffeys AJ, Wilson V. Neumann R, Kyte J: Amplification of human minisatellites by the polymerase chain reaction: Towards DNA fingerprinting of single cells. Nucleic Acids Res 1988;16:10953–10971.

Jeffreys AJ, Wilson V, Thein SL: Individual-specific 'fingerprints' of human DNA. Nature 1985;316:76–79.

Jeffreys AJ, Wilson V, Thein SL: Hypervariable 'minisatellite' regions in human DNA. Nature 1985:314:67–73.

Kidd KK, Bowcock AM, Schmidtke J, et al: Report of the DNA committee and catalogs of cloned and mapped genes and DNA polymorphisms: Human Gene Mapping 10 (1989): Tenth International Workshop on Human Gene Mapping. Cytogenet Cell Genet 1989;51:622–947.

Kirby LT: DNA Fingerprinting: An introduction. New York, Stockton Press, 1990, pp 135–147.

Litt M, Luty JA: A hypervariable microsatellite revealed by in vitro amplification of a dinucleotide repeat within the cardiac muscle action gene. Am J Hum Genet 1989; 44:397–401.

Nakamura Y, Leppert M, O'Connell P, et al: Variable number of tandem repeat (VNTR) markers for human gene mapping. Science 1987;235:1616–1622.

Richardson BJ, Baverstock PR, Adams M: Allozyme Electrophoresis: A Handbook for Animal Systematics and Population Studies. Sydney, Academic Press, 1986.

Righetti PG: Isoelectric Focusing: Theory, Methodology and Application. Laboratory Techniques in Biochemistry and Molecular Biology. Amsterdam, Elsevier Biomedical Press, 1983, vol 11.

Saiki RK, Scharf S, Faloona F, et al: Enzymatic amplification of β-globin genomic sequences and restriction site analysis for diagnosis of sickle cell anemia. Science 985;230:1350–1354.

Sambrook J, Fritsch EF, Maniatis T: Molecular cloning: A laboratory manual, ed 2. New York, Cold Spring Harbor Laboratory Press, 1989.

Schäfer R, Zischler H, Birsner U, Becker A, Epplen JT: Optimized oligonucleotide probes for DNA-fingerprinting. Electrophoresis 1988:9:369–374.

Tautz D: Hypervariability of simple sequences as a general source for polymorphic DNA markers. Nucleic Acid Res 1989;17:6463–6471.

VandeBerg JL, Aivaliotis MJ, Williams LE, et al: Biochemical genetic markers of squirrel monkeys and their use for pedigree validation. Biochem Genet 1990;28:41–56.

Vassart G, Georges M, Monsieur R, Brocas H. Lequarre AS, Christophe D: A sequence in M13 phage detects hypervariable minisatellites in human and animal DNA. Science 1987;235:683–684.

Washio K, Misawa S, Ueda S: Probe walking: Development of novel probes for DNA fingerprinting. Hum Genet 1989;83:223–226.

Weber JL: Informativeness of human $(dC\text{-}dA)_n \cdot (dT\text{-}dG)_n$ polymorphisms. Genomics 1990;7:524–530.

Weiss ML, Turner TR: Hypervariable minisatellites and VNTRs: in Devore E (ed): Molecular Approaches to Physical Anthropology. Cambridge, Cambridge University Press, in press.

Weiss ML, Wilson V, Chan C, Turner T, Jeffreys AJ: Application of DNA fingerprinting probes to Old World monkeys. Am J Primatol 1988;16:73–79.

Williams JG, Kubelik AR, Livak KJ, et al: DNA polymorphisms amplified by arbitrary primers are useful as genetic markers. Nucleic Acids Res 1991;18:6531–6535.

Wong Z, Wilson V, Patel I, et al: Characterization of a panel of highly variable minisatellites cloned from human DNA. Ann Hum Genet 1987;51:269–288.

Subject Index